Wearable and Portable Devices in Sport Biomechanics and Training Science

Wearable and Portable Devices in Sport Biomechanics and Training Science

Editors

Felipe García-Pinillos
Alejandro Pérez-Castilla
Diego Jaén-Carrillo

Basel • Beijing • Wuhan • Barcelona • Belgrade • Novi Sad • Cluj • Manchester

Editors
Felipe García-Pinillos
University of Granada
Granada
Spain

Alejandro Pérez-Castilla
University of Almería
Almería
Spain

Diego Jaén-Carrillo
Universität Innsbruck
Innsbruck
Austria

Editorial Office
MDPI AG
Grosspeteranlage 5
4052 Basel, Switzerland

This is a reprint of articles from the Special Issue published online in the open access journal *Sensors* (ISSN 1424-8220) (available at: https://www.mdpi.com/journal/sensors/special_issues/0R8NSC8S66).

For citation purposes, cite each article independently as indicated on the article page online and as indicated below:

Lastname, A.A.; Lastname, B.B. Article Title. *Journal Name* **Year**, *Volume Number*, Page Range.

ISBN 978-3-7258-1769-6 (Hbk)
ISBN 978-3-7258-1770-2 (PDF)
doi.org/10.3390/books978-3-7258-1770-2

Contents

About the Editors

Felipe García-Pinillos

Felipe García-Pinillos gained a degree in Sports Sciences at the University of Granada (2010) and a master's degree in Research in Physical Activity and Health at the University of Jaen (2011). His academic background also includes a master's degree in Injury Prevention and Return to Sport (2017, University of Jaen) and a master's degree in High Performance in Endurance Sports (2020, University of Murcia).

He started his PhD project as a fellow researcher at the University of Jaen (2013–2016). Thereafter, his main research interest was related to performance optimization in endurance athletes by determining the metabolic, physiological, neuromuscular, and biomechanical impact of different programs. Related to this topic, he stayed at different laboratories (e.g., Liverpool John Moores University, the Sport & Health Institute at the University of Granada, and the Sport Physiology Laboratory at the University of La Frontera) and finally attained his PhD degree (2016) at the University of Jaen (awarded with Extraordinary Mention by the University of Jaen, 2019).

When he finished his PhD studies, he started working for the University of La Frontera (Temuco, Chile) as a postdoctoral researcher. He stayed there for 2.5 years, investigating and teaching about training and its application to different contexts (from performance optimization to injury management). Finally, in February 2020, he started teaching as an Associate Professor at the Faculty of Sport Sciences (University of Granada).

From 2013 to today, he has published 206 papers, including 188 documents published in journals indexed in the JCR: 69 papers published in Q1 journals, 41 in Q2, 36 in Q3, and 42 in Q4. In 146 out of those 188 papers, he is the first or last author. According to Google Scholar's citation index, these papers have been cited 4381 times, and Felipe has an H-INDEX of 35.

Alejandro Pérez Castilla

Alejandro Pérez Castilla obtained a bachelor's degree in Sport Sciences in 2015, followed by two master's degrees: one in Physical Education in 2016 and another in Research in Sport and Physical Activity in 2017, all from the University of Granada. He then completed his doctoral studies in Biomedicine at the University of Granada between 2017 and 2020.

Alejandro secured various postdoctoral research contracts for two years at the Sports Sciences Faculty of the University of Granada. He then began working as an Assistant Professor at the Faculty of Education of the University of Almería. During this time, he published over 130 papers in international journals indexed in the Journal Citation Reports and has presented numerous works at national and international conferences. His main research areas focus on sports training and sports biomechanics. More specifically, most studies have addressed methodological issues related to velocity-based resistance training.

Diego Jaén-Carrillo

Diego Jaén-Carrillo was born in Murcia, Spain and completed his undergraduate degree in Sport Science (BSc) at the Catholic University San Antonio of Murcia (UCAM), Spain, in 2009. After that, Diego completed an MA in Teaching Proficiency at the University of Granada, Spain (2010) and a BA in English Studies at the University of Murcia, Spain, in 2014. During this time, he also obtained a funded scholarship at the UCAM Research and High-Performance Center as an MSc student to complete an MSc in High Performance Sport: Strength and Conditioning. His work during that time focused on the mechanisms of ACL rupture during landing after jumping in different sports. After spending some time working in Spain and Germany as a Sport Teacher for secondary school, Diego started a Graduate Teaching Assistant position at Universidad San Jorge, Zaragoza, (2018–2019) and completed his PhD in Health Sciences in 2020 at the same university. Immediately after that, he took the role of Assistant Professor of Sport Biomechanics and Sport Therapy at Universidad San Jorge, Zaragoza, until December 2022. Since January 2023, Diego has acted as a Postdoc Researcher at the Department of Sports Science of the University of Innsbruck, Austria, currently working in the "Training Science Research Group" under the leadership of Univ. Prof. Dr. Justin Lawley.

Since 2021, Diego has also collaborated as a Visiting Assistant Professor of Sport Biomechanics at the University of Murcia, Spain, in his MSc in High Performance in Cyclic Sports: running, swimming and cycling.

Diego's research focuses on human locomotion and endurance performance. He also has ongoing projects on Running with Power that aim to assess the validity and reliability of running power-related variables to quantify training load in order to prescribe training and racing accurately for both road and trail running based on scientific evidence.

Preface

Sport biomechanics and training have traditionally been tested under laboratory conditions, requiring specific settings and expensive equipment. The novel use of wearable devices addresses the lack of ecological validity in such measures and offers an affordable, user-friendly option for biomechanical assessments. Recently, wearable sensors have enabled the quantification of performance and workload by providing mechanical and physiological parameters, leading to their exponential growth in popularity. Many wearable sensors are now commercially available and capable of delivering both kinetic and kinematic data, thus improving the feasibility and efficiency of assessments and making them a viable alternative for sports practitioners and researchers. Additionally, wearable devices allow for real-time monitoring and biofeedback. This reprint aims to provide current information about the use and application of wearable sensors in sport biomechanics and training science.

Felipe García-Pinillos, Alejandro Pérez-Castilla, and Diego Jaén-Carrillo
Editors

Editorial

Wearable and Portable Devices in Sport Biomechanics and Training Science

Diego Jaén-Carrillo [1,*], Alejandro Pérez-Castilla [2,3] and Felipe García-Pinillos [4,5,6]

1   Department of Sport Science, University of Innsbruck, 6020 Innsbruck, Austria

2   Department of Education, Faculty of Education Sciences, University of Almería, 04120 Almería, Spain; alexperez@ual.es

3   SPORT Research Group (CTS-1024), CIBIS (Centro de Investigación para el Bienestar y la Inclusión Social) Research Center, University of Almería, 04120 Almería, Spain

4   Department of Physical Education and Sport, University of Granada, Carretera de Alfacar 21, 18011 Granada, Spain; fgpinillos@ugr.es

5   Sport and Health University Research Center (iMUDS), Parque Tecnologico de la Salud, Av. Del Conocimiento s/n, 18007 Granada, Spain

6   Department of Physical Education, Sport and Recreation, Universidad de La Frontera, Temuco 01145, Chile

*   Correspondence: diego.jaen@uibk.ac.at

Citation: Jaén-Carrillo, D.; Pérez-Castilla, A.; García-Pinillos, F. Wearable and Portable Devices in Sport Biomechanics and Training Science. *Sensors* **2024**, *24*, 4616. https://doi.org/10.3390/s24144616

Received: 21 June 2024

Accepted: 16 July 2024

Published: 17 July 2024

Sport biomechanics and training have traditionally been tested under laboratory conditions, requiring specific settings and expensive equipment [1,2]. The novel use of wearable devices addresses the lack of ecological validity in such measures and offers an affordable, user-friendly option for biomechanical assessments [3–5]. Recently, wearable sensors have enabled the quantification of performance and workload by providing mechanical and physiological parameters, leading to their exponential growth in popularity [3,5]. Many wearable sensors are now commercially available and capable of delivering both kinetic and kinematic data, thus improving the feasibility and efficiency of assessments and making them a viable alternative for sports practitioners and researchers [5]. Additionally, wearable devices allow for real-time monitoring and biofeedback [6]. This Special Issue of *Sensors* aims to provide current information about the use and application of wearable sensors in sport biomechanics and training science.

This Special Issue features twelve original articles (contributions 1–12) and a systematic review (contribution 13).

Contribution Summaries

Biomechanical Changes in Running Shoes: A study (contribution 1) investigated biomechanical changes in the lower limbs when transitioning from level ground to an uphill slope with different levels of longitudinal bending stiffness (LBS) in running shoes. The authors concluded that running uphill with high LBS shoes improves lower limb efficiency but increases knee joint energy absorption, potentially raising the risk of knee injuries. Amateurs should choose running shoes with optimal stiffness.

Metabolic Power and Energy Cost in Sprinting: Contribution 2 compared methods for calculating the metabolic power (MP) and energy cost (EC) of sprinting using GPS metrics and electromyography (EMG). The study found that GPS-based calculations underestimated neuromuscular and metabolic engagement, suggesting that EMG-derived methods are more accurate for MP and EC calculations during sprints.

Assessment of Torque–Velocity Profile in Cycling: One study (contribution 3) aimed to determine the feasibility, test–retest reliability, and long-term stability of a novel method for assessing the torque–velocity profile and maximal dynamic force (MDF) during leg-pedaling. Using a friction-loaded isoinertial cycle ergometer and a high-precision power meter, the data supported the method's validity, reliability, and long-term stability for cyclists.

Motor Development and Training: This empirical study (contribution 4) tested the assumed hierarchy for learning skills in motor development and training literature. The authors concluded that the proposed learning hierarchy is valid for some tasks, although the underlying reasons remain unknown.

Reliability of Wearable Systems: Four contributions (5, 10, 11, and 12) examined the reliability and agreement levels of different wearable systems:

- Contribution 5 evaluated the reliability of Xsens-based lower extremity joint angles during running on stable and unstable surfaces. The system captured within-day kinematic adaptations but showed less reliability for between-day measurements, particularly for ankle and hip joint angles in the frontal plane.
- Contribution 10 compared a markerless motion capture system (MotionMetrix) with an optoelectronic MCS (Qualisys) for kinematic evaluations during walking and running. The agreement varied by variable and speed, with some variables showing high agreement and others poor agreement.
- Contribution 11 assessed the validity of the three-sensor RunScribe Sacral Gait Lab™ IMU for measuring pelvic kinematics compared to the Qualisys system. The IMU did not meet the validity criteria for any variables or velocities tested.
- Contribution 12 analyzed the test–retest and between-device reliability of the Vmaxpro IMU for estimating vertical jump. The Vmaxpro was deemed unreliable for measuring vertical jumps.

Effects of Pressurization Modes on Muscle Activation: Contribution 6 examined the effects of different pressurization modes during high-load bench press training on muscle activation and subjective fatigue in bodybuilders. High-intensity bench press training with either continuous or intermittent pressurization significantly increased muscle activation, with continuous pressurization resulting in higher perceived fatigue.

Real-Time Monitoring in Fencing: Contribution 7 assessed the efficacy of a novel system for real-time monitoring of fencers' balance and movement control, enhanced by visual and haptic feedback modules. The findings suggest that integrating the Internet of Things (IoT) and real-time sensory feedback improves performance in fencing.

Thermoregulation in Athletes: A study (contribution 8) described bilateral variations in skin temperature of the anterior thigh and patellar tendon in healthy athletes, providing a model of baseline thermoregulation following a unilateral isokinetic fatigue protocol. The thermal challenge produced homogeneous changes in the quadriceps but not in the tendon areas, indicating that metabolic and blood flow changes depend on the tissue's physical and mechanical properties.

Discriminatory Power of Physical Tests: Contribution 9 examined whether specific physical tests can differentiate players with similar anthropometric characteristics but different playing levels. The authors concluded that a combination of the specific performance test and the force development standing test effectively identifies talent and differentiates between elite and sub-elite players.

Systematic Review on Beach Invasion Sports: The systematic review (contribution 13) aimed to (1) characterize internal and external loads during beach invasion sports, (2) identify monitoring technologies and metrics, (3) compare demands with indoor sports, and (4) explore differences by competition level, age, sex, and beach sport. Key findings demonstrated that beach sports involve moderate-to-high-intensity bouts with lower-intensity recovery, the unstable sand surface and variable outdoor conditions increase perceptual effort despite lower external load volumes compared to indoor sports, and substantial variability exists in acceleration, impact, and internal load intensity zones in the limited beach sports research.

Conclusions

In summary, this Special Issue provides valuable insights into sports biomechanics and training science, particularly the application of wearable devices for real-time monitoring and biofeedback.

Author Contributions: All authors have equally contributed to each part of the elaboration of the manuscript. All authors have read and agreed to the published version of the manuscript.

Conflicts of Interest: The authors declare no conflicts of interest.

List of Contributions

1. Lu, R.; Chen, H.; Huang, J.; Ye, J.; Gao, L.; Liu, Q.; Quan, W.; Gu, Y. Biomechanical Investigation of Lower Limbs during Slope Transformation Running with Different Longitudinal Bending Stiffness Shoes. *Sensors* **2024**, *24*, 3902. https://doi.org/10.3390/s24123902.
2. Grassadonia, G.; Alcaraz, P.E.; Freitas, T.T. Comparison of Metabolic Power and Energy Cost of Submaximal and Sprint Running Efforts Using Different Methods in Elite Youth Soccer Players: A Novel Energetic Approach. *Sensors* **2024**, *24*, 2577. https://doi.org/10.3390/s24082577.
3. Rodríguez-Rielves, V.; Barranco-Gil, D.; Buendía-Romero, Á.; Hernández-Belmonte, A.; Higueras-Liébana, E.; Iriberri, J.; Sánchez-Redondo, I.R.; Lillo-Beviá, J.R.; Martínez-Cava, A.; de Pablos, R.; et al. Torque–Cadence Profile and Maximal Dynamic Force in Cyclists: A Novel Approach. *Sensors* **2024**, *24*, 1997. https://doi.org/10.3390/s24061997.
4. Ribeiro, M.T.S.; Conceição, F.; Pacheco, M.M. Proficiency Barrier in Track and Field: Adaptation and Generalization Processes. *Sensors* **2024**, *24*, 1000. https://doi.org/10.3390/s24031000.
5. Debertin, D.; Wargel, A.; Mohr, M. Reliability of Xsens IMU-Based Lower Extremity Joint Angles during In-Field Running. *Sensors* **2024**, *24*, 871. https://doi.org/10.3390/s24030871.
6. He, K.; Sun, Y.; Xiao, S.; Zhang, X.; Du, Z.; Zhang, Y. Effects of High-Load Bench Press Training with Different Blood Flow Restriction Pressurization Strategies on the Degree of Muscle Activation in the Upper Limbs of Bodybuilders. *Sensors* **2024**, *24*, 605. https://doi.org/10.3390/s24020605.
7. Niță, V.-A.; Magyar, P. Improving Balance and Movement Control in Fencing Using IoT and Real-Time Sensorial Feedback. *Sensors* **2023**, *23*, 9801. https://doi.org/10.3390/s23249801.
8. Cabizosu, A.; Marín-Pagán, C.; Martínez-Serrano, A.; Alcaraz, P.E.; Martínez-Noguera, F.J. Myotendinous Thermoregulation in National Level Sprinters after a Unilateral Fatigue Acute Bout—A Descriptive Study. *Sensors* **2023**, *23*, 9330. https://doi.org/10.3390/s23239330.
9. Chirosa-Ríos, L.J.; Chirosa-Ríos, I.J.; Martínez-Marín, I.; Román-Montoya, Y.; Vera-Vera, J.F. The Role of the Specific Strength Test in Handball Performance: Exploring Differences across Competitive Levels and Age Groups. *Sensors* **2023**, *23*, 5178. https://doi.org/10.3390/s23115178.
10. Jaén-Carrillo, D.; García-Pinillos, F.; Chicano-Gutiérrez, J.M.; Pérez-Castilla, A.; Soto-Hermoso, V.; Molina-Molina, A.; Ruiz-Alias, S.A. Level of Agreement between the MotionMetrix System and an Optoelectronic Motion Capture System for Walking and Running Gait Measurements. *Sensors* **2023**, *23*, 4576. https://doi.org/10.3390/s23104576.
11. Ruiz-Malagón, E.J.; García-Pinillos, F.; Molina-Molina, A.; Soto-Hermoso, V.M.; Ruiz-Alias, S.A. RunScribe Sacral Gait Lab™ Validation for Measuring Pelvic Kinematics during Human Locomotion at Different Speeds. *Sensors* **2023**, *23*, 2604. https://doi.org/10.3390/s23052604.
12. Villalon-Gasch, L.; Jimenez-Olmedo, J.M.; Olaya-Cuartero, J.; Pueo, B. Test–Retest and Between–Device Reliability of Vmaxpro IMU at Hip and Ankle for Vertical Jump Measurement. *Sensors* **2023**, *23*, 2068. https://doi.org/10.3390/s23042068.
13. Vaccaro-Benet, P.; Gómez-Carmona, C.D.; Marzano-Felisatti, J.M.; Pino-Ortega, J. Internal and External Load Profile during Beach Invasion Sports Match-Play by Electronic Performance and Tracking Systems: A Systematic Review. *Sensors* **2024**, *24*, 3738. https://doi.org/10.3390/s24123738.

References

1. Verheul, J.; Nedergaard, N.J.; Vanrenterghem, J.; Robinson, M.A. Measuring biomechanical loads in team sports—From lab to field. *Sci. Med. Footb.* **2020**, *4*, 246–252. [CrossRef]
2. Watkins, J. *Laboratory and Field Exercises in Sport and Exercise Biomechanics*; Routledge: London, UK, 2017; ISBN 9781315306315.
3. Taborri, J.; Keogh, J.; Kos, A.; Santuz, A.; Umek, A.; Urbanczyk, C.; Van der Kruk, E.; Rossi, S. Sport biomechanics applications using inertial, force, and EMG sensors: A literatura overview. *Appl. Bionics. Biomech.* **2020**, *2020*, 2041549. [CrossRef] [PubMed]
4. McDevitt, S.; Hernandez, H.; Hicks, J.; Lowell, R.; Bentahaikt, H.; Burch, R.; Ball, J.; Chander, H.; Freeman, C.; Taylor, C.; et al. Wearables for biomechanical performance optimization and risk assessment in industrial and sports applications. *Bioengineering* **2022**, *9*, 33. [CrossRef] [PubMed]

5.   Adesida, Y.; Papi, E.; McGregor, A.H. Exploring the role of wearable technology in sport kinematics and kinetics: A systematic review. *Sensors* **2019**, *19*, 1597. [CrossRef] [PubMed]

6.   Lloyd, D. The future of in-field sports biomechanics: Wearables plus modelling compute real-time in vivo tissue loading to prevent and repair musculoskeletal injuries. *Sports Biomech.* **2021**, 1–29. [CrossRef] [PubMed]

sensors

Article

Biomechanical Investigation of Lower Limbs during Slope Transformation Running with Different Longitudinal Bending Stiffness Shoes

Runhan Lu [1], Hairong Chen [2,3], Jialu Huang [1], Jingyi Ye [1], Lidong Gao [4], Qian Liu [1,2,3], Wenjing Quan [1,*] and Yaodong Gu [1,*]

1   Faculty of Sports Science, Ningbo University, Ningbo 315211, China; lurunhan1@163.com (R.L.); hjialu@126.com (J.H.); yejingyi1999@gmail.com (J.Y.); liuqian19981008@163.com (Q.L.)
2   Doctoral School on Safety and Security Sciences, Óbuda University, 1034 Budapest, Hungary; chenhairong233@163.com
3   Faculty of Engineering, University of Szeged, 6724 Szeged, Hungary
4   Department of Material Science and Technology, Audi Hungaria Faculty of Automotive Engineering, Széchenyi István University, 9026 Győr, Hungary; gaolidong1997@hotmail.com
*   Correspondence: quanwenjing@nbu.edu.cn (W.Q.); guyaodong@nbu.edu.cn (Y.G.)

Abstract: Background: During city running or marathon races, shifts in level ground and up-and-down slopes are regularly encountered, resulting in changes in lower limb biomechanics. The longitudinal bending stiffness of the running shoe affects the running performance. Purpose: This research aimed to investigate the biomechanical changes in the lower limbs when transitioning from level ground to an uphill slope under different longitudinal bending stiffness (LBS) levels in running shoes. Methods: Fifteen male amateur runners were recruited and tested while wearing three different LBS running shoes. The participants were asked to pass the force platform with their right foot at a speed of 3.3 m/s ± 0.2. Kinematics data and GRFs were collected synchronously. Each participant completed and recorded ten successful experiments per pair of shoes. Results: The range of motion in the sagittal of the knee joint was reduced with the increase in the longitudinal bending stiffness. Positive work was increased in the sagittal plane of the ankle joint and reduced in the keen joint. The negative work of the knee joint increased in the sagittal plane. The positive work of the metatarsophalangeal joint in the sagittal plane increased. Conclusion: Transitioning from running on a level surface to running uphill, while wearing running shoes with high LBS, could lead to improved efficiency in lower limb function. However, the higher LBS of running shoes increases the energy absorption of the knee joint, potentially increasing the risk of knee injuries. Thus, amateurs should choose running shoes with optimal stiffness when running.

Keywords: level; uphill; running; kinematics; kinetics; longitudinal bending stiffness

Citation: Lu, R.; Chen, H.; Huang, J.; Ye, J.; Gao, L.; Liu, Q.; Quan, W.; Gu, Y. Biomechanical Investigation of Lower Limbs during Slope Transformation Running with Different Longitudinal Bending Stiffness Shoes. *Sensors* **2024**, *24*, 3902. https://doi.org/10.3390/s24123902

Academic Editor: Christian Peham

Received: 14 May 2024
Revised: 6 June 2024
Accepted: 12 June 2024
Published: 16 June 2024

1. Introduction

Recently, running has gained significant popularity as an accessible and inclusive sport. Running performance derives from a combination of anatomical, physiological, and behavioral traits that are uniquely evolved in humans [1]. As a result, research on the physiology and biomechanics of running never stops. This research has the potential to significantly improve human running performance. Most studies have focused on level running; however, in a marathon or city run, the route includes not just level ground but also uphill and downhill sections. For example, the Comrades Marathon, a world-famous ultra-marathon that has been held in South Africa since 1921, is a 90-km marathon with a course that incorporates a variety of road conditions (switching between uphill, level, and downhill). There was research discovered that uphill running leads to a decrease in both the maximal capacity to store and release elastic energy, in comparison to running on level ground. As a result, the body needs to generate additional mechanical energy to make up for the reduced elastic energy storage [1]. Much of this additional mechanical energy is generated by the hip [2]. During uphill walking, the ankle joint needs to release more energy to propel the body forward, while the absorption of energy by the ankle remains essentially constant [3]. For the knee joint, the angle of activity was greater when running uphill than when running on level ground [4]. In addition, when running downhill, the negative work absorbed by the knee increases, thereby increasing the risk of injury. In uphill running, the hip joint releases more energy and has an increased range of motion [5]. When jogging uphill at a consistent pace, the joints focus on releasing energy rather than using it up. It is vital to research how to release more energy to increase a runner's exercise performance.

The selection of appropriate running shoes is crucial for enhancing exercise performance and reducing the risk of sports-related injuries [6]. The longitudinal bending stiffness (LBS) of running shoes has been recognized as a key consideration in the development of performance footwear as part of the design to improve runner's exercise performance as well as minimize sports injuries [7,8]. The longitudinal bending stiffness of running shoes reduces the runner's walking frequency, prolongs the swing phase, and increases the peak vertical ground reaction and vertical impulse for each step [9]. Running economy is an important factor in determining long-distance running, and increasing the longitudinal bending stiffness of shoes could reduce the energy cost of running and effectively improve the running economy [10,11]. In addition, improving the LBS of a shoe has a significant impact on the ankle as well as the metatarsophalangeal joints. Properly fitted LBS running shoes play a key role in comfort as well as in the improvement of exercise performance [12]. Research has indicated that the metatarsophalangeal joints experience significant extension and make negative work while running, and this energy absorption has a negative impact on exercise performance. Increased LBS inhibits the extension of the metatarsophalangeal joint and reduces the time the supporting metatarsophalangeal joint goes from dorsiflexion to plantarflexion, thereby reducing the energy loss of the metatarsophalangeal joint during the running stance phase [13,14]. As Cigoja et al. found that increasing the LBS shortened the peak contraction velocity of the tendon unit of the calf muscles, these changes will result in a decrease in energy expenditure in the ankle plantarflexors and an increase in energy return to the Achilles tendon [15]. Additionally, an increase in LBS shifts the center of gravity of lower limb joint work from the knee to the metatarsophalangeal joints. Research has shown that as the LBS increases, there is a decrease in positive work at the knee joint and a significant increase in positive work at the metatarsophalangeal joints. This indicates a redistribution of work within the lower limb joints, with more work being redistributed from the knee to the ankle [16,17]. Furthermore, the increase in LBS results in a heightened rigidity of the shoe, providing enhanced support to the joints, exhibiting less mobility in the ankle and knee joints, and resulting in increased joint stability [18]. However, the transition from a level surface to an uphill slope in running shoes with different LBS has not been well studied.

Therefore, the research aimed to examine the variations in lower limb biomechanics during the transition from level to uphill running while wearing different LBS running shoes. The knee, ankle, and metatarsophalangeal joints of amateur runners transitioning from level to uphill running were analyzed in different LBS running shoes, helping to identify the most acceptable stiffness for them.

2. Materials and Methods

2.1. Participants

The sample size was calculated using G*Power 3.1 (Franz Faul, Germany) for univariate analysis of variance for detecting the number of groups = 1, number of measurements = 3, and power = 0.8 [19]. Based on these parameters, it was estimated that a minimum of 15 participants would be required for this research. A total of 15 male amateur runners were recruited (shoe size: 41, age: 22.5 ± 1.43, height: 175.3 ± 1.64 cm, weight: 65.25 ± 1.59 kg). Amateur runners are defined as running at least 3 times a week for 45 min or 10 km [20]. The participants' dominant leg is the right leg (the leg of the football goalkeeper). During the year prior to participation in the experiment, the subjects did not have severe lower limb injuries and surgeries, and before the experiment began, participants were asked to avoid any intense physical activity for 48 h to eliminate the possibility of fatigue. A healthy diet and full rest were required before the experiment. Alcohol or caffeine of any kind was also prohibited within 24 h of the start of the experiment [21]. All participants had to sign a written informed consent. The experiment was approved by the Ethics Committee of the University of Ningbo (2024RBG3272).

2.2. Experimental Processes

Prior to the official start of the experiment, participants were asked to warm up by walking at 2.2 m/s for 1 min on the treadmill [22]. We uniformly provided the participants with clothing and shoes to avoid experimental differences and maintain consistency. In the official experiment, the participants were asked to apply 38 reflective markers (diameter: 14 mm) [23]. The specific points are shown in Figure 1a. The participants were asked to step with their right foot on the force platform with both eyes looking forward until the complete static coordinates were captured. During the test, the participants were wearing the supplied clothing as well as shoes. In this research, based on the longitudinal bending stiffness selected for shoes in previous research (S-1: 5.0 Nm/rad, S-2: 6.3 Nm/rad, S-3: 8.6 Nm/rad), except for the difference in longitudinal bending stiffness, the other mechanical properties of the three pairs of running shoes were the same [13,24,25]. The LBS values of the shoes were measured by a rotational axis material-testing machine (Instron ElectroPuls E1000, Norwood, MA, USA). Then the participants passed the slope at a speed of 3.3 m/s $\pm$ 0.2 m/s [26]. The shoes are shown in Figure 1b. The longitudinal bending stiffness of the running shoes tested in this study was modeled on Willwacher's study, which was closer to previous studies and had not been tested in previous studies of transitioning from level to uphill running. The participants were asked to run with their right leg stepping on the force platform every time they switched from the plane to the uphill (Figure 1c). Each pair of shoes was tested 10 times and valid data were obtained. After each pair of shoes was tested, the participants rested for five minutes to prevent fatigue. The running data from the standing phase of an amateur runner were captured using the Vicon motion capture system (Vicon Metrics Ltd., Oxford, UK). Vertical ground reaction forces were used to identify the amateur runner's toe contact to toe-off. To control experimental variables, we used a Brauer timing light (Brower Timing System, Draper, UT, USA) to control running speed.

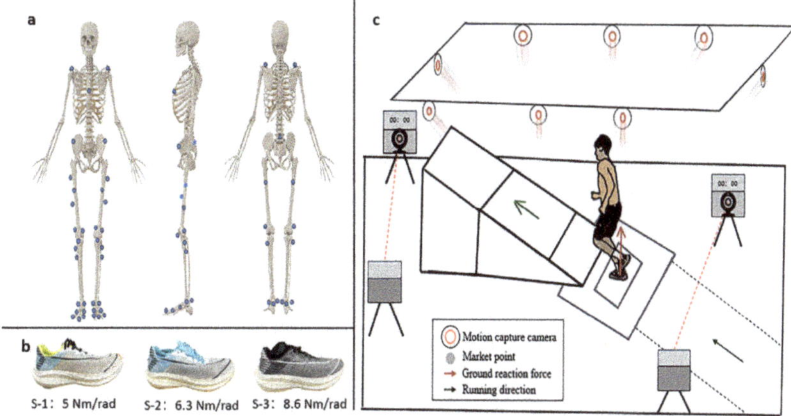

Figure 1. (**a**) The front, side, and back positions of markers. Blue dots: markers. (**b**) Three pairs of experimental shoes. (**c**) Display of experiment design for collecting the biomechanical data during the running stance phase.

2.3. Data Collection and Processing

This research analyzed the biomechanical properties of amateur runners' limbs during the transition from running on a level surface to running uphill. We converted the C3D file data exported by Vicon into '.trc' and '.mot' files. Then it was imported into OpenSim for the next step [27]. A musculoskeletal model from the OpenSim website (Gait 2392) was used [28,29]. The models in OpenSim were scaled to utilize the participant's marker point location and weights. Until the error value between the experimental and virtual markers was less than 0.02, the scaled model was applied to the data calculation. The inversion kinematics (IK) calculation tool was used in OpenSim to calculate the joint angle and optimize the results using the minimum binary method to minimize the error between experimental and virtual markings. Inversion Dynamics (ID) calculates the net moment of the knee, ankle, and thigh joints. The ID tool performs inversion dynamic analysis by applying the data of the kinematics of a given model description and the partial kinetics that may be applied to the model. The ID tool deals with the equations of force and acceleration in classical mechanics in the inversion dynamic sense, obtaining the net moment and torque of each joint that produces motion, from the data of IK and ID, the joint power and work performed are then calculated [30].

2.4. Statistical Analysis

The knee, ankle, metatarsophalangeal joint angles, peak moment, power, and work on the part of the plane were analyzed in SPSS 26.0 (IBM, Albany, NY, USA). The data obtained from the experiment used mean $\pm$ standard differential representation [31]. First, tests for normality and homogeneity of variances (Shapiro–Wilk and Levene's, respectively) were conducted on all data before the analysis. A one-way repeated measure ANOVA was utilized to analyze the impact of running in running shoes with different LBS on lower limb biomechanics, in accordance with tests for normality and homogeneity of variance. The significance level was set to $p < 0.05$. Post hoc tests were compared using the Bonferroni method to determine which of the two LBS running shoe conditions had significant differences in the range of motion, peak moments, peak positive and negative power, and positive and negative work at the hip, knee, ankle, and metatarsophalangeal joints based on p-values. The p-values require 3 comparisons, and the probability of 0.05 is divided by the number of comparisons to be made, 3, so the level of significance is set at $\alpha = 0.017$, and a difference of $p < 0.017$ is considered statistically significant.

3. Results

3.1. Kinematics

The changes in knee, ankle, and metatarsophalangeal joint angles during the stance phase of running in amateur runners in three different LBS running shoe conditions are shown in Figure 2; the range of joint motion is shown in Table 1. In three different LBS running shoe conditions, significant differences were seen in the motion angle of the knee adduction/abduction, as well as the metatarsophalangeal joint plantarflexion/dorsiflexion. As the LBS increased, joint angles decreased. There was no significant difference in the motion angle of the different planes of the other joints.

Table 1. Motion range, peak moment, peak positive, and negative power and positive and negative work of knee, ankle, and thigh joints in three running shoe conditions.

Index		Joint Motion	S-1	S-2	S-3	F	p
Motion angle (°)	Knee	Flexion/Extension	28.05 $\pm$ 5.41	28.59 $\pm$ 4.86	25.48 $\pm$ 5.41	2.259	0.112
		Adduction/Abduction	4.34 $\pm$ 1.24 [b]	4.92 $\pm$ 1.25 [c]	3.57 $\pm$ 0.93 [c]	7.779	0.001
		Intorsion/Extorsion	14.34 $\pm$ 6.69	14.25 $\pm$ 7.37	13.58 $\pm$ 7.34	0.075	0.927
	Ankle	Plantarflexion/Dorsiflexion	31.68 $\pm$ 8.60	32.50 $\pm$ 9.15	29.65 $\pm$ 9.14	0.677	0.511
		Inversion/Eversion	17.52 $\pm$ 4.04	18.24 $\pm$ 4.70	16.89 $\pm$ 4.33	0.556	0.576
	MTP	Plantarflexion/Dorsiflexion	13.03 $\pm$ 4.25 [ab]	11.84 $\pm$ 3.30 [ac]	8.56 $\pm$ 1.48 [bc]	11.156	0.001

Table 1. *Cont.*

Index		Joint Motion	S-1	S-2	S-3	F	*p*
Peak moment (Nm/kg)	Knee	Flexion/Extension	3.51 ± 0.99 [b]	3.33 ± 1.17 [c]	2.24 ± 0.56 [bc]	8.632	0.001
		Adduction/Abduction	-2.59 ± 1.21 [ab]	-1.43 ± 0.31 [a]	-1.85 ± 0.94 [b]	7.362	0.001
		Intorsion/Extorsion	-1.05 ± 0.57	-0.86 ± 0.86	-0.59 ± 0.26	2.604	0.082
	Ankle	Plantarflexion/Dorsiflexion	-3.22 ± 1.81	-3.46 ± 0.64	-2.79 ± 0.66	1.901	0.158
		Inversion/Eversion	2.16 ± 1.04 [b]	1.72 ± 0.68	1.52 ± 0.34 [b]	4.1	0.021
	MTP	Plantarflexion/Dorsiflexion	0.56 ± 0.13	0.55 ± 0.18	0.51 ± 0.15	0.44	0.646
Peak positive power (W/kg)	Knee	Flexion/Extension	1.68 ± 0.33 [ab]	1.37 ± 0.22 [a]	1.17 ± 0.20 [b]	8.868	0.001
		Adduction/Abduction	1.92 ± 0.70 [ab]	1.37 ± 0.52 [a]	1.26 ± 0.32 [b]	5.608	0.007
		Intorsion/Extorsion	0.15 ± 0.09	0.13 ± 0.11	0.11 ± 0.15	0.668	0.517
	Ankle	Plantarflexion/Dorsiflexion	6.89 ± 0.70 [b]	6.20 ± 1.01 [c]	4.79 ± 0.85 [bc]	13.92	0.001
		Inversion/Eversion	0.54 ± 0.11	0.52 ± 0.15	0.49 ± 0.12	0.355	0.704
	MTP	Plantarflexion/Dorsiflexion	1.65 ± 0.74	2.04 ± 0.96	2.16 ± 0.85	1.579	0.217
Peak negative power (W/kg)	Knee	Flexion/Extension	-2.14 ± 0.59	-2.20 ± 0.38	-2.13 ± 0.4	0.071	0.932
		Adduction/Abduction	-1.76 ± 0.50 [ab]	-0.76 ± 0.55 [a]	-1.17 ± 0.70 [b]	9.562	0.001
		Intorsion/Extorsion	-0.22 ± 0.25	-0.13 ± 0.07	-0.14 ± 0.06	1.575	0.217
	Ankle	Plantarflexion/Dorsiflexion	-4.30 ± 0.93 [ab]	-3.36 ± 0.74 [a]	-3.07 ± 0.58 [b]	6.284	0.006
		Inversion/Eversion	-0.44 ± 0.12	-0.45 ± 0.35	-0.46 ± 0.11	0.014	0.986
	MTP	Plantarflexion/Dorsiflexion	-1.80 ± 0.89	-1.52 ± 0.61	-1.79 ± 0.85	0.656	0.524
Positive work (J/kg)	Knee	Flexion/Extension	0.24 ± 0.13	0.32 ± 0.12	0.29 ± 0.08	2.387	0.102
		Adduction/Abduction	0.12 ± 0.09 [b]	0.16 ± 0.09 [c]	0.06 ± 0.01 [bc]	9.469	0.001
		Intorsion/Extorsion	0.09 ± 0.08	0.06 ± 0.06	0.07 ± 0.10	0.879	0.42
	Ankle	Plantarflexion/Dorsiflexion	0.54 ± 0.36	0.41 ± 0.09 [c]	0.64 ± 0.23 [c]	3.409	0.041
		Inversion/Eversion	0.31 ± 0.16 [a]	0.20 ± 0.07 [a]	0.23 ± 0.07	4.598	0.015
	MTP	Plantarflexion/Dorsiflexion	0.05 ± 0.02 [b]	0.07 ± 0.05	0.09 ± 0.04 [b]	5.005	0.009
Negative work (J/kg)	Knee	Flexion/Extension	-0.24 ± 0.13 [b]	-0.34 ± 0.15	-0.42 ± 0.11 [b]	7.848	0.001
		Adduction/Abduction	-0.13 ± 0.10	-0.09 ± 0.09	-0.12 ± 0.19	0.491	0.614
		Intorsion/Extorsion	-0.07 ± 0.06 [b]	-0.09 ± 0.02 [c]	-0.03 ± 0.02 [bc]	14.783	0.001
	Ankle	Plantarflexion/Dorsiflexion	-0.28 ± 0.23	-0.21 ± 0.05 [c]	-0.37 ± 0.10 [c]	4.726	0.013
		Inversion/Eversion	-0.28 ± 0.13	-0.25 ± 0.17	-0.17 ± 0.06	2.881	0.065
	MTP	Plantarflexion/Dorsiflexion	-0.09 ± 0.08	-0.07 ± 0.07	-0.06 ± 0.04	1.283	0.283

Notes: (a) represents a statistically significant difference between S-1 shoe and S-2 shoe data; (b) represents a statistically significant difference between S-1 shoe and S-3 shoe data; (c) represents a statistically significant difference between S-2 shoe and S-3 shoe data.

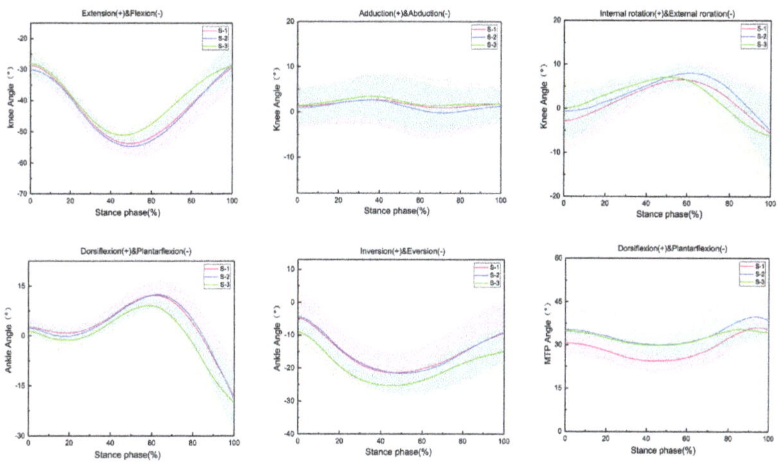

Figure 2. Changes in stance phase and knee, ankle, and metatarsophalangeal joint angles in three running shoe conditions.

3.2. Kinetics

The changes in the knee, ankle, and metatarsophalangeal joint peak moment during the stance phase of running in amateur runners in three different LBS running shoe conditions are shown in Figure 3; the peak joint moment is shown in Table 1. In three different LBS running shoe conditions, significant differences were seen in the peak moments of knee flexion/extension, adduction/abduction, and ankle inversion/eversion. As the LBS increased, the peak joint moment decreased. There was no significant difference in the peak moment of the different surfaces of the other joints.

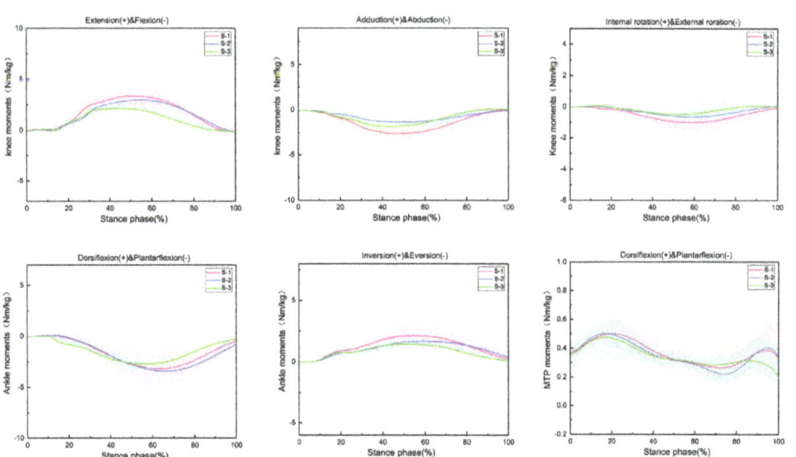

Figure 3. Changes in knee, ankle, and metatarsophalangeal joint moments during the support period in the three running shoe conditions.

The changes in knee, ankle, and metatarsophalangeal joint peak positive and negative power during the stance phase of running in amateur runners in three different LBS running shoe conditions are shown in Table 1. In three different LBS running shoe conditions, significant differences were seen in the peak positive power of knee flexion/extension, adduction/abduction, and ankle plantarflexion/dorsiflexion, and significant differences were seen in the peak negative power of knee adduction/abduction as well as ankle plantarflexion/dorsiflexion. The peak positive and negative power of joints decreased with increasing LBS. There was no significant difference in the peak positive and negative power of the different surfaces of the other joints.

The changes in knee, ankle, and metatarsophalangeal joint positive and negative work during the stance phase of running in amateur runners in three different LBS running shoe conditions are shown in Table 1. In three different LBS running shoe conditions, significant differences were seen in the positive work of knee adduction/abduction, plantarflexion/dorsiflexion with the ankle and metatarsophalangeal joint, and inversion/eversion with the ankle, and significant differences were seen in the negative work of knee flexion/extension, internal-external, and ankle plantarflexion/dorsiflexion. As the LBS increased, the positive work of knee adduction/abduction and ankle inversion/eversion decreased, while the plantarflexion/dorsiflexion of the ankle and metatarsophalangeal joint increased, the negative work of knee flexion/extension and ankle plantarflexion/dorsiflexion increased, and the internal-external of the knee decreased. There was no significant difference in the positive and negative work of the different surfaces of the other joints.

4. Discussion

This research aims to investigate the kinematics and kinetics of the knee, ankle, and metatarsophalangeal joints in amateur runners transitioning from level to uphill running while wearing different LBS running shoes. When transitioning from level to uphill running, our research observed that, as the LBS increased, there was a decrease in the range of motion and positive work of the knee, an increase in the positive work of the ankle, and a decrease in the range of motion and negative work of the metatarsophalangeal joint, together with an increase in positive work.

From a kinematic point of view, there was a significant difference in the knee joint angle when transitioning from level to uphill running in different LBS running shoes, and the S-2 shoe showed a higher range of knee adduction/abduction compared to the S-3 shoe. Thorsten Sterzing's research found that the characteristics of running shoes affect the knee and ankle range of motion during running, which impacts the control of knee and ankle stabilization [18]; a decreased knee and ankle range of motion indicates that higher LBS improves knee stability in amateur runners transitioning from level to uphill running. Moreover, the rise in LBS enhances the shoe's stiffness, resulting in improved foot support. Shoes with low stiffness offer less foot support than shoes with high stiffness, leading to amateur runners having difficulty controlling the stability of the knee joint. Additionally, significant differences in the range of motion of the metatarsophalangeal joints were found, and the range of motion of the metatarsophalangeal joints of amateur runners decreases as the LBS increases [32]. Shoes with higher LBS will limit the range of motion of the joint. Hoogkamer et al. discovered that high-LBS shoes decreased peak metatarsophalangeal joint dorsiflexion angles by 6° and 12° [13]. Stefanyshyn et al.'s research identified restricted activity in the plantarflexion/dorsiflexion of the metatarsophalangeal joint caused by heightened LBS in running shoes [33,34]. Not only that, shoes with a higher LBS increase the dorsiflexion of the metatarsophalangeal joint, while at the same time enhancing the foot's ability to rotate outward as a potential compensatory strategy [35].

In terms of moment, this research found that the peak moments of knee flexion/ extension, adduction/abduction, and ankle inversion/eversion were significantly different between S-1 and S-3 shoes in running shoes with different LBS transitions from level to uphill and that an increase in LBS led to a decrease in the peak moments at the knee and ankle joints. Zhou and Debelle's research demonstrated an increased moment at the ankle joint and improved strength of the peroneus longus and peroneus brevis muscles when runners wore unstable shoes [36,37]. Amateur runners experience decreased ankle stabilization due to the inadequate support provided by low-stiffness running shoes. To ensure ankle stabilization while running, the muscles responsible for ankle motion must generate greater forces in the frontal plane to meet the demands for ankle stabilization. Wearing a running shoe with low stiffness has been shown to result in higher peak moments in ankle inversion and eversion. Willwacher also demonstrated a decrease in the mean ankle moment when running in shoes with medium stiffness compared to lower stiffness shoes [38]. Therefore, there was a higher possibility of injury when running in low-stiffness shoes. Researchers have usually assumed that there would be an increase in peak moments in the plantarflexion/dorsiflexion of the ankle. Only Roy and Stefanyshyn, however, have shown that there is an increase in the peak ankle moment as the LBS increases while jogging at a constant speed on a treadmill inclined at 1% [39]. Cigoja's research demonstrated that wearing high-LBS shoes led to the metatarsophalangeal joints transitioning into the plantarflexion phase sooner. In amateur runners, the decrease in negative work generation at the time of metatarsophalangeal joints was accompanied by a corresponding increase in positive work, although this change did not significantly change the moment reduction [40]. This is consistent with our experimental results.

Hoogkamer's research also found an increase in positive metatarsophalangeal joint plantarflexion/dorsiflexion work during the stance phase of running as the LBS of the running shoe increased [13]. From an energy loss perspective, increasing the LBS of a shoe stiffens the shoe, limiting the dorsiflexion of the metatarsophalangeal joints, slowing angular velocity, and decreasing energy loss at the metatarsophalangeal joints [41]. Therefore, wearing higher LBS running shoes can increase the exercise performance of amateur runners. In addition, the research found that, as the LBS increased, peak positive power and positive work of the knee in adduction/abduction decreased, but the positive work of the ankle in plantarflexion/dorsiflexion increased. Cigoja found that as LBS increased, there was a decrease in positive work at the knee joint. Additionally, the distribution of positive work shifts from the knee joint to the ankle joint in the experiment [34]. The muscle tendons at the proximal joints have lower energy storage and return capabilities compared to those at the distal joints, resulting in higher energy expenditure to produce the same amount of work [42,43]. Therefore, reallocating the workload among the lower limb joints benefits amateur runners in terms of enhancing their performance. Furthermore, the research found significant differences in ankle plantarflexion/dorsiflexion as well as knee flexion and extension negative work. In uphill running, negative work was mainly absorbed by the knee [44]. Willwacher found that the amount of negative work absorbed by the knee and ankle increased as the LBS increased when an amateur runner ran uphill with three different LBS shoes [45]. Thus, greater energy absorption is required at the knee joint to provide cushioning during the stance phase while running in high-LBS running shoes.

In summary, in this experiment, the biomechanical data of the lower limbs of amateur runners wearing three pairs of running shoes with different longitudinal bending stiffness were compared, and the shoe with the highest longitudinal bending stiffness in the sole was more suitable for amateur runners transitioning from level to uphill running. Increasing the LBS of running shoes can increase the running performance of amateur runners, promote safety, and decrease the risk of joint problems. However, excessive LBS in running shoes is not ideal for amateur runners. If the longitudinal bending stiffness of shoes is excessive, amateur runners may be hindered in terms of exercise performance [46]. Selecting the appropriate LBS running shoe based on amateur runner conditions and the level of running intensity was crucial to prevent avoidable injuries.

5. Conclusions

Enhancing the LBS of running shoes can impact the biomechanics of the lower limb when transitioning from level to uphill running. The phenomenon of positive work shifting from proximal to distal joints in the lower extremity joints is observed, which would be beneficial in improving the efficient work performed in the lower limb. Amateur runners must absorb additional energy for cushioning in the knee joint when wearing high-LBS running shoes, which raises the risk of knee injury. In high-LBS running shoes, the metatarsophalangeal joint does more positive work and reduces negative work absorption. The angle of activity of the joints was reduced when wearing high longitudinal bending stiffness running shoes, and amateur runners were able to better control the stability of their joints. Therefore, high longitudinal bending stiffness running shoes were more suitable for amateur athletes to transition from level to uphill running. Furthermore, amateur runners could use these research findings to receive valuable suggestions when choosing running shoes with optimal LBS.

Author Contributions: Conceptualization, R.L.; methodology, R.L.; software, H.C.; validation, H.C. and J.H.; investigation, J.Y.; resources, L.G.; data curation, Q.L.; writing—original draft preparation, R.L.; writing—review and editing, W.Q. and Y.G.; visualization, W.Q.; supervision, Y.G. All authors have read and agreed to the published version of the manuscript.

Funding: This study was sponsored by the Zhejiang Provincial Natural Science Foundation of China for Distinguished Young Scholars (LR22A020002), Zhejiang Provincial Key Research and Development Program of China (2021C03130), Zhejiang Provincial Natural Science Foundation (LTGY23H040003), Ningbo key R&D Program (2022Z196), Research Academy of Medicine Combining Sports, Ningbo (No.2023001), the Project of NINGBO Leading Medical and Health Discipline (No.2022-F15, No.2022-F22), Ningbo Natural Science Foundation (20221JCGY010532, 20221JCGY010607), Public Welfare Science and Technology Project of Ningbo, China (2021S134), and K. C. Wong Magna Fund in Ningbo University, Zhejiang Rehabilitation Medical Association Scientific Research Special Fund (ZKKY2023001).

Institutional Review Board Statement: The study was conducted according to the guidelines of the Declaration of Helsinki and approved by the Ethics Committee of Ningbo university (2024RBG3272).

Informed Consent Statement: Informed consent was obtained from all participants involved in study.

Data Availability Statement: The data that support the findings of this study are available on reasonable request from the corresponding author. The data are not publicly available due to privacy or ethical restrictions.

Conflicts of Interest: The authors declare no conflicts of interest.

References

1. Mattson, M.P. Evolutionary aspects of human exercise—Born to run purposefully. *Ageing Res. Rev.* **2012**, *11*, 347–352. [CrossRef]
2. Snyder, K.L.; Kram, R.; Gottschall, J.S. The role of elastic energy storage and recovery in downhill and uphill running. *J. Exp. Biol.* **2012**, *215*, 2283–2287. [CrossRef] [PubMed]
3. Montgomery, J.R.; Grabowski, A.M. The contributions of ankle, knee and hip joint work to individual leg work change during uphill and downhill walking over a range of speeds. *R. Soc. Open Sci.* **2018**, *5*, 180550. [CrossRef] [PubMed]
4. Okudaira, M.; Willwacher, S.; Kuki, S.; Yamada, K.; Yoshida, T.; Tanigawa, S. Three-dimensional CoM energetics, pelvis and lower limbs joint kinematics of uphill treadmill running at high speed. *J. Sports Sci.* **2020**, *38*, 518–527. [CrossRef] [PubMed]
5. Vernillo, G.; Giandolini, M.; Edwards, W.B.; Morin, J.-B.; Samozino, P.; Horvais, N.; Millet, G.Y. Biomechanics and physiology of uphill and downhill running. *Sports Med.* **2017**, *47*, 615–629. [CrossRef]
6. Tenforde, A.; Hoenig, T.; Saxena, A.; Hollander, K. Bone Stress Injuries in Runners Using Carbon Fiber Plate Footwear. *Sports Med.* **2023**, *53*, 1499–1505. [CrossRef]
7. Park, S.-K.; Lam, W.-K.; Yoon, S.; Lee, K.-K.; Ryu, J. Effects of forefoot bending stiffness of badminton shoes on agility, comfort perception and lower leg kinematics during typical badminton movements. *Sports Biomech.* **2017**, *16*, 374–386. [CrossRef]
8. Stefanyshyn, D.J.; Wannop, J.W. The influence of forefoot bending stiffness of footwear on athletic injury and performance. *Footwear Sci.* **2016**, *8*, 51–63. [CrossRef]
9. Roberts, T.J.; Belliveau, R.A. Sources of mechanical power for uphill running in humans. *J. Exp. Biol.* **2005**, *208*, 1963–1970. [CrossRef]
10. Hoogkamer, W.; Kram, R.; Arellano, C.J. How Biomechanical Improvements in Running Economy Could Break the 2-hour Marathon Barrier. *Sports Med.* **2017**, *47*, 1739–1750. [CrossRef]
11. Rodrigo-Carranza, V.; González-Mohíno, F.; Santos-Concejero, J.; González-Ravé, J.M. The effects of footwear midsole longitudinal bending stiffness on running economy and ground contact biomechanics: A systematic review and meta-analysis. *Eur. J. Sport Sci.* **2022**, *22*, 1508–1521. [CrossRef] [PubMed]
12. Jiang, C. The effect of basketball shoe collar on ankle stability: A systematic review and meta-analysis. *Phys. Act. Health* **2020**, *4*, 11–18. [CrossRef]
13. Hoogkamer, W.; Kipp, S.; Kram, R. The Biomechanics of Competitive Male Runners in Three Marathon Racing Shoes: A Randomized Crossover Study. *Sports Med.* **2019**, *49*, 133–143. [CrossRef] [PubMed]
14. Cigoja, S.; Asmussen, M.J.; Firminger, C.R.; Fletcher, J.R.; Edwards, W.B.; Nigg, B.M. The Effects of Increased Midsole Bending Stiffness of Sport Shoes on Muscle-Tendon Unit Shortening and Shortening Velocity: A Randomised Crossover Trial in Recreational Male Runners. *Sports Med. Open* **2020**, *6*, 9. [CrossRef] [PubMed]
15. Cigoja, S.; Fletcher, J.R.; Esposito, M.; Stefanyshyn, D.J.; Nigg, B.M. Increasing the midsole bending stiffness of shoes alters gastrocnemius medialis muscle function during running. *Sci. Rep.* **2021**, *11*, 749. [CrossRef]
16. Ortega, J.A.; Healey, L.A.; Swinnen, W.; Hoogkamer, W. Energetics and Biomechanics of Running Footwear with Increased Longitudinal Bending Stiffness: A Narrative Review. *Sports Med.* **2021**, *51*, 873–894. [CrossRef] [PubMed]
17. Cigoja, S.; Fletcher, J.R.; Nigg, B.M. Can changes in midsole bending stiffness of shoes affect the onset of joint work redistribution during a prolonged run? *J. Sport Health Sci.* **2022**, *11*, 293–302. [CrossRef]
18. Sterzing, T.; Thomsen, K.; Ding, R.; Cheung, J.T.-M. Running shoe crash-pad design alters shoe touchdown angles and ankle stability parameters during heel–toe running. *Footwear Sci.* **2015**, *7*, 81–93. [CrossRef]
19. Bartlett, J. Introduction to sample size calculation using G* Power. *Eur. J. Soc. Psychol.* **2019**. Available online: https://www.studocu.com/row/document/sukkur-institute-of-business-administration/business-research/g-power-guide/12418441 (accessed on 20 May 2024).
20. Hespanhol Junior, L.C.; Pena Costa, L.O.; Lopes, A.D. Previous injuries and some training characteristics predict running-related injuries in recreational runners: A prospective cohort study. *J. Physiother.* **2013**, *59*, 263–269. [CrossRef]
21. Gao, L.; Ye, J.; Bálint, K.; Radak, Z.; Mao, Z.; Gu, Y. Biomechanical effects of exercise fatigue on the lower limbs of men during the forward lunge. *Front. Physiol.* **2023**, *14*, 1182833. [CrossRef] [PubMed]
22. Chen, H.; Song, Y.; Liu, Q.; Ren, F.; Bíró, I.; Gu, Y. Gender Effects on Lower Limb Biomechanics of Novice Runners before and after a 5 km Run. *J. Men's Health* **2022**, *18*, 176. [CrossRef]
23. Cen, X.; Lu, Z.; Baker, J.S.; István, B.; Gu, Y. A comparative biomechanical analysis during planned and unplanned gait termination in individuals with different arch Stiffnesses. *J. Men's Health* **2021**, *11*, 1871. [CrossRef]

24. Willwacher, S.; Kurz, M.; Menne, C.; Schrödter, E.; Brüggemann, G.-P. Biomechanical response to altered footwear longitudinal bending stiffness in the early acceleration phase of sprinting. *Footwear Sci.* **2016**, *8*, 99–108. [CrossRef]
25. Day, E.; Hahn, M. Optimal footwear longitudinal bending stiffness to improve running economy is speed dependent. *Footwear Sci.* **2019**, *12*, 3–13. [CrossRef]
26. Jiang, X.; Yang, X.; Zhou, H.; Baker, J.S.; Gu, Y. Prolonged Running Using Bionic Footwear Influences Lower Limb Biomechanics. *Healthcare* **2021**, *9*, 236. [CrossRef] [PubMed]
27. Zhou, H.; Xu, D.; Quan, W.; Liang, M.; Ugbolue, U.C.; Baker, J.S.; Gu, Y. A pilot study of muscle force between normal shoes and bionic shoes during men walking and running stance phase using opensim. *Actuators* **2021**, *10*, 274. [CrossRef]
28. Mei, Q.; Fernandez, J.; Xiang, L.; Gao, Z.; Yu, P.; Baker, J.S.; Gu, Y. Dataset of lower extremity joint angles, moments and forces in distance running. *Heliyon* **2022**, *8*, e11517. [CrossRef]
29. Rajagopal, A.; Dembia, C.L.; DeMers, M.S.; Delp, D.D.; Hicks, J.L.; Delp, S.L. Full-body musculoskeletal model for muscle-driven simulation of human gait. *IEEE Trans. Biomed. Eng.* **2016**, *63*, 2068–2079. [CrossRef]
30. Butler, A.B.; Caruntu, D.I.; Freeman, R.A. Knee joint biomechanics for various ambulatory exercises using inverse dynamics in OpenSim. In Proceedings of the ASME 2017 International Mechanical Engineering Congress and Exposition, Tampa, FL, USA, 3–9 November 2017; p. V003T004A045.
31. Lee, D.K.; In, J.; Lee, S. Standard deviation and standard error of the mean. *Korean J. Anesthesiol.* **2015**, *68*, 220–223. [CrossRef]
32. Chen, H.; Shao, E.; Sun, D.; Xuan, R.; Baker, J.S.; Gu, Y. Effects of footwear with different longitudinal bending stiffness on biomechanical characteristics and muscular mechanics of lower limbs in adolescent runners. *Front. Physiol.* **2022**, *13*, 907016. [CrossRef] [PubMed]
33. Madden, R.; Sakaguchi, M.; Wannop, J.; Stefanyshyn, D. Forefoot bending stiffness, running economy and kinematics during overground running. *Footwear Sci.* **2015**, *7*, S11–S13. [CrossRef]
34. Flores, N.; Rao, G.; Berton, E.; Delattre, N. The stiff plate location into the shoe influences the running biomechanics. *Sports Biomech.* **2021**, *20*, 815–830. [CrossRef] [PubMed]
35. Ortega, J.A. The Effects of Footwear Longitudinal Bending Stiffness on the Energetics and Biomechanics of Uphill Running. Master's Thesis, University of Massachusetts Amherst, Amherst, MA, USA, 2022. [CrossRef]
36. Debelle, H.; Maganaris, C.N.; O'Brien, T.D. Role of Knee and Ankle Extensors' Muscle-Tendon Properties in Dynamic Balance Recovery from a Simulated Slip. *Sensors* **2022**, *22*, 3483. [CrossRef]
37. Huiyu, Z.; Datao, X.; Wenjing, Q.; Ugbolue, U.C.; Sculthorpe, N.F.; Baker, J.S.; Yaodong, G. A foot joint and muscle force assessment of the running stance phase whilst wearing normal shoes and bionic shoes. *Acta Bioeng. Biomech.* **2022**, *24*. [CrossRef]
38. Willwacher, S.; Konig, M.; Braunstein, B.; Goldmann, J.P.; Bruggemann, G.P. The gearing function of running shoe longitudinal bending stiffness. *Gait Posture* **2014**, *40*, 386–390. [CrossRef]
39. Roy, J.P.; Stefanyshyn, D.J. Shoe midsole longitudinal bending stiffness and running economy, joint energy, and EMG. *Med. Sci. Sports Exerc.* **2006**, *38*, 562–569. [CrossRef]
40. Cigoja, S.; Firminger, C.R.; Asmussen, M.J.; Fletcher, J.R.; Edwards, W.B.; Nigg, B.M. Does increased midsole bending stiffness of sport shoes redistribute lower limb joint work during running? *J. Sci. Med. Sport* **2019**, *22*, 1272–1277. [CrossRef]
41. Riiser, K.; Ommundsen, Y.; Småstuen, M.C.; Løndal, K.; Misvær, N.; Helseth, S. The relationship between fitness and health-related quality of life and the mediating role of self-determined motivation in overweight adolescents. *Scand. J. Public Health* **2014**, *42*, 766–772. [CrossRef]
42. Biewener, A.A.; Roberts, T.J. Muscle and tendon contributions to force, work, and elastic energy savings: A comparative perspective. *Exerc. Sport Sci. Rev.* **2000**, *28*, 99–107.
43. Alexander, R.M. Tendon elasticity and muscle function. *Comp. Biochem. Physiol. Part A Mol. Integr. Physiol.* **2002**, *133*, 1001–1011. [CrossRef] [PubMed]
44. Nuckols, R.W.; Takahashi, K.Z.; Farris, D.J.; Mizrachi, S.; Riemer, R.; Sawicki, G.S. Mechanics of walking and running up and downhill: A joint-level perspective to guide design of lower-limb exoskeletons. *PLoS ONE* **2020**, *15*, e0231996. [CrossRef] [PubMed]
45. Willwacher, S.; Lichtwark, G.; Cresswell, A.G.; Kelly, L.A. Effects of midsole bending stiffness and shape on lower extremity joint work per distance in level, incline and decline running. *Footwear Sci.* **2021**, *13*, S41–S42. [CrossRef]
46. Flores, N.; Rao, G.; Berton, E.; Delattre, N. Increasing the longitudinal bending stiffness of runners' habitual shoes: An appropriate choice for improving running performance? *Proc. Inst. Mech. Eng. Part P J. Sports Eng. Technol.* **2021**, *237*, 121–133. [CrossRef]

 *sensors*

Article

Comparison of Metabolic Power and Energy Cost of Submaximal and Sprint Running Efforts Using Different Methods in Elite Youth Soccer Players: A Novel Energetic Approach

Gabriele Grassadonia [1,2,3,4], Pedro E. Alcaraz [1,5,6] and Tomás T. Freitas [1,5,6,7,*]

[1] UCAM Research Center for High Performance Sport, Universidad Católica de Murcia (UCAM), 30107 Murcia, Spain; gabriele.grassadonia@gmail.com (G.G.); palcaraz@ucam.edu (P.E.A.)
[2] UPSS—International Department of Motor Arts, Popular University of Sport Sciences, 00122 Rome, Italy
[3] UPM—Department of Medical Sciences, Popular University of Milan, 20122 Milan, Italy
[4] MIU—Department of Sport Sciences, Miami International University, Miami, FL 33131, USA
[5] Faculty of Sport, Universidad Católica de Murcia (UCAM), 30107 Murcia, Spain
[6] Strength and Conditioning Society, 30008 Murcia, Spain
[7] NAR—Nucleus of High Performance in Sport, São Paulo 04753-060, Brazil
* Correspondence: tfreitas@ucam.edu

Abstract: Sprinting is a decisive action in soccer that is considerably taxing from a neuromuscular and energetic perspective. This study compared different calculation methods for the metabolic power (MP) and energy cost (EC) of sprinting using global positioning system (GPS) metrics and electromyography (EMG), with the aim of identifying potential differences in performance markers. Sixteen elite U17 male soccer players (age: 16.4 ± 0.5 years; body mass: 64.6 ± 4.4 kg; and height: 177.4 ± 4.3 cm) participated in the study and completed four different submaximal constant running efforts followed by sprinting actions while using portable GPS-IMU units and surface EMG. GPS-derived MP was determined based on GPS velocity, and the EMG-MP and EC were calculated based on individual profiles plotting the MP of the GPS and all EMG signals acquired. The goodness of fit of the linear regressions was assessed by the coefficient of determination (R^2), and a repeated measures ANOVA was used to detect changes. A linear trend was found in EMG activity during submaximal speed runs ($R^2 = 1$), but when the sprint effort was considered, the trend became exponential ($R^2 = 0.89$). The EMG/force ratio displayed two different trends: linear up to a 30 m sprint ($R^2 = 0.99$) and polynomial up to a 50 m sprint ($R^2 = 0.96$). Statistically significant differences between the GPS and EMG were observed for MP splits at 0–5 m, 5–10 m, 25–30 m, 30–35 m, and 35–40 m and for EC splits at 5–10 m, 25–30 m, 30–35 m, and 35–40 m ($p \leq 0.05$). Therefore, the determination of the MP and EC based on GPS technology underestimated the neuromuscular and metabolic engagement during the sprinting efforts. Thus, the EMG-derived method seems to be more accurate for calculating the MP and EC in this type of action.

Keywords: football; maximum velocity; maximal running; GPS; EMG/force ratio

Citation: Grassadonia, G.; Alcaraz, P.E.; Freitas, T.T. Comparison of Metabolic Power and Energy Cost of Submaximal and Sprint Running Efforts Using Different Methods in Elite Youth Soccer Players: A Novel Energetic Approach. *Sensors* **2024**, *24*, 2577. https://doi.org/10.3390/s24082577

Academic Editor: Georg Fischer

Received: 18 March 2024
Revised: 12 April 2024
Accepted: 15 April 2024
Published: 17 April 2024

1. Introduction

The energy cost (EC) and kinematics of various forms of locomotion (e.g., running) have been analyzed in numerous investigations [1–7] with the aim of elucidating the main mechanisms of different movements. These studies have practical applications and allow for evaluating the metabolic energy expenditure or predicting the "ideal" performance [8–14] based on the relationship between mechanics and energetics [7,15–19], which is one of the most crucial and extensively researched domains of human movement [3,4,16,20–27].

For example, di Prampero et al. [22] estimated the EC of the first 30 m of a sprint running from a standing position to overcome the challenges of directly measuring effort during dynamic actions. In brief, the method relied on the equivalence between acceleration on flat ground and ascent at constant speed, with an equivalent slope defined by forward acceleration. Since the EC of constant speed on a varied range of slopes is well known [1–3,5,13,28], estimating the EC of the run is possible when the equivalence between forward acceleration and the slope is known. Therefore, di Prampero et al.'s [22] model has been suggested to redefine the concept of "high intensity". Nevertheless, despite the new possibilities that arise from this approach [22] in terms of workload quantification and physical performance evaluation during training and competition [18,29–34], more evidence is still needed to determine the feasibility of EC estimation in applied scenarios.

Some researchers tried to evaluate neuromuscular and metabolic engagement during training and match events by using portable technology to understand muscle activation thresholds. A first attempt to characterize the profile of neuromuscular activation during a soccer match was proposed by Montini et al. [35], with the intention to integrate, in competition, more traditional laboratory-based approaches (e.g., electromyography [EMG]) to help better understand the physiological demands of competitive soccer. The authors analyzed different intensity zones to create a relative performance model and suggested that this approach could be used to improve the understanding of the physiological requirements of competitive soccer [35]. However, the EC and metabolic power (MP) calculated by EMG were not determined; thus, additional research is still necessary to consolidate measurements of economy and neuromuscular activation during performance activities that involve high-intensity running. This type of methodological approach is important for practitioners since, by using portable technologies, it is possible to collect data on more ecologically valid conditions than in laboratory settings.

The literature has explored the behavior of EMG during sprints and submaximal runs since Mero & Komi [36]; however, to the authors' knowledge, it has not been utilized for the calculation of the MP and EC until Colli's work (unpublished data retrieved from laltrametodologia.com). Thus, this remains a topic that needs further investigation to better understand the main mechanistic–energetic needs and, consequently, make meaningful methodological choices. Currently, there are numerous existing studies evaluating MP and energy expenditure, utilizing global positioning systems (GPS) and inertial measurement units (IMU) [24,32,33,37–43], but there is a complete absence of studies calculating these parameters from EMG technology. Analyzing submaximal and maximal sprint behavior with the aim of determining the MP and EC calculated by EMG and the EMG and force relationship could help clarify actual metabolic and neuromuscular engagement during linear running actions. The comparison of two distinct technologies (i.e., EMG and GPS-IMU) has the potential to provide precise estimates of relative effort for actions such as sprints, yielding hypothetical benefits.

Therefore, the aims of this study were to (1) analyze submaximal running efforts at various constant speeds to investigate possible differing mechanical–energetic demands when compared to sprinting; (2) examine the behavior of the EMG activity-to-force ratio (EMG/F) in linear sprints over 30 m and 50 m and their corresponding 5 m sections; and (3) determine the EC and MP of sprinting assessed by GPS-IMU and EMG by creating an ad hoc neuromuscular profile utilizing muscle activation patterns. The present study may have significant implications for the establishment and structuring of training objectives.

2. Materials and Methods

2.1. Study Design

A cross-sectional study design was used (Figure 1). Data were collected during the 2020/2021 competitive season, during the months of September through November, with players from the under-17 (U17) age category of a professional soccer club academy. To avoid a potential source of bias, de-identified data were analyzed by a researcher not directly involved in data collection. After a careful theoretical explanation accompanied by a practical demonstration, players completed four different submaximal constant running efforts followed by sprinting actions while using portable GPS-IMU units and surface EMG. All athlete measures were taken in a single testing session for each player during the pre-season period. The warm-up included mobility and running-based exercises for a duration of ~15 min. All warm-up exercises had been previously used by all the players, as they were applied in daily training.

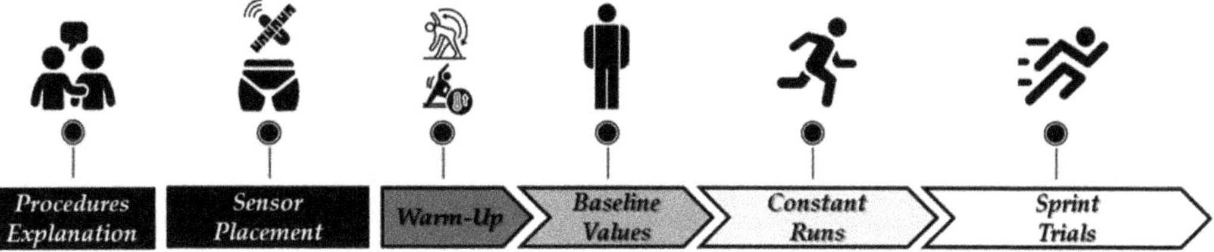

Figure 1. Overview of the study design.

2.2. Participants

A convenience sample of sixteen U17 football players (age: 16.4 ± 0.5 years; body mass: 64.6 ± 4.4 kg; height: 177.4 ± 4.3 cm; and BMI: 20.5 ± 1.3) of the "Elite Italian Championship" volunteered to participate in this study. A normal team practice and competition schedule, consisting of at least four training sessions and one match per week, was maintained during the investigation period. Only players who were free from recent injuries or medical conditions that could limit their maximum effort were included in the study. Detailed information regarding all testing and training procedures was provided to the subjects and their legal guardians before the latter signed a written informed consent. The Local Human Subjects Ethics Committee approved the study in compliance with the Declaration of Helsinki.

2.3. Procedures

2.3.1. Constant Running and Sprint Testing

Four incremental constant ($C_{1,2,3,4}$) running speeds (over 50 m, at theoretical required times of ~22.5, ~15, ~11.3, and ~9 s in "C_1", "C_2", "C_3", and "C_4", respectively) and a sprint effort (where only the split of the maximum speed phase was taken) were used for the construction of an individual profile (detailed below; coded with "S_5"). Timing adherence was manually controlled using stopwatches during the constant runs in the trials. All tests were conducted on the training and match field, and each player was given the appropriate technical clothing to maintain their running characteristics (ecological field test). As mentioned, the players started by performing the constant runs with the objective of having an approximate constant difference between runs rather than a set datum (impossible for a field test that does not take place on an ergometer); thus, they were asked to maintain the same running characteristics during each trial. After the constant runs and a rest period (2 min), the players performed a total of three all-out sprints over 50 m. To establish the zone of maximum sprinting speed, a plateau with a delta of no more than 3 km·h^{-1} in the GPS data was selected to objectively determine when athletes reached their peak speed. A 5 min passive rest period was provided between trials to minimize fatigue effects on performance. Participants were encouraged to perform each sprint trial as fast as possible.

2.3.2. Electromyography Recording and Analysis

During the trials, EMG shorts equipped with textile electrodes (Myonear Pro, Myontec, Kuopio, Finland) were used to collect muscle activation data (Figure 2). The conductive electrodes and the associated wires were integrated into the fabric. These electrodes covered three main muscle groups, bilaterally, with 6 differential EMG biosignal channels: quadriceps, hamstrings, and glutes. Two sizes of shorts were available (medium and large), and the best fit was chosen for each participant. The proper size of the shorts is essential to establish necessary contact between electrodes and skin and to minimize or avoid any movement artifacts during dynamic activities [44]. Additionally, a small amount of water was applied to the electrodes before the participant put on the shorts to ensure adequate signal conduction, as previously recommended [45]. EMG signals were transmitted to a laptop and analyzed and collected at 1000 Hz with the Myontec "Muscle Monitor" software version 3.1.0.4 (Myontec Ltd., Kuopio, Finland). Textile electrodes embedded in shorts appeared to provide comparable lower limb muscle activation data to traditional surface EMG [44]. Each trial was firstly filtered with a second-order Butterworth band pass (a bandwidth of 40–200 Hz, derived through an exploration of the frequency domain with a signal voltage and -3 dB cutoff frequency) filter, before being rectified and averaged over 100 Hz. In accordance with Kyröläinen et al. [46], who criticized the use of voluntary maximum isometric contractions (MVICs) for evaluating neuromuscular activation during running, the EMG data were normalized using the peak EMG activity (EMGpeak) detected during the sprint, thus allowing for greater repeatability of the measurements. In addition, the EMG signals during the runs were segmented into subphases to enable a detailed analysis not only of the overall trend (total EMG recording [EMGTOT_ecf], comprehensive of ground contact, eccentric and concentric, and of the flight phases) but also of the characteristics of each phase (i.e., eccentric [EMGe], concentric [EMGc], and flight phase) utilizing the LagalaColli software (version 1.0.2.218, Spinitalia S.R.L., Rome, Italy). The EMG/F ratio was determined with an arbitrary unit, consisting of the ratio of the normalized EMG signal with peak values and expressing it as a percentage and the resulting force (in N·kg^{-1}), which was calculated by integrating the accelerations from the three axes (x, y, and z) using IMU technology.

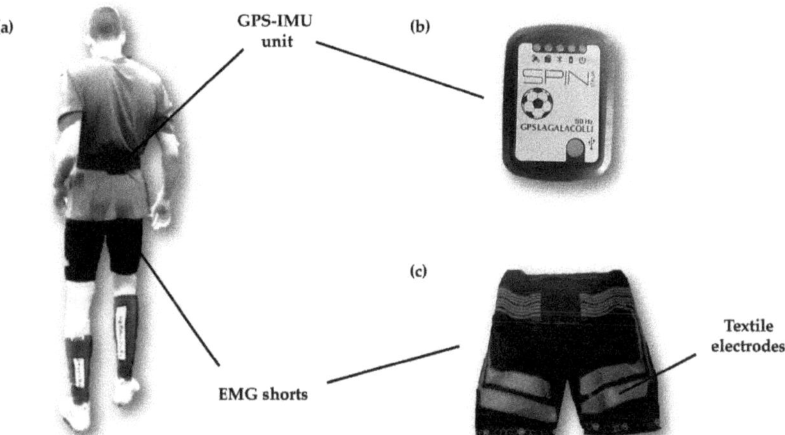

Figure 2. (**a**) Back view of the sensor placement; (**b**) GPS unit; (**c**) EMG shorts equipped with textile electrodes with 6 differential EMG biosignal channels.

2.3.3. Ad Hoc Profiling and Metabolic Power Calculation

Prior to the sprint analysis, an individual linear profile (including slope and intercept) was constructed for each athlete by plotting the MP of the GPS and muscle load (ML) from all the EMG signals acquired. The profile was individualized and made it possible to recalculate the MP from the EMG by a simple method that consisted of multiplying by the slope and then adding intercept (Equation (1)). Then, the EC was calculated by dividing the obtained value of MP by the speed achieved (Equation (2)).

$$\mathbf{MP}_{EMG} = (\mathbf{ML}_{EMG} \cdot \mathbf{SLOPE}) + \mathbf{INTERCEPT} \tag{1}$$

$$\mathbf{EC}_{EMG} = \frac{\mathbf{MP}_{EMG}}{\mathbf{SPEED}_{IMU}} \tag{2}$$

The EMG data were integrated with GPS-IMU signals to permit us to temporally and kinematically differentiate phases. Sprint analyses were conducted utilizing personalized spreadsheets. The GPS MP was based on the GPS velocity, and the integrated GPS-IMU velocity was utilized to determine the EMG MP. The GPS data were recorded at 50 Hz and the IMU at 100 Hz in accordance with the manufacturer's instructions.

2.4. Statistical Analysis

Statistical analyses were conducted using the Statistical Package for Social Sciences (SPSS) software, version 25.0 (Chicago, IL, USA), Microsoft Excel 2019 (Redmond, WA, USA), and the free Statistical Software Jamovi 2.3.28. Data are presented as means and standard deviations. The goodness of fit of the linear regressions was assessed by the coefficient of determination (R^2) and the confidence interval (CI, set at 95%). The Shapiro–Wilk test was used to verify if the values were normally distributed, and the Wilcoxon signed rank nonparametric test was used for data not normally distributed. A repeated measures ANOVA was used to detect changes, with a two-sample F-test for variances. The effect size (ES, Cohen's d) of the intervention was calculated using Cohen's guidelines [47,48]. The threshold values for the ES were small ($\geq$0.2), medium ($\geq$0.5), and large ($\geq$0.8). For all procedures, a level of $p \leq 0.05$ was selected to indicate statistical significance.

3. Results

During the experimental period, no injuries were sustained by any of the players, and the compliance with the assessments and degree to which the participants adhered to the study protocol and accepted the interventions and assessments were maximal, as there were no dropouts. Regarding the study results, these include the EMG signal and speed mean values, detected with the EMG signal in a bipodal static ($2.4 \pm 0.6\%$ relative to the EMG_{peak}) and obtained during the four incremental constant running and sprint efforts over 50 m. The first constant running (C_1) exercise was completed at $6.9 \pm 0.8\ km\cdot h^{-1}$, with an EMG_e of $16.2 \pm 8.3\%$, an EMG_c of $14.0 \pm 3.8\%$, and an EMG_{TOT_ecf} of $13.9 \pm 5.3\%$. In the second constant running (C_2) exercise, the speed was $10.2 \pm 0.8\ km\cdot h^{-1}$ with an EMG_e of $21.6 \pm 11.8\%$, an EMG_c of $17.7 \pm 6.4\%$, and an EMG_{TOT_ecf} of $18.5 \pm 8.3\%$. C_3 was completed at $13.3 \pm 1.5\ km\cdot h^{-1}$, with EMG_e, EMG_c, and EMG_{TOT_ecf} values of $27.4 \pm 11.8\%$, $23.0 \pm 5.9\%$, and $23.1 \pm 7.8\%$, respectively. Finally, the speed reached in C_4 was $17.4 \pm 1.3\ km\cdot h^{-1}$, with an EMG_e of $33.1 \pm 8.7\%$, an EMG_c of $30.4 \pm 12.6\%$, and EMG_{TOT_ecf} of $28.6 \pm 8.0\%$. Regarding the sprint effort (S_5), the speed achieved was $27.3 \pm 1.8\ km\cdot h^{-1}$, the EMG_e was $62.3 \pm 8.6\%$, the EMG_c was $57.4 \pm 8.6\%$, and the EMG_{TOT_ecf} $63.1 \pm 8.7\%$. Figure 3 displays the corresponding total EMG patterns in relation to running speed.

During sprinting, the EMG/F ratio data were processed for each 5 m interval (Figure 4) and presented a double behavior interpolated with two types of fit. The EMG/F ratio was linear up to 30 m ($R^2 = 0.99$) and polynomial (fourth degree) up to the completion of 50 m ($R^2 = 0.96$).

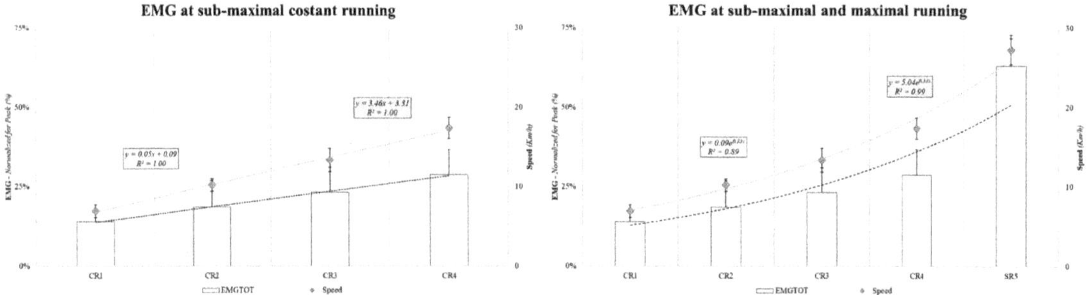

Figure 3. EMG at four different submaximal running speeds (**left** panel) and with sprint efforts (**right** panel).

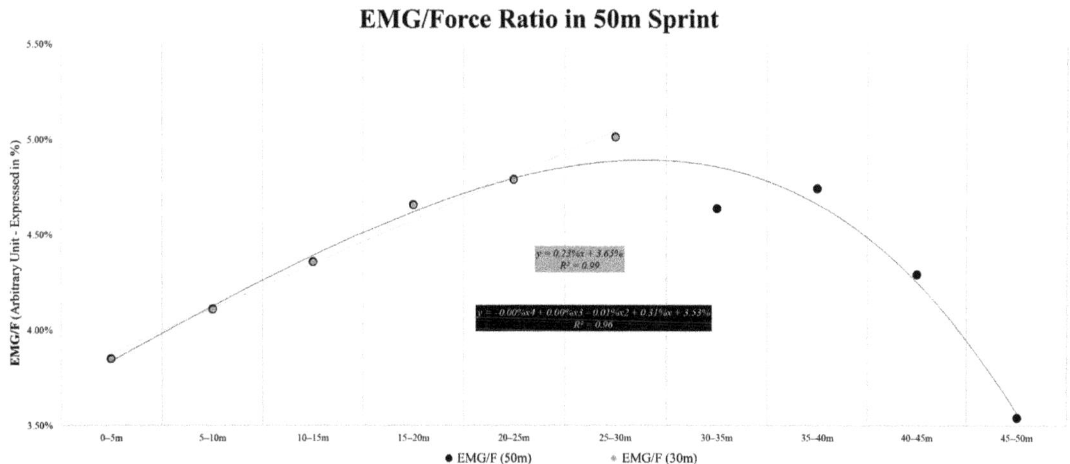

Figure 4. Linear (gray) and curvilinear (polynomial, in black) fits of the EMG/F ratio over 50 m sprints.

Comparisons of the MP for the different split distances obtained with the GPS and EMG are presented in Figure 5. The data were not normally distributed at 20–25 m and 35–40 m distances. In the 0–5 m and 5–10 m splits, the MP calculated with the GPS was significantly higher than with EMG (0–5 m: $p = 0.03$, F = 1.13, ES = 0.81; 5–10 m: $p = 0.02$, F = 1.67, ES = 0.89). Conversely, in the 10–15 m and 15–20 m ranges, no significant differences were observed between both methods (10–15 m: $p = 0.31$, F = 1.42, ES = 0.37; 15–20 m: $p = 0.58$, F = 1.76, ES = 0.20; 20–25 m: $p = 0.39$, F = 1.66, ES = 0.10). In the 20–25 m, 25–30 m, 30–35 m, and 35–40 m splits, the MP was significantly lower when determined via the GPS rather than EMG (20–25 m: $p = 0.01$, F = 1.66, ES = 0.31; 25–30 m: $p \leq 0.001$, F = 1.17, ES = 1.42; 30–35 m: $p = 0.002$, F = 0.48; ES = 1.19; and 35–40 m: $p = 0.02$, F = 1.15, ES = 0.98). Lastly, no differences between the MP determined via the GPS and EMG were found in the 40–45 m ($p = 0.14$; F = 0.94, ES = 0.54) and 45–50 m splits ($p = 0.53$; F = 1.38, ES = 0.22). Table S1 (in Supplementary Files) illustrates the values obtained after adjustments, applying the nonparametric statistical test.

The EC estimated through the GPS and EMG is displayed in Figure 6. The data were not normally distributed at the distances 15–20 m, 20–25 m, 25–30 m, 30–35 m, and 35–40 m. No differences were found between both approaches in the 0–5 m split ($p = 0.30$, F = 0.68, ES = 0.38), which contrasts with the 5–10 m split, in which the EC determined via the GPS was significantly greater ($p = 0.001$, F = 1.03, ES = 1.33). In the 10–15 m ($p = 0.09$, F = 1.03, ES = 0.63), 15–20 m ($p = 0.09$, F = 1.32, ES = 0.40), and 20–25 m ($p = 0.54$, F = 1.66, ES = 0.20) ranges, no differences in EC were identified. Regarding the 25–30 m ($p = 0.02$; F = 0.96, ES = 1.35), 30–35 m ($p = 0.003$, F = 0.61, ES = 0.95), and 35–40 m ($p = 0.05$, F = 1.14, ES = 0.76) ranges, the EC estimated through EMG was significantly higher. Finally, in the 40–45 m ($p = 0.30$, F = 1.57, ES = 0.37) and 45–50 m splits ($p = 0.12$, F = 2.78, ES = 0.56), no differences were observed in the EC estimated with the GPS and EMG. Table S2 (in Supplementary Files) illustrates the values obtained after adjustments, applying the nonparametric statistical test.

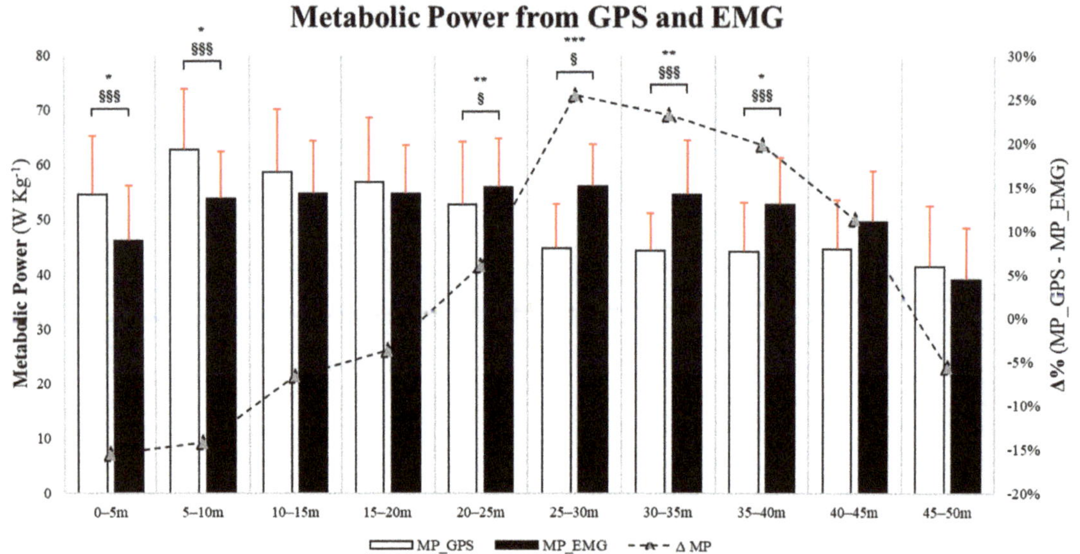

Figure 5. Metabolic power calculated via GPS and EMG methods during the linear 50 m sprint. * p-value ≤ 0.05, ** $p \leq 0.01$, *** $p \leq 0.001$; § ES ≥ 0.20, §§ ES ≥ 0.50, §§§ ES ≥ 0.80.

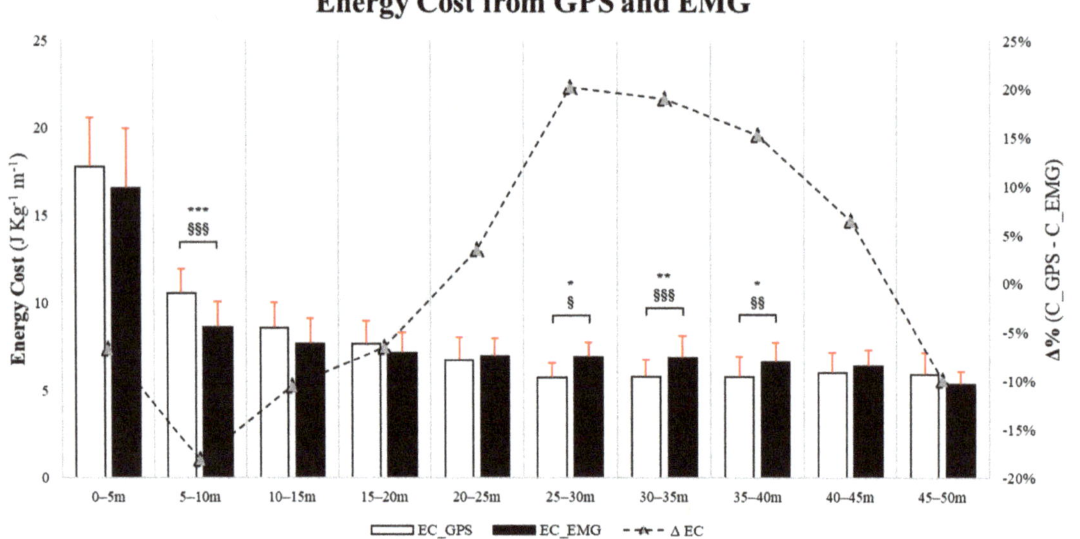

Figure 6. Energy cost calculated via GPS and EMG methods during the linear 50 m sprint. * p-value ≤ 0.05, ** $p \leq 0.01$, *** $p \leq 0.001$; § ES ≥ 0.20, §§ ES ≥ 0.50, §§§ ES ≥ 0.80.

4. Discussion

The present study aimed to investigate the EMG activity and the EMG/F ratio during different submaximal and maximal runs and to study the differences between GPS-IMU and EMG technologies in the calculation of the MP and EC during constant running and sprinting efforts. The main findings indicate a linear increase in EMG values with running speed during the submaximal runs, which becomes exponential when considering the inclusion of sprinting. Moreover, the current results demonstrate the existence of a linear increase in the EMG/F ratio in sprints up to a breaking point (i.e., observed at 30 m) when an alteration in the overall trend is observed (i.e., considering the whole 50 m). In addition, differences were found at certain splits between the MP and EC calculated from the GPS-IMU and EMG, which indicates that these technologies cannot be used interchangeably to determine these metrics. Taking this into account, the present findings suggest that EMG seems to be a more precise technology for accurately estimating the MP and EC, showing a higher EC for sprinting, especially at greater speeds.

Notably, in the constant-speed runs, the EMG activity increased linearly with increasing speed. However, when also considering the sprint actions, the best-fitting trend becomes exponential (Figure 3). This might have an important implication for the study of the metabolic engagement of running efforts, as it supports the idea that sprint situations may cause an important increase in the energy expenditure of soccer players [36,46,49,50]. However, it should be considered that the present study did not specifically consider the EC of acceleration, high-speed running, and deceleration efforts, although from previous studies [3,12,22,33], we could already hypothesize significant differences between these types of actions. From a coaching perspective, the results herein could be used to determine the effective energetic and neuromuscular engagement needed for different types of actions and better understand performance models to develop the best training methodologies.

Another important parameter to be considered when investigating the interaction between external and internal loads during actions such as linear sprinting is the EMG/F ratio [51–55]. The current data highlight that this ratio appears to linearly increase until a 'breakpoint', where a decrease occurs (i.e., at 30 m, as evidenced by the data interpolation in Figure 4), which may have significant practical applications. In brief, it indicates that the expression of neuromuscular parameters likely varies across different distances and sports contexts, providing practitioners with an ideal range of distances that could be used in sprint training. For example, for youth soccer players, from an energetic and neuromuscular perspective, it may not be optimal to perform linear sprints greater than 30 m due to the observed decline in EMG/F. In our analyzed sample, there appears to be difficulty in maintaining the neuromuscular engagement characteristics indicated by the EMG/F marker over longer distances. This may be due to poor sprinting habits for longer distances, particularly under static start conditions. Future studies should verify the behavior of this parameter in other populations of athletes (e.g., sprinters), assuming that the drop in the EMG/F ratio should be postponed as much as possible for those who must perform linear sprints.

Finally, based on the construction of an ad hoc profile that allowed for the calculation of the MP and EC from EMG, it appears that the GPS-IMU approach may systematically underestimate the actual cost of sprinting in a statistically significant manner, especially for sprint actions between 25 and 40 m, when compared to EMG. This may be explained, at least in part, by the fact that at higher speeds, acceleration rates are considerably lower [49,56] but more costly; hence, the GPS-IMU may not be the most appropriate approach to quantify energy expenditure. Of note, EMG technology seems to display different MP and EC engagement with a much more "curvilinear pattern" (i.e., a fourth-degree polynomial relationship) during sprints, thus emphasizing a different, realistically more accurate engagement in some splits. These considerations could be useful for coaches and physical trainers to understand actual energy engagement and neuromuscular parameters in soccer, knowing more about its limitations and potential [56,57]. However, further research is required to determine the practical applications of this area of study in different populations with different purposes. In accordance with Van Hooren et al. [6], the calculation of the markers would be important for optimizing energetic and mechanical efficiency, possibly minimizing injury occurrence resulting from internal (i.e., physiology) and external (i.e., environment) sources. All these findings seem to be important in characterizing sprint action. The MP and EC analyses using EMG in comparison to the GPS may provide more precise results for evaluating neuromuscular and metabolic activity. This approach can be advantageous for optimizing mechanical–energetic requirements and warrants further investigation, including cognitive engagement during exercises involving a ball.

This study has several limitations that should be considered when interpreting the results. Firstly, the cross-sectional design used prevents us from drawing any causal inferences regarding the examined variables. Secondly, only isolated and "decontextualized" linear sprints without a ball were assessed when it is known that, in soccer, most physical capacities are expressed along with technical–tactical elements with the ball [32,38]. Thus, the data here should not be directly extrapolated to sprinting during soccer matchplay. Thirdly, other important running-based actions, such as accelerations and decelerations, were not assessed and compared in detail. Additionally, the EC and MP were estimated through a GPS-IMU and EMG, and the use of a portable gas analyzer could have en hanced the study's accuracy, providing practical assistance and a better understanding when comparing data. This was demonstrated by Savoia et al. [33] when comparing the GPS algorithm based upon di Prampero's theoretical model in elite soccer players with a measure obtained with a portable gas analyzer. Nevertheless, the methodological approach here is more practical and easier to apply in real-world contexts, which is an important point worth highlighting. Further research is necessary to determine whether and how the current findings may be affected by training adaptations.

The outcomes of this study may be useful for strength and conditioning coaches to plan their sessions more effectively. Our data examined the EC of running at different speeds and identified the EMG trends indicative of actual neuromuscular demands. The analysis of an internal-to-external load ratio, such as the EMG/F ratio, may be useful in determining appropriate distances for training. In addition, the differences between the MP and EC calculated by the GPS-IMU and EMG suggested an important underestimation of the actual demands of high-speed actions by the former (which must be considered when developing training exercises). However, it is important to note that the current data were collected from a sample of U17 soccer players from a Mediterranean context and that the generalization of the results to other populations should be made cautiously. Further research should be conducted to investigate these aspects and potential disparities in game scenarios.

5. Conclusions

In summary, this study presents a new perspective for characterizing running activities in soccer, utilizing parameters such as the EMG/F ratio and using the MP and EC calculated from EMG, and just a GPS-IMU. Defining and characterizing the specifics of physical engagement are strategic factors for designing a novel approach [58–60] to study neuromuscular and metabolic activity to continue development [3,33,49,56]. Although additional research is necessary, these indicators appear suitable for accurately studying workload, improving performance, examining the dose–response relationship of exercise, and identifying the onset and modification of fatigue during competitions. In the future, a GPS-IMU and EMG should be validated against direct measures of energy expenditure, both external and internal, to determine their relationship with direct measures of fitness and performance.

Supplementary Materials: The following supporting information can be downloaded at https://www.mdpi.com/article/10.3390/s24082577/s1: Table S1: MP; Table S2: EC.

Author Contributions: Conceptualization, G.G.; data curation, G.G.; formal analysis, G.G. and T.T.F.; investigation, G.G.; methodology, G.G., P.E.A. and T.T.F.; project administration, G.G., P.E.A. and T.T.F.; supervision, P.E.A. and T.T.F.; visualization, G.G., P.E.A. and T.T.F.; writing—original draft, G.G. and T.T.F.; writing—review and editing, G.G., P.E.A. and T.T.F. All authors have read and agreed to the published version of the manuscript.

Funding: This research received no external funding.

Institutional Review Board Statement: The study was conducted in accordance with the Declaration of Helsinki, and approved by the Institutional Review Board of UCAM—Universidad Católica San Antonio de Murcia (protocol code CE022106 and date of approval 26/02/2021).

Informed Consent Statement: Informed consent was obtained from all subjects involved in the study.

Data Availability Statement: Data are contained within the article.

Acknowledgments: First, the leading author would like to thank Roberto Colli for the many exchanges and teachings received during these years. Specifically, this article discusses a profiling method developed by Roberto Colli, although with some modifications. The author believes, however, that in the future, it is possible to still improve the profiling, as suggested in the article and already discussed privately with Roberto Colli. Also, the leading author would like to thank Antonio Buglione for sharing and developing, several years ago, important insights into the energetics of human locomotion (in particular, with the metabolic power and energy cost approach).

Conflicts of Interest: Authors P.E.A. and T.T.F. was employed by the company Strength and Conditioning Society, 30008 Murcia, Spain. The remaining authors declare that the research was conducted in the absence of any commercial or financial relationships that could be construed as a potential conflict of interest.

References

1. Minetti, A.E.; Ardigò, L.P.; Saibene, F. Mechanical determinants of the minimum energy cost of gradient running in humans. *J. Exp. Biol.* **1994**, *195*, 211–225. [CrossRef]
2. Minetti, A.E.; Ardigò, L.P.; Saibene, F. The transition between walking and running in humans: Metabolic and mechanical aspects at different gradients. *Acta Physiol. Scand.* **1994**, *150*, 315–323. [CrossRef]
3. Minetti, A.E.; Moia, C.; Roi, G.S.; Susta, D.; Ferretti, G. Energy cost of walking and running at extreme uphill and downhill slopes. *J. Appl. Physiol. (1985)* **2002**, *93*, 1039–1046. [CrossRef]
4. Zamparo, P.; Pavei, G.; Nardello, F.; Bartolini, D.; Monte, A.; Minetti, A.E. Mechanical work and efficiency of 5 + 5 m shuttle running. *Eur. J. Appl. Physiol.* **2016**, *116*, 1911–1919. [CrossRef]
5. Minetti, A.E.; Pavei, G. Update and extension of the e'quivalent slope' of speed-changing level locomotion in humans: A computational model for shuttle running. *J. Exp. Biol.* **2018**, *221 Pt 15*, jeb182303. [CrossRef]
6. Van Hooren, B.; Meijer, K.; McCrum, C. Attractive Gait Training: Applying Dynamical Systems Theory to the Improvement of Locomotor Performance Across the Lifespan. *Front. Physiol.* **2019**, *9*, 1934. [CrossRef]
7. Zamparo, P.; Pavei, G.; Monte, A.; Nardello, F.; Otsu, T.; Numazu, N.; Fujii, N.; Minetti, A.E. Mechanical work in shuttle running as a function of speed and distance: Implications for power and efficiency. *Hum. Mov. Sci.* **2019**, *66*, 487–496. [CrossRef]
8. Furusawa, K.; Hill, A.V.; Parkinson, J.L. The Dynamics of Sprint Running. *Proc. R. Soc. B Biol. Sci.* **1927**, *102*, 29–42.
9. Fenn, W.O. Frictional and kinetic factors in the work of sprint running. *Am. J. Physiol. Leg. Content* **1930**, *92*, 583–611.
10. Fenn, W.O. Work against gravity and work due to velocity changes in running: Movements of the center of gravity within the body and foot pressure on the ground. *Am. J. Physiol. Leg. Content* **1930**, *93*, 433–462.
11. Hill, A.V. The heat of shortening and the dynamic constants of muscle. *Proc. R. Soc. Lond. Ser. B-Biol. Sci.* **1938**, *126*, 136–195.
12. Margaria, R. Sulla fisiologia e specialmente sul consumo energetico della marcia e della corsa a varia velocità ed inclinazione del terreno. *Atti Accad. Nazionale dei Lincei* **1938**, *7*, 299–368.
13. Margaria, R.; Cerretelli, P.; Aghemo, P.; Sassi, G. Energy cost of running. *J. Appl. Physiol.* **1963**, *18*, 367–370. [CrossRef] [PubMed]
14. Margaria, R. La fisiologia della locomozione. *Le Scienze* **1970**, *V*, 11–21. Available online: http://vivicitta.uisp.it/home/wp-content/uploads/1970_025_1.pdf (accessed on 17 March 2024).
15. Marey, E.J.; Demeny, G. Variations de travail mécanique dépensé dans les différentes allures de l'homme. *Comptes Rendus Hebd. Séances L'académie Sci.* **1885**, *101*, 910–915.
16. Cavagna, G.A.; Kaneko, M. Mechanical work and efficiency in level walking and running. *J. Physiol.* **1977**, *268*, 467–481. [CrossRef] [PubMed]
17. Pinnington, H.C.; Dawson, B. The energy cost of running on grass compared to soft dry beach sand. *J. Sci. Med. Sport* **2001**, *4*, 416–430. [CrossRef] [PubMed]
18. Colli, R.; Buglione, A.; Introini, E.; D'Ottavio, S. L'allenamento intermittente tra scienza e prassi. *SdS Sc. Dello Sport* **2007**, *72*, 45–52.
19. di Prampero, P.E.; Botter, A.; Osgnach, C. The energy cost of sprint running and the role of metabolic power in setting top performances. *Eur. J. Appl. Physiol.* **2015**, *115*, 451–469. [CrossRef]
20. Deutsch, F. Analytic posturology. *Psychoanal. Q.* **1952**, *21*, 196–214. [CrossRef]
21. Cavagna, G.A.; Komarek, L.; Mazzoleni, S. The mechanics of sprint running. *J. Physiol.* **1971**, *217*, 709–721. [CrossRef] [PubMed]
22. di Prampero, P.E.; Fusi, S.; Sepulcri, L.; Morin, J.B.; Belli, A.; Antonutto, G. Sprint running: A new energetic approach. *J. Exp. Biol.* **2005**, *208 Pt 14*, 2809–2816. [CrossRef] [PubMed]
23. Buglione, A.; di Prampero, P.E. The energy cost of shuttle running. *Eur. J. Appl. Physiol.* **2013**, *113*, 1535–1543. [CrossRef]

24. Piras, A.; Raffi, M.; Atmatzidis, C.; Merni, F.; Di Michele, R. The Energy Cost of Running with the Ball in Soccer. *Int. J. Sports Med.* **2017**, *38*, 877–882. [CrossRef]

25. Monte, A.; Zamparo, P. Correlations between muscle-tendon parameters and acceleration ability in 20 m sprints. *PLoS ONE* **2019**, *14*, e0213347. [CrossRef]

26. Pavei, G.; Zamparo, P.; Fujii, N.; Otsu, T.; Numazu, N.; Minetti, A.E.; Monte, A. Comprehensive mechanical power analysis in sprint running acceleration. *Scand. J. Med. Sci. Sports* **2019**, *29*, 1892–1900. [CrossRef]

27. Monte, A.; Baltzopoulos, V.; Maganaris, C.N.; Zamparo, P. Gastrocnemius Medialis and Vastus Lateralis in vivo muscle-tendon behavior during running at increasing speeds. *Scand. J. Med. Sci. Sports* **2020**, *30*, 1163–1176. [CrossRef] [PubMed]

28. Minetti, A.E. A model equation for the prediction of mechanical internal work of terrestrial locomotion. *J. Biomech.* **1998**, *31*, 463–468. [CrossRef] [PubMed]

29. Osgnach, C.; Poser, S.; Bernardini, R.; Rinaldo, R.; di Prampero, P.E. Energy cost and metabolic power in elite soccer: A new match analysis approach. *Med. Sci. Sports Exerc.* **2010**, *42*, 170–178. [CrossRef]

30. Buglione, A.; di Prampero, P.E. Energy Cost of Elite Soccer Players Before and During Season. In Proceedings of the XXXII World Congress of Sports Medicine, Rome, Italy, 27–30 September 2012.

31. Osgnach, C.; Paolini, E.; Roberti, V.; Vettor, M.; di Prampero, P.E. Metabolic Power and Oxygen Consumption in Team Sports: A Brief Response to Buchheit et al. *Int. J. Sports Med.* **2016**, *37*, 77–81. [CrossRef]

32. Licciardi, A.; Grassadonia, G.; Monte, A.; Ardigò, L.P. Match metabolic power over different playing phases in a young professional soccer team. *J. Sports Med. Phys. Fitness* **2020**, *60*, 1170–1171. [CrossRef] [PubMed]

33. Savoia, C.; Padulo, J.; Colli, R.; Marra, E.; McRobert, A.; Chester, N.; Azzone, V.; Pullinger, S.A.; Doran, D.A. The Validity of an Updated Metabolic Power Algorithm Based upon di Prampero's Theoretical Model in Elite Soccer Players. *Int. J. Environ. Res. Public Health* **2020**, *17*, 9554. [CrossRef] [PubMed]

34. Spyrou, K.; Alcaraz, P.E.; Marín-Cascales, E.; Herrero-Carrasco, R.; Cohen, D.D.; Freitas, T.T. Neuromuscular Performance Changes in Elite Futsal Players Over a Competitive Season. *J. Strength Cond. Res.* **2023**, *37*, 1111–1116. [CrossRef] [PubMed]

35. Montini, M.; Felici, F.; Nicolò, A.; Sacchetti, M.; Bazzucchi, I. Neuromuscular demand in a soccer match assessed by a continuous electromyographic recording. *J. Sports Med. Phys. Fitness* **2017**, *57*, 345–352. [CrossRef] [PubMed]

36. Mero, A.; Komi, P.V. Electromyographic activity in sprinting at speeds ranging from sub-maximal to supra-maximal. *Med. Sci. Sports Exerc.* **1987**, *19*, 266–274. [CrossRef] [PubMed]

37. Savoia, C.; Iellamo, F.; Caminiti, G.; Doran, D.A.; Pullinger, S.; Innaurato, M.R.; Annino, G.; Manzi, V. Rethinking training in elite soccer players: Comparative evidence of small-sided games and official match play in kinematic parameters. *J. Sports Med. Phys. Fitness* **2021**, *61*, 763–770. [CrossRef]

38. Ju, W.; Doran, D.; Hawkins, R.; Gómez-Díaz, A.; Martin-Garcia, A.; Ade, J.; Laws, A.; Evans, M.; Bradley, P. Contextualised peak periods of play in English Premier League matches. *Biol. Sport* **2022**, *39*, 973–983. [CrossRef] [PubMed]

39. Rodriguez-Barbero, S.; González Ravé, J.M.; Juárez Santos-García, D.; Rodrigo-Carranza, V.; Santos-Concejero, J.; González-Mohíno, F. Effects of a Regular Endurance Training Program on Running Economy and Biomechanics in Runners. *Int. J. Sports Med.* **2023**, *44*, 1059–1066. [CrossRef] [PubMed]

40. di Prampero, P.E.; Osgnach, C.; Morin, J.B.; Zamparo, P.; Pavei, G. Mechanical and Metabolic Power in Accelerated Running-PART I: The 100-m dash. *Eur. J. Appl. Physiol.* **2023**, *123*, 2473–2481. [CrossRef]

41. Osgnach, C.; di Prampero, P.E.; Zamparo, P.; Morin, J.B.; Pavei, G. Mechanical and metabolic power in accelerated running-Part II: Team sports. *Eur. J. Appl. Physiol.* **2024**, *124*, 417–431. [CrossRef]

42. Van Hooren, B.; Jukic, I.; Cox, M.; Frenken, K.G.; Bautista, I.; Moore, I.S. The Relationship between Running Biomechanics and Running Economy: A Systematic Review and Meta-Analysis of Observational Studies. *Sports Med.* **2024**, 1–48. [CrossRef] [PubMed]

43. Van Hooren, B.; Willems, P.; Plasqui, G.; Meijer, K. Changes in running economy and running technique following 6 months of running with and without wearable-based real-time feedback. *Scand. J. Med. Sci. Sports* **2024**, *34*, e14565. [CrossRef] [PubMed]

44. Colyer, S.L.; McGuigan, P.M. Textile Electrodes Embedded in Clothing: A Practical Alternative to Traditional Surface Electromyography when Assessing Muscle Excitation during Functional Movements. *J. Sports Sci. Med.* **2018**, *17*, 101–109. [PubMed]

45. Finni, T.; Hu, M.; Kettunen, P.; Vilavuo, T.; Cheng, S. Measurement of EMG activity with textile electrodes embedded into clothing. *Physiol. Meas.* **2007**, *28*, 1405–1419. [CrossRef] [PubMed]

46. Kyröläinen, H.; Avela, J.; Komi, P.V. Changes in muscle activity with increasing running speed. *J. Sports Sci.* **2005**, *23*, 1101–1109. [CrossRef] [PubMed]

47. Cohen, J. A power primer. *Psychol. Bull.* **1992**, *112*, 155–159. [CrossRef] [PubMed]

48. Cohen, J. *Statistical Power Analysis for the Behavioral Sciences*, 2nd ed.; Lawrence Erlbaum Associates: Hillside, NJ, USA, 1988.

49. Sonderegger, K.; Tschopp, M.; Taube, W. The Challenge of Evaluating the Intensity of Short Actions in Soccer: A New Methodological Approach Using Percentage Acceleration. *PLoS ONE* **2016**, *11*, e0166534. [CrossRef]

50. Hoppe, M.W.; Baumgart, C.; Slomka, M.; Polglaze, T.; Freiwald, J. Variability of Metabolic Power Data in Elite Soccer Players During Pre-Season Matches. *J. Hum. Kinet.* **2017**, *58*, 233–245. [CrossRef] [PubMed]

51. Bosco, C. *La Forza Muscolare*; Società Stampa Sportiva: Roma, Italy, 1997.

52. Grant, K.A.; Habes, D.J. An electromyographic study of strength and upper extremity muscle activity in simulated meat cutting tasks. *Appl. Ergon.* **1997**, *28*, 129–137. [CrossRef]

53. Hautier, C.A.; Arsac, L.M.; Deghdegh, K.; Souquet, J.; Belli, A.; Lacour, J.R. Influence of fatigue on EMG/force ratio and cocontraction in cycling. *Med. Sci. Sports Exerc.* **2000**, *32*, 839–843. [CrossRef]
54. Madeleine, P.; Bajaj, P.; Søgaard, K.; Arendt-Nielsen, L. Mechanomyography and electromyography force relationships during concentric, isometric and eccentric contractions. *J. Electromyogr. Kinesiol.* **2001**, *11*, 113–121. [CrossRef] [PubMed]
55. Oksanen, A.; Pöyhönen, T.; Ylinen, J.J.; Metsähonkala, L.; Anttila, P.; Laimi, K.; Hiekkanen, H.; Aromaa, M.; Salminen, J.J.; Sillanpää, M. Force production and EMG activity of neck muscles in adolescent headache. *Disabil. Rehabil.* **2008**, *30*, 231–239. [CrossRef]
56. Polglaze, T.; Dawson, B.; Peeling, P. Gold Standard or Fool's Gold? The Efficacy of Displacement Variables as Indicators of Energy Expenditure in Team Sports. *Sports Med.* **2016**, *46*, 657–670. [CrossRef] [PubMed]
57. Hoppe, M.W.; Baumgart, C.; Polglaze, T.; Freiwald, J. Validity and reliability of GPS and LPS for measuring distances covered and sprint mechanical properties in team sports. *PLoS ONE* **2018**, *13*, e0192708. [CrossRef] [PubMed]
58. Rumpf, M.C.; Lockie, R.G.; Cronin, J.B.; Jalilvand, F. Effect of Different Sprint Training Methods on Sprint Performance Over Various Distances: A Brief Review. *J. Strength Cond. Res.* **2016**, *30*, 1767–1785. [CrossRef] [PubMed]
59. Matusiński, A.; Gołas, A.; Zajac, A.; Maszczyk, A. Acute effects of resisted and assisted locomotor activation on sprint performance. *Biol. Sport* **2022**, *39*, 1049–1054. [CrossRef] [PubMed]
60. Zabaloy, S.; Carlos-Vivas, J.; Freitas, T.T.; Pareja-Blanco, F.; Loturco, I.; Comyns, T.; Gálvez-González, J.; Alcaraz, P.E. Muscle Activity, Leg Stiffness, and Kinematics During Unresisted and Resisted Sprinting Conditions. *J. Strength Cond. Res.* **2022**, *36*, 1839–1846. [CrossRef] [PubMed]

Article

Torque–Cadence Profile and Maximal Dynamic Force in Cyclists: A Novel Approach

Víctor Rodríguez-Rielves [1,†], David Barranco-Gil [2,†], Ángel Buendía-Romero [1,3], Alejandro Hernández-Belmonte [1], Enrique Higueras-Liébana [1], Jon Iriberri [4], Iván R. Sánchez-Redondo [2], José Ramón Lillo-Beviá [1], Alejandro Martínez-Cava [1], Raúl de Pablos [2], Pedro L. Valenzuela [5,6], Jesús G. Pallarés [1,*,‡] and Lidia B. Alejo [2,‡]

1 Human Performance and Sports Science Laboratory, Faculty of Sport Sciences, University of Murcia, 30720 Murcia, Spain; victor@entrenamiento.pro (V.R.-R.); angel.buendiaromero@uclm.es (Á.B.-R.); alejandro.hernandez7@um.es (A.H.-B.); enriquehiglie@gmail.com (E.H.-L.); jr.lillo@ua.es (J.R.L.-B.); alejandro.martinez12@um.es (A.M.-C.)

2 Faculty of Sport Sciences, European University of Madrid, 28670 Madrid, Spain; david.barranco@universidadeuropea.es (D.B.-G.); ivanrodriguez.trainer@gmail.com (I.R.S.-R.); raulbq98@gmail.com (R.d.P.); lidia.brea@universidadeuropea.es (L.B.A.)

3 GENUD Toledo Research Group, Faculty of Sports Sciences, University of Castilla-La Mancha, 45071 Toledo, Spain

4 Jumbo Visma Professional Cycling Team, 5215 MV Den Bosch, The Netherlands; j-iriberri@euskadi.eus

5 Physical Activity and Health Research Group (PaHerg), Research Institute of Hospital 12 de Octubre (Imas12), 28041 Madrid, Spain; pedro.valenzuela92@gmail.com

6 Department of Systems Biology, University of Alcalá, 28871 Madrid, Spain

* Correspondence: jgpallares@um.es

† These authors contributed equally to this work.

‡ These authors contributed equally to this work.

Citation: Rodríguez-Rielves, V.; Barranco-Gil, D.; Buendía-Romero, Á.; Hernández-Belmonte, A.; Higueras-Liébana, E.; Iriberri, J.; Sánchez-Redondo, I.R.; Lillo-Beviá, J.R.; Martínez-Cava, A.; de Pablos, R.; et al. Torque–Cadence Profile and Maximal Dynamic Force in Cyclists: A Novel Approach. *Sensors* **2024**, *24*, 1997. https://doi.org/10.3390/s24061997

Academic Editor: Valentina Agostini

Received: 7 March 2024
Revised: 15 March 2024
Accepted: 19 March 2024
Published: 21 March 2024

Abstract: We aimed to determine the feasibility, test–retest reliability and long-term stability of a novel method for assessing the force (torque)-velocity (cadence) profile and maximal dynamic force (MDF) during leg-pedaling using a friction-loaded isoinertial cycle ergometer and a high-precision power-meter device. Fifty-two trained male cyclists completed a progressive loading test up to the one-repetition maximum (1RM) on a cycle ergometer. The MDF was defined as the force attained at the cycle performed with the 1RM-load. To examine the test–retest reliability and long-term stability of torque–cadence values, the progressive test was repeated after 72 h and also after 10 weeks of aerobic and strength training. The participants' MDF averaged 13.4 ± 1.3 N·kg^{-1}, which was attained with an average pedal cadence of 21 ± 3 rpm. Participants' highest power output value was attained with a cadence of 110 ± 16 rpm ($52 \pm 5\%$ MDF). The relationship between the MDF and cadence proved to be very strong ($R^2 = 0.978$) and independent of the cyclists' MDF ($p = 0.66$). Cadence values derived from this relationship revealed a very high test–retest repeatability (mean SEM = 4 rpm, 3.3%) and long-term stability (SEM = 3 rpm, 2.3%); despite increases in the MDF following the 10-week period. Our findings support the validity, reliability and long-term stability of this method for the assessment of the torque–cadence profile and MDF in cyclists.

Keywords: assessment; cycling; force; testing; laboratory

1. Introduction

Force–velocity evaluations, usually performed by incremental loading tests, enable a comprehensive evaluation of muscle mechanical capabilities [1]. For resistance (e.g., in kg), an athlete has to overcome increases, and so does the force he/she has to apply to move it. However, the difference between this force applied by the athlete and that represented by the resistance becomes smaller and smaller along the incremental test, which decreases the resulting velocity (whether linear or radial). Thus, a relationship by plotting the main two variables (i.e., force and velocity) or others derived (e.g., power) can be modeled [2,3].

Although these tests are typically used in the context of strength exercises (e.g., knee extension, squat, bench press), they can also be applied in other muscle tasks such as leg pedaling [4]. To this effect, several methods have been used for evaluating the force (torque)-velocity (cadence) profile in bicycle ergometer exercises, including the completion of a series of isokinetic efforts [5] or a single isoinertial effort after a stationary start [6,7]. These approaches, however, have some limitations, including low specificity or a small number of resulting data (in the case of isokinetic and single-sprint tests, respectively) [4].

Several studies have assessed the torque–cadence profile during bicycling through short-duration maximal efforts performed against different resistive forces [4,8–11]. These methods show a high test–retest reliability for estimating the theoretical maximal values of torque (T_0 or F_0), pedal cadence and power output, as well as the ('optimal') cadence associated with the maximal power output produced [10,11]. Of note, these indicators have been positively associated with cycling performance [8] and the torque–cadence profile can be used to identify imbalances in the lower-limb mechanical capacities, thereby potentially allowing training programs to be prescribed based on each individual's needs (e.g., high-load or high-cadence training targeting improvements in the maximal levels of torque or cadence, respectively) [12]. Despite the practical applicability of the torque–cadence profile, its long-term stability in relative terms (i.e., the cadence associated with the percentage of the individual maximal torque value) is unknown. If present, long-term stability, which has been already studied for other upper- and lower-limb isoinertial exercises [13,14], can reinforce the applicability of the torque–cadence profile as an evaluation and training prescription method in the sport of cycling.

Another indicator commonly determined during the assessment of the force–velocity profile in muscle strength exercises is the maximal dynamic force (MDF, usually represented by the one-repetition maximum, 1RM) [15]. However, to the best of our knowledge, the MDF has not yet been assessed during pedaling, even if this parameter might provide potentially useful information. Indeed, the determination of MDF enables the force applied in each pedaling stroke to be expressed during training or racing (i.e., as a percentage of the MDF). For instance, during training sessions aimed at improving torque production capacity, a training modality that has gained popularity in recent years [16–18], researchers and coaches could also prescribe and monitor training loads based on the percentages of MDF in addition to other 'classical' indicators such as relative intensity based on heart rate or power output 'zones' that do not accurately identify the actual medium-to-high (>50% MDF) intensity efforts [19].

Considering all the above, the present study aimed to determine the feasibility, test–retest reliability, and long-term stability of a novel methodological procedure for determining the torque–cadence profile and the MDF during leg pedaling using a high-precision power-meter sensor mounted on a friction-loaded isoinertial cycle ergometer in trained cyclists.

2. Materials and Methods

2.1. Participants

Fifty-two male cyclists volunteered to participate in the study (age [mean $\pm$ standard deviation (SD)] 29.3 ± 8.3 years; training experience, 17.5 ± 7.3 years; height, 174 ± 5 cm; body mass, 71.9 ± 6.9 kg). As reflected by the results of previous testing in our laboratory [19] (maximum oxygen uptake = 63.8 ± 6.7 mL·kg^{-1}·min^{-1}, peak power output = 5.4 ± 0.7 W·kg^{-1}), cyclists were considered highly trained [20]. All subjects were instructed to maintain their normal diet during the study period, as well as to refrain from performing high-intensity exercise or ingesting caffeine or other stimulants 48 h before each testing session. They were informed of the study procedures and provided written informed consent. The study was approved by the Ethical Committee of the Local University (ID: 4135/2022), and all procedures were conducted following the standards established by the Declaration of Helsinki and its later amendments.

2.2. Experimental Approach

All the participants performed three familiarization sessions with the testing procedures. To study whether the torque–cadence profile was dependent on individual MDF levels, the participants were ranked according to their MDF and divided into three tertile groups as follows: low ($n = 17$), medium ($n = 18$) and high MDF ($n = 18$), respectively. On the other hand, seven subjects from each group (i.e., total = 21 participants) were randomly selected to analyze the test–retest reliability ($n = 10$) and long-term stability ($n = 11$), as described below.

2.3. Procedures

Feasibility. All incremental pedaling tests were conducted in a friction-loaded isoinertial cycle ergometer (Monark© 874E; Varberg, Sweden) equipped with a 175 mm crank. The position of the saddle (height and setback) and handlebar (reach and drop) was individually adjusted to replicate the participant's own bike (Figure 1A). The test started with the crank of the preferred leg at 45° relative to the vertical position. The initial load (2 kp) was progressively increased by 0.5 to 3 kp in each trial through the addition of calibrated disks (Eleiko, Sport AB; Halmstad, Sweden) until the heaviest load was reached above which the cyclist could no longer properly perform a whole (360°) pedaling cycle (i.e., 1RM; the precision of 0.5 kp). Load increments were individualized so that participants reached 1RM in less than 8 attempts, interspersed by 5-min rests (i.e., 2-min, free-cadence active recovery against 1 kp followed by 3-min passive recovery). Participants were required to perform a 5-s all-out effort with each load. Only the pedal cycle (i.e., a complete cycle with both legs) associated with the highest cadence was used for the subsequent analyses. In addition, the force (N), torque ($N \cdot m^{-1}$, considering the crack length), and power output (W) achieved during the highest cadence cycle were registered. The MDF was defined as the force attained with the 1RM-load. In order to directly measure the pedaling force and crank position during each pedaling cycle and trial, a recently validated high-precision power meter (Rotor 2INpower, Madrid, Spain; 50 Hz) [21] was adapted to the bottom bracket of the cycle ergometer (Figure 1B). The power meter was calibrated at the beginning of each testing session following the manufacturer's instructions. A specific software (Rotor INPower Software 2.2) was used for the analysis of force, torque, cadence and power output data (Figure 2).

Test–retest reliability. The above-described incremental test (using the same absolute loads in kp) was repeated after 72 h (test–retest reliability).

Long-term stability. The above-described incremental test (same absolute loads) was also repeated after a 10-week combined endurance and resistance training program. In addition to their habitual cycling endurance training (10.4 ± 0.8 h per week), participants underwent a standardized resistance training intervention 3 days per week (5 sets of 7 free-weight squat repetitions at 70% of 1RM per session) during the aforementioned program. The cyclist's 1RM in the full squat exercise was accurately estimated by the lifting velocity as detailed elsewhere [14]. During the training program, both relative intensity (70% 1RM) and intra-set volume (half of the possible repetitions per set) were programmed using the level of effort strategy, which has been proven to be a precise, reliable, and practical alternative to velocity-based training [22].

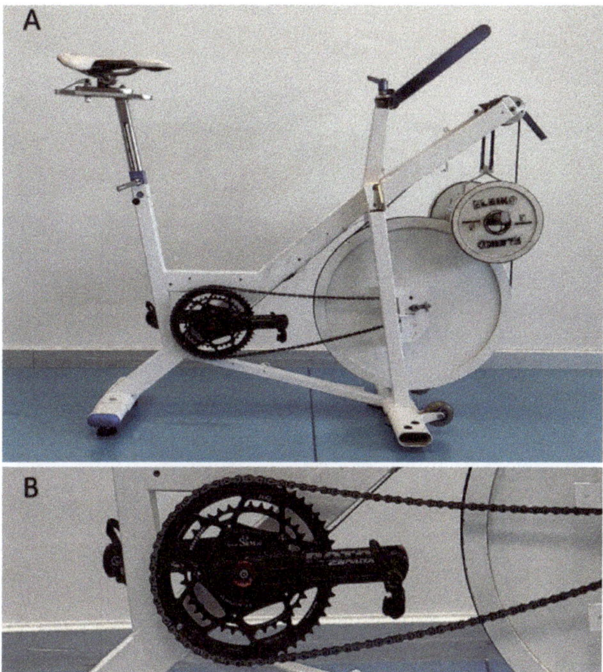

Figure 1. Cycle-ergometer (**A**) and power-meter used in the study (**B**).

2.4. Statistical Analyses

Standard statistical methods were used for the calculation of the mean, standard deviation (SD), coefficient of determination (R^2), standard error of the estimate (SEE), and 95% confidence interval (CI). Relationships between variables were studied by fitting second-order polynomials to the data. The standard error of measurement (SEM) was calculated in absolute and relative ([100 × SEM]/mean) terms from the square root of the mean square error in a repeated-measures ANOVA test. The normality of the data was verified using the Shapiro–Wilk test. Cross-sectional differences between MDF-tertile groups were examined through a one-way ANOVA test with Scheffé's post hoc comparisons. Differences between the test–retest results (test–retest reliability) and pre- and post-training results (long-term stability) were analyzed with paired *t*-tests. The level of significance was set at 0.05. Analyses were performed using SPSS software version 20.0 (IBM Corporation; Armonk, NY, USA).

Test–retest reliability. The above-described incremental test (using the same absolute loads in kp) was repeated after 72 h (test–retest reliability).

Long-term stability. The above-described incremental test (with the same absolute loads) was also repeated after a 10-week combined endurance and resistance training program. In addition to their habitual cycling endurance training (10.4 ± 0.8 h per week), participants underwent a standardized resistance training intervention 3 days per week (5 sets of 7 free-weight squat repetitions at 70% of 1RM per session) during the aforementioned.

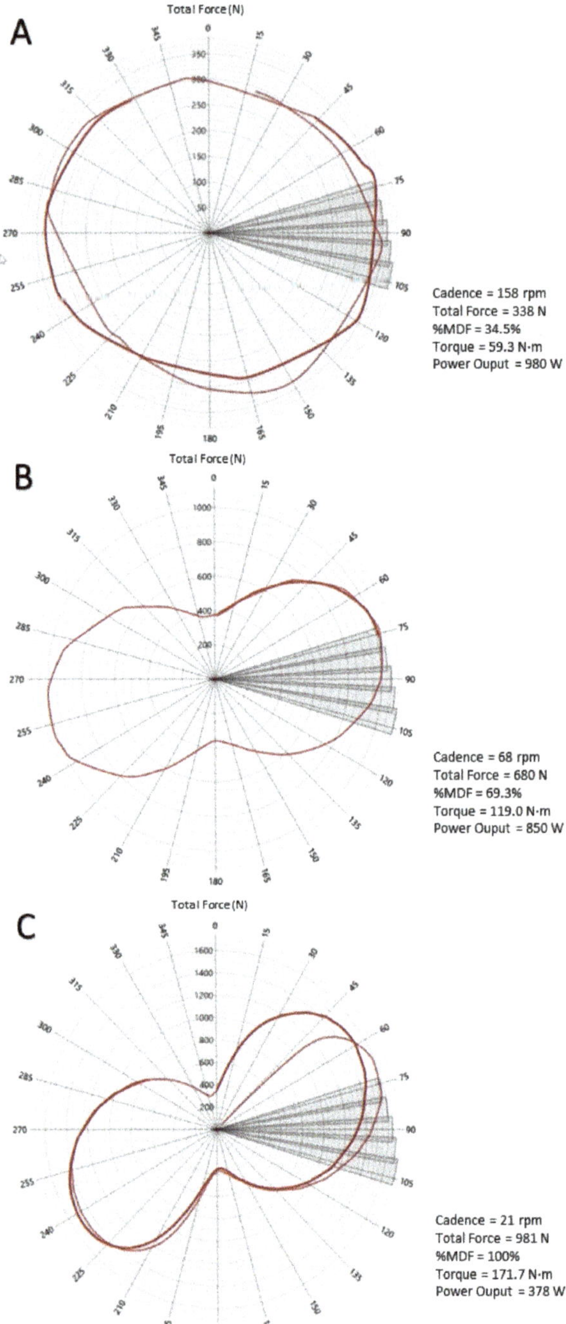

Figure 2. Screenshots of the power-meter software analysis with examples of the applied total force attained by a study participant at each crank angle (50 Hz) for a pedaling cycle against a low (~35% of maximal dynamic force [MDF]) (**A**), moderate (~70% of MDF) (**B**) or maximum resistive force (100 of MDF, i.e., one-repetition maximum) (**C**).

3. Results

3.1. Feasibility

On average, participants performed 7 ± 1 attempts until reaching their MDF, which was successfully determined in all of them. The loads used during the incremental pedaling test ranged between 2.0 and 21.5 kp. No adverse events were noted during the tests. After plotting pedaling cadence, on the one hand, against the % of MDF and, on the other, fitting a second-order polynomial to all data points, a very close relationship between these two variables was found ($R^2 = 0.978$; SEE = 9 rpm; Figure 3). Individual curve fits for each test yielded an R^2 value of 0.980 ± 0.013 (95% confidence interval, 0.976 to 0.983). A prediction equation to estimate the relative torque (% of MDF) from cadence (rpm) could be obtained ($R^2 = 0.975$; SEE = 4.5% of MDF) as follows:

$$\text{the \% of MDF} = (0.0007595 \times \text{rpm}^2) - (0.6163 \times \text{rpm}) + 111.4$$

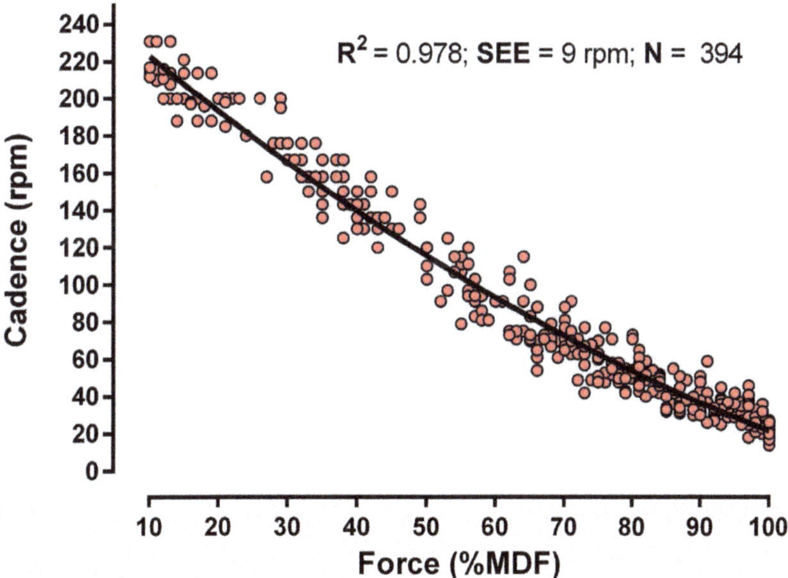

Figure 3. Force (expressed relative to the maximum dynamic force [MDF])–cadence relationship ($n = 52$ participants).

The torque–power output (panels A and B), force–cadence (panels C and D) and torque–cadence (panels E and F) relationships are shown in Figure 4. Participants' MDF (961 ± 108 N or 13.4 ± 1.3 N·kg^{-1}) was achieved with a load of 17 ± 2 kp and a cadence of 21 ± 3 rpm. Participants attained the highest power output with a cadence of 110 ± 16 rpm, corresponding to $52 \pm 5\%$ of their MDF. The polynomial equations showed a good fit ($R^2 \geq 0.893$) for the force–cadence and torque–cadence relationships.

Finally, no significant differences were observed for the average (including the whole force–cadence spectrum, $p = 0.528$, F-value = 0.301) or minimum cadence (i.e., that attained at the MDF, $p = 0.487$, F-value = 0.287) across participants with different MDF levels (Table 1).

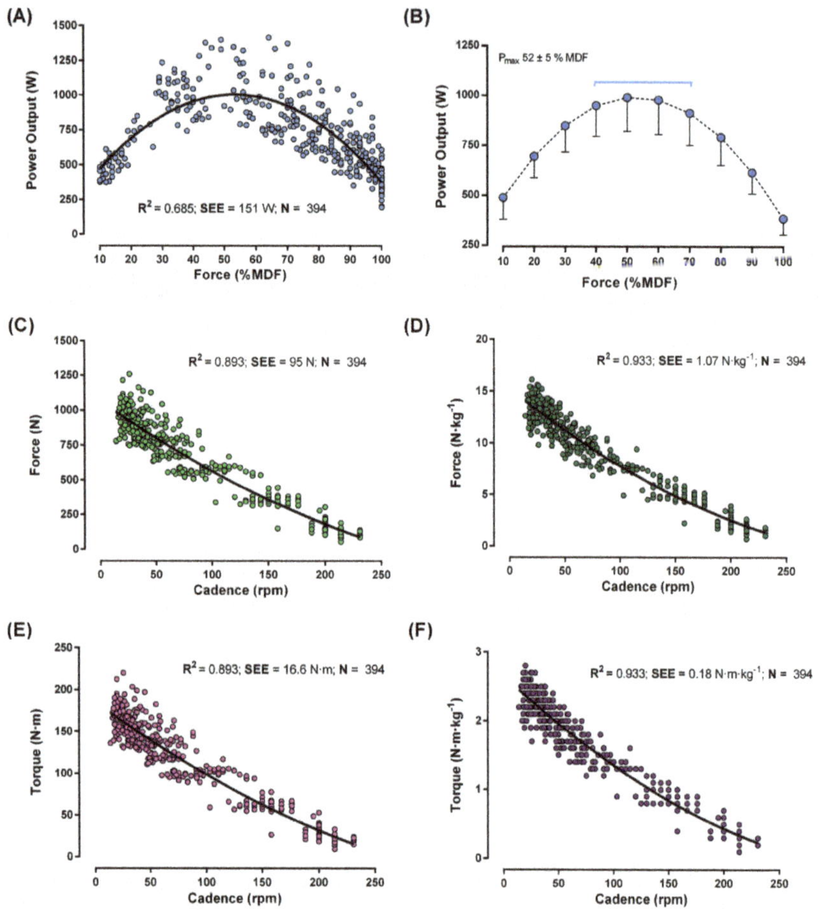

Figure 4. Power output to force (expressed relative to the maximum dynamic force [MDF]) relationship (individual and average data shown in (**A,B**)), force–cadence relationship (**C,D**), and torque–cadence relationship (**E,F**) (n = 52 participants).

Table 1. Comparison of maximum torque values, average pedal cadence, and cadence associated with the maximum dynamic force (MDF) between subgroups of different MDF levels.

Subgroup	VO_{2max} (ml·kg^{-1}·min^{-1})	MAP (W·kg^{-1})	MDF (N·kg^{-1})	Max Torque (N·m·kg^{-1})	Average Cadence (rpm) †	Cadence at MDF (rpm)
G1 (n = 17)	61.5 ± 1.6	5.2 ± 0.2	12.0 ± 0.8 [#]	2.1 ± 0.1 [#]	109 ± 7	22 ± 5
G2 (n = 17)	63.0 ± 1.9	5.4 ± 0.2	13.4 ± 0.4 *	2.3 ± 0.1 *	112 ± 10	22 ± 5
G3 (n = 18)	67.9 ± 2.9	5.9 ± 0.3	14.8 ± 0.6 *	2.6 ± 0.1 *[#]	111 ± 5	21 ± 3

G1, G2 and G3 are the groups based on MDF-tertiles (from lowest to highest). Abbreviations: MAP: maximal aerobic power, rpm: revolutions per minute. Symbols. * Significantly different from G1 (p < 0.05); [#] significantly different from G2 (p < 0.05); † average of all the force–cadence spectrum (from 10 to 100% MDF).

3.2. Test–Retest Reliability

The MDF (969 ± 74 N vs. 965 ± 65 N, p > 0.05) and its associated cadence (23 ± 4 rpm vs. 22 ± 2 rpm, p > 0.05) were similar on days 1 and 2. When analyzing the cadence attained at different percentages of MDF, the results showed a very high test-rest repeatability (mean SEM = 4 rpm, 3.3%) (Table 2).

Table 2. Force–velocity profile (pedal cadence attained with each relative resistive force (% of maximal dynamic force [MDF], $n = 52$ participants) as follows: test–retest reliability ($n = 10$) and long-term stability after 10 weeks of combined aerobic and resistance (leg squat) training ($n = 11$).

Relative Resistive Force (% MDF)	General %MDF—Cadence Relationship		Cadence (rpm)									
			Test–Retest Reliability					Long-Term Stability				
			Day 1	Day 2	Difference (rpm)	SEM (rpm)	SEM (%)	Pre-Training	Post-Training	Difference (rpm)	SEM (rpm)	SEM (%)
	Mean ± SD	95% CI	Mean ± SD	Mean ± SD				Mean ± SD	Mean ± SD			
10	223 ± 15	219–227	222 ± 28	222 ± 23	0.9	7.5	3.4	223 ± 16	226 ± 11	−3.2	9.4	4.2
20	192 ± 12	189–196	192 ± 24	190 ± 21	1.1	5.3	2.8	194 ± 11	195 ± 7	−0.9	6.4	3.3
30	164 ± 11	161–167	163 ± 20	161 ± 19	1.4	4.4	2.7	167 ± 9	166 ± 7	0.6	4.3	2.6
40	138 ± 10	135–141	136 ± 17	135 ± 17	1.5	4.3	3.2	141 ± 10	139 ± 9	1.7	3.6	2.6
50	114 ± 10	111–117	112 ± 15	111 ± 14	1.0	4.6	4.1	117 ± 11	116 ± 10	1.8	3.6	3.1
60	92 ± 10	89–95	90 ± 13	89 ± 12	0.8	4.7	5.3	95 ± 11	93 ± 11	2.5	4.0	4.2
70	71 ± 9	69–74	70 ± 10	69 ± 9	0.8	4.5	6.5	74 ± 10	72 ± 10	2.8	4.0	5.5
80	53 ± 7	51–55	52 ± 8	51 ± 7	1.5	4.0	7.7	55 ± 9	53 ± 9	2.7	3.8	7.0
90	36 ± 5	35–38	36 ± 5	35 ± 4	1.2	3.2	9.0	38 ± 6	36 ± 6	2.2	3.2	8.6
100	22 ± 4	21–23	23 ± 4	22 ± 2	1.1	3.3	14.5	22 ± 6	21 ± 4	1.3	2.9	13.2

Abbreviations: CI, confidence interval; rpm, revolutions per minute; SD, standard deviation; SEM, standard error of the measurement.

3.3. Long-Term Stability

Although the MDF of the participants who underwent the 10-week resistance program significantly increased with training (966 ± 76 N vs. 1001 ± 92 N at pre- and post-training, respectively, $p = 0.013$), values from the %MDF–cadence relationship remained stable from pre- to post-training (SEM = 4 rpm, 2.3%) (Table 2).

4. Discussion

In the present study, we propose a novel test for the assessment of the torque–cadence profile and MDF during pedaling. Through the proposed equation, the relative resistive load (expressed as a % of MDF) produced during a given effort could be estimated by attending to the attained cadence. Our findings support an overall high test–retest reliability of the force–velocity profile, as well as high stability in the face of performance changes or different levels of cyclists' MDF. Thus, this test might be useful for prescribing or identifying relative intensities and for monitoring training-induced changes in different zones of the force–velocity curve.

Previous studies have implemented methodological procedures for the assessment of the force–cadence profile during cycling. Rudsits et al. [4] determined this profile through six sprints against increasing external loads, eliciting torque values from 0 to 4 $N \cdot m \cdot kg^{-1}$, which resulted in cadences ranging between ~41 and ~214 rpm. Of note, the authors concluded that a robust assessment of the torque–velocity profile during pedaling required recording a large number of pedal cycles completed over a wide range of cadences [4]. However, these authors mostly focused on how the testing and modeling procedures can influence the torque–cadence profile and did not assess the test–retest reliability or long-term stability as we did here.

García-Ramos et al. [10] also determined the force–velocity profile through 5–6 sprints against increasing external resistive forces between 0.4 $N \cdot kg^{-1}$ (172 rpm) and 1.3 $N \cdot kg^{-1}$ (83 rpm). Interestingly, the authors observed a higher test–retest reliability for the cadence associated with the lightest compared to the heaviest loads, respectively, which was confirmed in the present study. For this reason, these authors recommended using two distant but relatively light loads (corresponding to >110 rpm) when applying the so-called 'two-point method' for the estimation of the force–velocity profile [10]. In a subsequent study, the same research group confirmed that the two-point method, using 180–200 rpm and 110–125 rpm, could be a reliable procedure for assessing the force–velocity profile during pedaling [11].

In the present study, we assessed the torque–velocity profile using the widest range of cadences assessed to date (from ~22 to ~220 rpm). In this regard, although in line with García-Ramos et al., we found lower reliability with the heavier loads (e.g., SEM > 5% with cadences < 90 rpm), and the overall force–velocity profile appeared highly reliable (SEM of 2 to 3%). Moreover, we observed a very high consistency in this profile, with a given cadence representing a similar relative load despite between- and within-subject variations for the MDF (as shown in Tables 1 and 2, respectively). These results might support the validity of the torque–cadence profile for identifying potential limitations or weaknesses in an individual's profile and for the assessment of training-induced changes. For instance, García-Ramos et al. [12] found that 6 weeks of both heavy- and light-load sprint training induced a shift in the slope of the torque–cadence profile. However, those individuals who trained with heavy loads improved their maximum torque levels to a greater extent than those who trained using a light load, whereas the opposite trend was observed for the highest cadences.

In the present study, we also propose a novel indicator, such as the MDF during pedaling, defined as the maximum force that can be produced during a whole pedaling cycle. This parameter seems to correspond, at least in the present cohort, to a cadence of 21–22 rpm regardless of individuals' MDF levels, as confirmed in both within- and between-subject analyses. Thus, our findings might support the validity of the percentage of the MDF as an indicator of the relative load of efforts as performed during training or competition, as well as during specific strength training stimuli (e.g., the so-called 'torque' training) [16–18]. This could be of particular relevance for research purposes, allowing to match relative training loads during cycling using the velocity of muscle contractions (i.e., pedal cadence), similar to what is typically performed in resistance exercises such as leg squat, bench press, or prone bench pulls (i.e., velocity-based training) [14,23]. This practical application for cycling can be exemplified by recent studies assessing the effects of the so-called 'torque' training (i.e., performing short-duration bouts at low cadences [40–60 rpm] to increase torque production capacity) [16–18]. However, the authors of these studies [16–18] could not quantify the relative loads of these bouts with respect to the participants' MDF (Figure 2), and therefore, whether these training sessions actually elicited high individual torque levels remains unknown.

It is worth emphasizing that we propose a relatively simple and economical procedure for the assessment of the force–velocity profile during cycling, as well as for the determination of a novel parameter such as the MDF. Our results showed that the % of the MDF–cadence relationship is reliable and stable over time, regardless of the changes or different levels of cyclists' MDF. In practice, the very close adjustment we found for this relationship would allow cyclists: (i) to determine the percentage of the MDF that is being used during every training or competition effort and (ii) to program the target cadence to train at a planned percentage of the MDF. Moreover, the cadence achieved against the same load (in kp) pre-and post-training could be measured (iii) for practically quantifying changes in cyclist's performance—e.g., the pre-post training differences of ~12 rpm at intensities $\leq$ 50% of MDF or ~8 rpm at intensities > 50% of MDF would represent a performance change of ~5%.

This research is not exempt from limitations. Firstly, only trained male cyclists were included. Although it is hypothesized that the fit of the % of the MDF–cadence relationship would also be very strong in recreational and female cyclists, this aspect should be verified. Secondly, only one training stimulus (squat exercise) was used to examine the long-term stability of the % of the MDF–cadence relationship, so this stability should be examined after applying other stimuli like the so-called "torque" training. Finally, future studies should examine the mechanisms that could be behind the reduction in the linearity of the % of the MDF–cadence relationship at high intensities. Among others, changes in aspects like muscle recruitment and/or pedaling technique when pedaling at the maximal voluntariness against high resistances could explain this fact.

5. Practical Applications

We propose a relatively simple and economical procedure for the assessment of the force–velocity profile during cycling, as well as for the determination of a novel parameter such as the MDF. Our results showed that the % of the MDF–cadence relationship is reliable and stable over time, regardless of the changes in or different levels of cyclists' MDF. In practice, the very close adjustment we found for this relationship would allow cyclists: (i) to determine the percentage of the MDF that is being used during every training or competition effort and (ii) to program the target cadence to train at a planned percentage of the MDF. The accuracy of these first two practical applications could even be maximized by using each cyclist's individual relationship, thus reducing the slight between-subject differences associated with each % of the MDF. Moreover, the cadence achieved against the same load (in kp) pre-and post-training could be measured (iii) for practically quantifying changes in cyclist's performance—e.g., the pre-post training differences of ~12 rpm at intensities < 50% of MDF or ~8 rpm at intensities > 50% of MDF would represent a performance change of ~5%.

Author Contributions: Conception and design: V.R.-R., D.B.-G., A.H.-B., J.I., P.L.V. and J.G.P.; Acquisition of data: D.B.-G., A.H.-B., E.H.-L., J.R.L.-B., A.M.-C., Á.B.-R., L.B.A., V.R.-R., I.R.S.-R. and R.d.P.; Analysis and interpretation of data: A.H.-B., E.H.-L., J.I., P.L.V. and J.G.P.; Drafting the article or revising it critically for important intellectual content: D.B.-G., A.H.-B., E.H.-L., P.L.V. and J.G.P.; Final approval of the version to be published and agreement to be accountable for all aspects of the work: D.B.-G., A.H.-B., J.I., J.R.L.-B., A.M.-C., Á.B.-R., L.B.A., V.R.-R., I.R.S.-R., R.d.P., P.L.V. and J.G.P. All authors have read and agreed to the published version of the manuscript.

Funding: PLV is funded by a postdoctoral contract granted by Instituto de Salud Carlos III (Sara Borrell, CD21/00138). D.B.-G., and P.L.V. are funded by the Spanish Ministry of Economy and Competitiveness and Fondos FEDER (PI18/00139). AHB is funded by the Spanish Ministry of Education, Culture and Sports.

Institutional Review Board Statement: The study was conducted in accordance with the Declaration of Helsinki and approved by the Ethical Committee of the University of Murcia (ID: 4135/2022), approval date: 22 September 2022.

Informed Consent Statement: Informed consent was obtained from all subjects involved in this study. Written informed consent was obtained from the participants to publish this paper.

Data Availability Statement: Data that support the findings of this study are available from the corresponding author upon reasonable request. The data are not publicly available due to privacy.

Acknowledgments: The authors thank the subjects for their invaluable contribution to this study.

Conflicts of Interest: The authors declare no conflicts of interest.

References

1. Alcazar, J.; Csapo, R.; Alegre, L.M. Editorial: The Force-Velocity Relationship: Assessment and Adaptations Provoked by Exercise, Disuse and Disease. *Front. Physiol.* **2022**, *13*, 1062041. [CrossRef]
2. Taylor, K.B.; Deckert, S.; Sanders, D. Field-Testing to Determine Power—Cadence and Torque—Cadence Profiles in Professional Road Cyclists. *Eur. J. Sport Sci.* **2023**, *23*, 1085–1093. [CrossRef] [PubMed]
3. García Ramos, A.; Jaric, S. Two-Point Method: A Quick and Fatigue-Free Procedure for Assessment of Muscle Mechanical Capacities and the One-Repetition Maximum. *Strength Cond. J.* **2017**, *40*, 54–66. [CrossRef]
4. Rudsits, B.L.; Hopkins, W.G.; Hautier, C.A.; Rouffet, D.M. Force-Velocity Test on a Stationary Cycle Ergometer: Methodological Recommendations. *J. Appl. Physiol.* **2018**, *124*, 831–839. [CrossRef] [PubMed]
5. Sargeant, A.J.; Hoinville, E.; Young, A. Maximum Leg Force and Power Output during Short-Term Dynamic Exercise. *J. Appl. Physiol. Respir. Environ. Exerc. Physiol.* **1981**, *51*, 1175–1182. [CrossRef] [PubMed]
6. Martin, J.C.; Wagner, B.M.; Coyle, E.F. Inertial-Load Method Determines Maximal Cycling Power in a Single Exercise Bout. *Med. Sci. Sports Exerc.* **1997**, *29*, 1505–1512. [CrossRef] [PubMed]
7. Coso, J.D.; Mora-Rodríguez, R. Validity of Cycling Peak Power as Measured by a Short-Sprint Test versus the Wingate Anaerobic Test. *Appl. Physiol. Nutr. Metab.* **2006**, *31*, 186–189. [CrossRef]
8. Dorel, S.; Hautier, C.A.; Rambaud, O.; Rouffet, D.; Van Praagh, E.; Lacour, J.R.; Bourdin, M. Torque and Power-Velocity Relationships in Cycling: Relevance to Track Sprint Performance in World-Class Cyclists. *Int. J. Sports Med.* **2005**, *26*, 739–746. [CrossRef]
9. Dorel, S.; Couturier, A.; Lacour, J.-R.; Vandewalle, H.; Hautier, C.; Hug, F. Force-Velocity Relationship in Cycling Revisited: Benefit of Two-Dimensional Pedal Forces Analysis. *Med. Sci. Sports Exerc.* **2010**, *42*, 1174–1183. [CrossRef]
10. García-Ramos, A.; Torrejón, A.; Morales-Artacho, A.J.; Pérez-Castilla, A.; Jaric, S. Optimal Resistive Forces for Maximizing the Reliability of Leg Muscles' Capacities Tested on a Cycle Ergometer. *J. Appl. Biomech.* **2018**, *34*, 47–52. [CrossRef]
11. García-Ramos, A.; Zivkovic, M.; Djuric, S.; Majstorovic, N.; Manovski, K.; Jaric, S. Assessment of the Two-Point Method Applied in Field Conditions for Routine Testing of Muscle Mechanical Capacities in a Leg Cycle Ergometer. *Eur. J. Appl. Physiol.* **2018**, *118*, 1877–1884. [CrossRef]
12. García-Ramos, A.; Torrejón, A.; Pérez-Castilla, A.; Morales-Artacho, A.J.; Jaric, S. Selective Changes in the Mechanical Capacities of Lower-Body Muscles After Cycle-Ergometer Sprint Training Against Heavy and Light Resistances. *Int. J. Sports Physiol. Perform.* **2018**, *13*, 290–297. [CrossRef]
13. Hernández-Belmonte, A.; Martínez-Cava, A.; Morán-Navarro, R.; Courel-Ibáñez, J.; Pallarés, J.G. A Comprehensive Analysis of the Velocity-Based Method in the Shoulder Press Exercise: Stability of the Load-Velocity Relationship and Sticking Region Parameters. *Biol. Sport* **2020**, *38*, 235–243. [CrossRef] [PubMed]
14. Pareja-Blanco, F.; Walker, S.; Häkkinen, K. Validity of Using Velocity to Estimate Intensity in Resistance Exercises in Men and Women. *Int. J. Sports Med.* **2020**, *41*, 1047–1055. [CrossRef]
15. Loturco, I.; McGuigan, M.R.; Pereira, L.A.; Pareja-Blanco, F. The Load-Velocity Relationship in the Jump Squat Exercise. *Biol. Sport* **2023**, *40*, 611–614. [CrossRef]
16. Nimmerichter, A.; Eston, R.; Bachl, N.; Williams, C. Effects of Low and High Cadence Interval Training on Power Output in Flat and Uphill Cycling Time-Trials. *Eur. J. Appl. Physiol.* **2012**, *112*, 69–78. [CrossRef] [PubMed]
17. Kristoffersen, M.; Gundersen, H.; Leirdal, S.; Iversen, V.V. Low Cadence Interval Training at Moderate Intensity Does Not Improve Cycling Performance in Highly Trained Veteran Cyclists. *Front. Physiol.* **2014**, *5*, 34. [CrossRef] [PubMed]

18. Paton, C.D.; Hopkins, W.G.; Cook, C. Effects of Low- vs. High-Cadence Interval Training on Cycling Performance. *J. Strength Cond. Res.* **2009**, *23*, 1758–1763. [CrossRef]
19. Pallarés, J.G.; Morán-Navarro, R.; Ortega, J.F.; Fernández-Elías, V.E.; Mora-Rodriguez, R. Validity and Reliability of Ventilatory and Blood Lactate Thresholds in Well-Trained Cyclists. *PLoS ONE* **2016**, *11*, e0163389. [CrossRef]
20. McKay, A.K.A.; Stellingwerff, T.; Smith, E.S.; Martin, D.T.; Mujika, I.; Goosey-Tolfrey, V.L.; Sheppard, J.; Burke, L.M. Defining Training and Performance Caliber: A Participant Classification Framework. *Int. J. Sports Physiol. Perform.* **2022**, *17*, 317–331. [CrossRef]
21. Rodríguez-Rielves, V.; Martínez-Cava, A.; Buendía-Romero, Á.; Lillo-Beviá, J.R.; Courel-Ibáñez, J.; Hernández-Belmonte, A.; Pallarés, J.G. Reproducibility of the Rotor 2INpower Crankset for Monitoring Cycling Power Output: A Comprehensive Analysis in Different Real-Context Situations. *Int. J. Sports Physiol. Perform.* **2021**, *1*, 120–125. [CrossRef] [PubMed]
22. Hernández-Belmonte, A.; Courel-Ibáñez, J.; Conesa-Ros, E.; Martínez-Cava, A.; Pallarés, J.G. Level of Effort: A Reliable and Practical Alternative to the Velocity-Based Approach for Monitoring Resistance Training. *J. Strength Cond. Res.* **2022**, *36*, 2992–2999. [CrossRef] [PubMed]
23. Grazioli, R.; Loturco, I.; Lopez, P.; Setuain, I.; Goulart, J.; Veeck, F.; Inácio, M.; Izquierdo, M.; Pinto, R.S.; Cadore, E.L. Effects of Moderate-to-Heavy Sled Training Using Different Magnitudes of Velocity Loss in Professional Soccer Players. *J. Strength Cond. Res.* **2023**, *37*, 629–635. [CrossRef] [PubMed]

Article

Proficiency Barrier in Track and Field: Adaptation and Generalization Processes

M. Teresa S. Ribeiro [1,2], Filipe Conceição [2] and Matheus M. Pacheco [2,3,*]

[1] Research Center in Sport Sciences, Health and Human Development (CIDESD), Physical Education and Sport Sciences Department, University of Maia, 4475-690 Maia, Portugal; teresaribeiro@umaia.pt
[2] Center for Investigation, Formation, Innovation and Intervention in Sports, Faculty of Sport, University of Porto, 4200-450 Porto, Portugal; filipe@fade.up.pt
[3] GEDEM, Department of Physical Education, Federal University of Rondônia, Porto Velho 78900-000, Brazil
* Correspondence: matheusp@fade.up.pt; Tel.: +351-22-04-25-217

Abstract: The literature on motor development and training assumes a hierarchy for learning skills—learning the "fundamentals"—that has yet to be empirically demonstrated. The present study addressed this issue by verifying (1) whether this strong hierarchy (i.e., the proficiency barrier) holds between three fundamental skills and three sport skills and (2) considering different transfer processes (generalization/adaptation) that would occur as a result of the existence of this strong hierarchy. Twenty-seven children/adolescents participated in performing the countermovement jump, standing long jump, leap, high jump, long jump, and hurdle transposition. We identified the proficiency barrier in two pairs of tasks (between the countermovement jump and high jump and between the standing long jump and long jump). Nonetheless, the transfer processes were not related to the proficiency barrier. We conclude that the proposed learning hierarchy holds for some tasks. The underlying reason for this is still unknown.

Keywords: transfer of performance; motor development; stages; skill acquisition

Citation: Ribeiro, M.T.S.; Conceição, F.; Pacheco, M.M. Proficiency Barrier in Track and Field: Adaptation and Generalization Processes. *Sensors* **2024**, *24*, 1000. https://doi.org/10.3390/s24031000

Academic Editors: Felipe García-Pinillos, Diego Jaén-Carrillo and Alejandro Pérez-Castilla

Received: 30 December 2023
Revised: 22 January 2024
Accepted: 2 February 2024
Published: 4 February 2024

1. Introduction

In the motor development and training literature (e.g., [1–4]), there is a long-standing assumption that maintains a high priority for learning the "fundamentals". In general terms, this is the assumption that simpler or basic skills must be practiced and learned before one can further specialize in or learn more complex skills. Examples of this assumption are found in descriptive models and guidelines in sports/physical education (see, for instance, [5]).

These fundamentals are usually related to two different ideas, depending on the area. In motor development, a strong case is made for the "fundamental movement patterns" [6]. The fundamental movement patterns are broadly defined [7] as those performed in "common" motor activities without imposed specific performance goals. Examples of these common activities are throwing, kicking, and running. In these descriptive models, it is implied that one would be unable to learn sport-specific movement patterns (e.g., dart throwing, specific kicking patterns in soccer, and 100 m sprint running) if these fundamental skills are not being performed skillfully [1,8,9].

In training, the expected relationship between fundamentals and performance is more subtle. Usually, the claim is in terms of a given physical ability (e.g., "reactive force", "explosive strength", and "agility") that must be developed so one can demonstrate good performances of sport-specific movement patterns (see [3]). Note, nonetheless, that the discussed physical ability is usually considered (and measured) within the context of a given "basic" skill (e.g., explosive strength can be measured by the countermovement jump). In some cases, the basic skill is used for practice, as it would also be the best way to develop such a physical ability.

In both cases, there is an implied hierarchy of the learning contents for a learner to perform well in sport skills. Vern Seefeldt [1,8] promoted a strong case for this hierarchy: he stated that those with low proficiency (a large difference from a ("gold"/optimum) standard movement pattern) in the fundamental movement skills would have great difficulty in learning more complex skills in motor development. He termed this considerable difficulty the "proficiency barrier". It should be emphasized that this hypothesis is a direct implication from many of the stage models of motor development [1,2,5,10] that encompass a hierarchy of learning.

In recent years, a debate has started concerning the ontological status of these fundamentals. Newell [11], for example, questioned whether the so-called fundamental movement skills meet the necessary conditions to be called "fundamental" to start with. Others have tried to provide new terminology (calling them foundational movement skills [12]), calling into question their centrality in developmental theories [13] or defending their centrality in motor development [6].

Notably, one still needs to demonstrate empirical evidence on the proposed hierarchy involving these skills (or the assumed hierarchy). Only recently did three studies directly support the proficiency barrier hypothesis [9,14,15]. These studies considered whether superior results in two fundamental movement patterns' assessments (e.g., running and bouncing a ball) were necessary for either demonstrating or learning a more complex (sport-specific) movement pattern that is supposed to be a combination of two simpler ones (e.g., dribbling). Nonetheless, it is not true that the hierarchy holds for all situations. The aforementioned supportive results contrast with other studies that failed to show such a dependency [16,17]. In the contrasting studies, either one could learn a sport-specific skill without showing good results in the fundamental skills [17] or the relationship among skills existed but showed no signs of a given proficiency barrier [16].

There are many potential reasons why the proficient barrier does not hold for all cases. The main one, we argue, is that one can expect more than a single relationship between skills: generalization and adaptation. Both can be said to be types of transfer (i.e., the effect that practice in task A has on performance of task B [18]). Generalization is simply the application of something learned in a specific context to others. This is usually what is supposedly studied in motor learning experiments when researchers implement experimental designs with transfer tests of 10 to 20 trials after practice (also, see the rationale behind some theories [19,20]). Adaptation, on the other hand, is a process of change that occurs *based* on what was learned in the original practice. This is what is usually implied in motor development—what can be learned is dependent on the previous experiences/tendencies of the individual (see, for instance, [21,22]). Few studies have used such terminology, but they demonstrated this process (see [23,24] for an approach based on such a process). Importantly, studies have only considered one or the other in their designs and, for this reason, the principles that differentiate when one or the other is observed are unexplored.

Our main purpose is to address a long-standing assumption on the necessity of learning fundamental movement patterns for learning sport-specific skills and the processes that would be based on such a necessity. Specifically, we investigate whether the appearance of a proficiency barrier is a matter of the type of transfer that occurred. Our hypothesis is that the proficiency barrier occurs when adaptation is required. This would be the case when what is learned in the fundamental skill is not what is performed for the sport skill but a necessary condition for the sport skill to be learned. Only after learning the required (i.e., fundamental) skill "components" can the more complex skill be learned, i.e., a nonlinear relationship (such as in [14,15]). On the other hand, if one can generalize learning from the fundamental skill to the sport skill, then the better one is at the fundamental skill, the better one is at the sport skill, i.e., a linear relation (such as in [16]).

To reach our goal, the present study investigates how three "fundamental" skills relate to three sport-specific skills in track and field. We selected the countermovement jump, the horizontal jump, and the obstacle jump as fundamental movement skills and the high jump, long jump, and hurdles as sport-specific skills. To investigate the relationships among the skills, we (1) assessed the potential existence of a proficiency barrier and (2) verified whether the proficiency barrier relates to generalization/adaptation processes. Different from previous studies on the topic [9,14–16] that implemented checklist-based assessments, we analyzed the hypothesis through motion tracking technology to identify movement pattern characteristics that differentiate individuals without an a priori assumption of them.

2. Methods

2.1. Sample

Twenty-seven young athletes (15 girls) participated in this study. Table 1 shows the sample's characteristics. The participants were, at that time, affiliated with a track and field outreach program of the Faculty of Sport. Participation in this study was voluntary. Legal guardians read and signed an informed consent form. The Ethics Committee of the Faculty of Sport, University of Porto, approved all procedures.

Table 1. Sample characteristics (age, height, weight, body fat percentage, and categories).

Age	Height (m)	Weight (kg)	Body Fat %
12.31 ± 3.19	1.51 ± 0.19	45.66 ± 15.60	19.40 ± 6.52
	Age Categories (n)		
Under 10		5	
Under 12		5	
Under 14		5	
Under 16		7	
Under 18		3	
Under 20		2	

2.2. Task and Materials

To characterize participants' anthropometry, we used a stadiometer (Seca 213 stadiometer, Hamburg, Germany) with a precision of 0.1 cm and a portable bioimpedance scale (Tanita BC-730, Tokyo, Japan). All measurements adhered to the protocols established by the International Working Group on Kineanthropometry standards [25].

Participants performed six tasks. Figure 1 shows the countermovement jump, standing long jump, leap, high jump, long jump, and hurdle transposition. The order of data collection was always the same for all participants: countermovement jump, standing long jump, leap, long jump, high jump, and hurdle transposition. We considered the first three tasks as "fundamental" in the sense that they are performed as general activities (not necessarily in the context of sports). The last three tasks are specific events in track and field (representing, thus, the sport-specific assessment). Eight motion capture cameras recorded all tasks at 100 Hz (Miqus Video, Qualisys AB, Gothenburg, Sweden). For all tasks, participants performed ten repetitions with a 30 s rest interval.

For the countermovement jump, the athlete starts in a standing position with feet parallel and shoulder-width distance. After the experimenter commands "go", the athlete was instructed to jump as high as possible after a fast flexion of the hips and knees.

In the standing long jump, the athlete begins in a standing position, with feet parallel and behind the starting line. The experimenter instructed the athlete that, upon the "go" signal, he/she should jump as far as possible forward. The jumps were measured with a measuring tape placed parallel to the jumping direction.

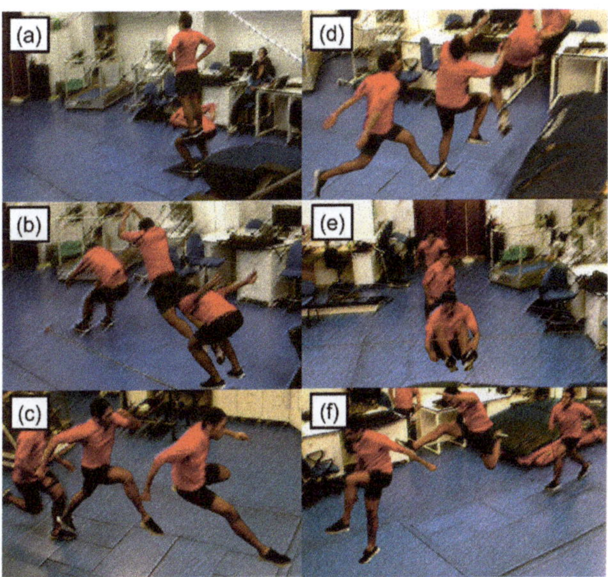

Figure 1. Exemplary participant performing the (**a**) countermovement jump; (**b**) standing long jump; (**c**) leap; (**d**) high jump; (**e**) long jump; and (**f**) hurdle transposition.

For the leap, we implemented an adapted version of the leap used in the Test of Gross Movements Development (TGMD-2 [26]). The experimenter instructed the athlete to run toward a "water puddle" and leap over it. Also, the athlete was supposed to continue to run after landing.

For the high jump test, the experimenter instructed participants to run and jump with their preferred foot, aiming to jump as high as possible (over an "imaginary" bar) and then land with their back on the mattress. If the preferred foot was not known, a few familiarization trials were permitted.

In the long jump test, participants were instructed to run and jump with their preferred foot, aiming to jump as far as possible and then land in a seated position. The landing was performed on a suitable mattress for this purpose.

For the hurdle transposition, the experimenter instructed athletes to run and jump over an imaginary hurdle (clear the hurdle) placed in the middle of the running path. After clearing the hurdle, the athletes were instructed to continue running.

2.3. Procedures

Each evaluation session was carried out with a single participant and in the same place (University of Porto Biomechanics Laboratory). Before each session, the space was calibrated while participants performed a warm-up (slow run and mobility exercises). Then, anthropometric measures were taken, and the participant received instructions on each motor task to be performed. For all tasks, the experimenters provided no demonstration to the participant beyond the general instructions described in the previous session.

2.4. Data Analysis

After data collection, we used Theia3D (v2023.1.0.3161, Theia Markerless, Inc., Kingston, ON, Canada) to extract the joint motion of each participant. Theia3D offers a markerless motion capture solution that relies on synchronized video data to generate precise and dependable 3D pose estimations of human subjects visible in the footage. The system employs advanced deep learning algorithms (deep convolutional neural networks) trained to recognize humans and accurately predict the 2D positions of over 100 landmarks on the human body for each frame in every camera's video. By applying a subject-specific inverse kinematic model scaled to the predicted landmarks, the 3D pose of the human is reconstructed and continuously tracked throughout their movements. This data-driven approach ensures a robust solution that is applicable across various environments and movements, enabling the efficient collection of high-quality 3D motion capture data. Theia3D has shown higher reliability than marker-based methods for lower limb kinematics (less than 3.5° of variability) [27] and, given the complexity of the upper limb kinematics, an error in the range of 8.1 to 23° for the upper limb joints (root mean squared error) [28].

The data were processed considering arms and feet as 6-degrees of freedom segments and with a low-pass filter at 20 Hz. The data were further processed using a designed Matlab script (all codes can be assessed in https://osf.io/bgxa8/, accessed on 3 February 2024) (Matlab R2023b Update 4 [23.2.0.2428915]).

For all joint motion analyses, 20 dimensions were considered: ankle flexion, knee flexion, hip flexion and abduction, thorax flexion, abduction and rotation, shoulder flexion and abduction, elbow flexion, and pelvis angle relative to the lab. We also considered some position measures of the center of mass, left foot center of gravity, and right foot center of gravity.

For all joint motions, we shifted the angles that showed discontinuities around ±180°. Then, we filled the potential missing data within a trial with the *spline* function in Matlab and filtered the data with a 10 Hz fourth-order Butterworth low-pass filter.

Provided some issues with near-static moments of the trial, the Theia3D software would create spurious 360° rotations around a given joint. The designed script would identify joint angles with rotations above 300° and identify the moment of rotation using the *findchangepts* function in Matlab (with a maximum of 2 changes: beginning and end of the spurious rotation). For trials in which this spurious rotation took less than 25% of the trial, the script cleared the frames in which the spurious rotation occurred, decreased the time series after the rotation by adding ±360° (depending on the direction of the spurious rotation), and filled the data with the *spline* function. Trials in which the spurious rotation took longer than 25% of the trial were not further considered in the analysis. This was not considered for pelvis rotations with reference to the laboratory (as this could occur in the high jump).

Considering potential issues in the Theia3D processing (such as not processing a given trial) or the aforementioned spurious rotations, we missed 48 trials (2.96% of the trials).

Before calculation of the performance measures or the movement patterns, we selected only the moment of interest in the whole recording. For the countermovement jump, we considered, as the beginning of the trial, the moment of the lower center of mass height before the center of the mass peak height and, as the end, the moment of the center of mass peak height. For the standing long jump, we considered, as the beginning of the trial, the moment of the maximum knee flexion before the peak velocity forward of the center of mass and, as the end of the trial, the first negative peak acceleration of the center of mass after the peak velocity forward. For the leap, long jump, and the hurdle transposition, as the beginning and end of the trial, respectively, we found the minima before and after the peak height of the center of mass. For the high jump, we considered, as the beginning of the trial, the second minimum before the center of the mass peak height and, as the end of the trial, the first minimum after the center of the mass peak height.

For all further analyses, the trials were time-normalized using the *spline* function in Matlab. Thus, all trials have 100 frames (1 to 100%).

2.4.1. Performance

The performance of each task was derived according to the task's demands. For the countermovement jump, we used the peak height of the center of mass as the performance. For the standing long jump, we used the landing distance of the individual. If the individual fell, we considered the local where the feet first touched the ground after the jump. For the leap, high jump, and hurdle transposition, we considered performance as the highest height the body achieved in the trial. For this, we considered the height of both the feet and center of mass over time and selected the lowest height "segment" over time. From this, we determined the maximum that the lowest height segment reached. For the long jump, we considered the distance reached derived from the velocity of the center of mass after the jump.

2.4.2. Movement Patterns

To characterize the movement pattern, we performed a principal component analysis for each trial using all 20 joint motions. From the outcome, we considered the first 2 principal component coefficients (accounting for, on average, 89.78% of the variance of the data) and compared them among skills using the normalized dot product. We also calculated the cumulative sum of variance accounted for each principal component and noted how many principal components were necessary to explain at least 90% of the data.

2.4.3. General Associations

As we have a large variety of ages and anthropometric characteristics, as well as six motor tasks being performed, we decided to characterize in general terms the associations among all variables before delving into our primary questions. For this reason, we first performed seven linear mixed effect models with performance outcomes for each condition as dependent variables and sex, age, fat percentage, and trials as independent variables.

For the number of components, we used the average number per condition per participant and performed a Friedman's ANOVA to understand whether these different tasks had a tendency toward qualitatively different movement patterns.

Additionally, as specific pairings were considered for the proficiency barrier and transfer processes assessments (see below), we determined the spearman's ρ correlations among all six tasks, for the sake of completeness.

2.4.4. Proficiency Barrier Assessment

The proficiency barrier is a phenomenon that is, primarily, tested longitudinally: if someone demonstrates low proficiency in a fundamental skill, the learning (a longitudinal process) of the sport skill becomes difficult, if not impossible. Pacheco et al. [15] pointed out that such a longitudinal process shows its signature in cross-sectional measurements and demonstrated it through a sigmoidal relationship between fundamental and sport skills.

It is important to note, however, that the relationship is demonstrated among the skills' movement patterns on a continuous scale. Despite the fact that movement patterns are categorical (i.e., one movement pattern is not intrinsically comparable to another by itself), studies got around this issue by employing criteria-based movement assessments. One can sum the achieved criteria for each skill and then see whether the relationship fits the expectation.

There are issues with these criteria-based assessments, nonetheless. First, this type of assessment implies that there is a given standard (i.e., a "champion model") that the performer must follow. To our knowledge, this is hardly justifiable a priori: there is no reason to believe that individuals, with their own previous experiences and biomechanical individualities, converge to the same optimal technique (see [29,30]). Further, it implies a *unique* pathway between fundamental and sport skills learning: only if the athlete performs the fundamental movement pattern in the way that the assessment considers, then the athlete can learn the sport skill. It also assumes that the movement pattern is sufficient for measuring skill level. As numerous studies argue (see [31,32]), an athlete can reach a high level of performance through different movement patterns. The second issue is that, even if there was a single movement pattern standard for fundamental and sport skills, one would need to validate this type of assessment for all skills in order to understand the phenomenon at stake here. Considering the two issues together creates an insurmountable problem.

Thus, to infer a proficiency barrier between the fundamental tasks and the sport-specific ones, we performed a two-step procedure. First, we performed the same procedure as in Pacheco et al. [15]: we compared the linear and sigmoid functions fit between fundamental and sport-specific performances. The rationale behind the comparison is that, if there is a minimum value of a fundamental skill performance on which an individual must demonstrate to learn a sport skill, then the curve between the fundamental skill and sport skills performance would be nonlinear: two relationships (i.e., regimes) separated by a threshold value (i.e., the proficiency barrier). Pacheco tested two different possibilities (piecewise and sigmoid functions) and the sigmoid function demonstrated the best fit.

For this, we compared the resultant corrected Akaike Information Criterion (AICc, see [33]) between the linear function

$$S = \alpha + \beta \, \text{FMS} \tag{1}$$

and the sigmoidal function

$$S = \alpha + (\beta - \alpha)/(1 + \exp(-\delta^*(\text{FMS} - \gamma))), \tag{2}$$

where S is the sport-specific task and FMS is the fundamental movement skill; and α, β, δ, and γ are free parameters. The AICc penalizes the number of required parameters provided the explained variance of the fit—also being appropriate for small sample sizes. For interpretation, smaller AICc values refer to a better fit.

In Pacheco et al. [15], for the S and FMS in the above functions, they used a score of the summed criteria achieved by the performance of the movement pattern. Given the lack of such measures here, we fitted the function considering the performance average for each task.

The sigmoidal fitting was performed with the nonlinear least squares method using the *fit* function from Matlab R2023b. The free parameters α and β were constrained to have minima and maxima values of 10% (of the range) below and above, respectively, the S variable values considered in the pairing, and δ was constrained from 0 to infinity and γ from 10% below up to 10% above the FMS variable values. The starting points were considered the 25th percentile of the S variable values, the 75th percentile of the S variable values, 1, and the median of the FMS variable values for α, β, δ, and γ. The linear function was fitted using the same function with the *poly1* option. The AICc was calculated as in [33].

The pairings of FMS and S were defined by an arbitrary "proximity" relationship between the skills. The pairings were defined as a countermovement jump and high jump, standing long jump and long jump, and leap and hurdle transposition.

Provided that the performances' relationships can be misleading (see [15]), we also considered the demonstrated movement pattern. The second step, then, considered whether participants with the best performances of the sport skills showed similar movement patterns at the fundamental skills. This would imply that those who reached higher levels of performance had to demonstrate something fundamental to reach these performances. For this, we performed, for each of the fundamental skills, the normalized dot product among participants (considering both the first and second principal components). The normalized dot product ranges from 0 to 1 and values above 0.9 are considered as high similarity (see [34]).

2.4.5. Generalization and Adaptation

After the assessment of a potential proficiency barrier, we investigated whether we would find signs of generalization/adaptation between the performed movement patterns in fundamental and sport skills. For all cases, we ordered the participants in terms of the demonstrated performance of the sport skill.

First, we tested whether individuals' "generalization" (maintenance of the fundamental movement pattern in the sport skill) was a function of the performance demonstrated in the sport skill. Generalization was measured using the normalized dot product between the same skill's pairings used for testing the proficiency barrier. We compared the skills using the first and second principal components. Then, we calculated the Spearman's ρ

correlation to evaluate whether there was any relationship between the performance of the sport skill and the relationship among movement patterns.

Second, we aimed to compare whether generalization/adaptation of the movement pattern occurred, in general, for tasks with/without the presence of a proficiency barrier. We considered the first and second principal components of each task as a single vector and performed the normalized dot product between the fundamental and sport skills pairings. Then, we performed a Friedman's ANOVA to establish whether tasks that showed a potential proficiency barrier were, indeed, the ones that required adaptation of the performed movement pattern.

Third, considering the same skill pairings, we tested whether the change in the number of components needed to perform the sport skills from the fundamental skill was a function of the skill reached for the sport skill. For this, we calculated the cumulative sum of variance accounted for each principal component and noted how many principal components were necessary to explain at least 90% of the data. Then, we subtracted the required components of the sport skills from the fundamental skills and correlated (using Spearman's ρ correlation) this with the performance achieved for the sport skill.

Considering the current sample size, our analysis had a sensitivity for an effect size of 0.49 for the correlations (*Point Biserial Model*, power of 0.80, α of 0.05, and two tails).

3. Results

3.1. General Associations

The linear mixed effect models for the performance of each motor task showed that performance was always affected by age: countermovement jump (estimate: 0.05; t [260] = 12.99; $p < 0.001$); standing long jump (estimate: 5.82; t [248] = 3.64; $p < 0.001$); leap (estimate: 0.02; t [257] = 2.52; $p = 0.012$); high jump (estimate: 0.06; t [255] = 4.91; $p < 0.001$); long jump (estimate: 0.11; t [262] = 3.94; $p < 0.001$); and hurdle transposition (estimate: 0.03; t [260] = 4.00; $p < 0.001$). Thus, in all cases, the older the individual, the better the performance.

In a few tasks, trial also showed a significant effect: standing long jump (estimate: -3.73; t [248] = 2.11; $p = 0.036$) and hurdle transposition (estimate: -0.004; t [260] = 2.53; $p = 0.012$). Thus, for these two tasks, individuals showed a tendency toward a decrease in performance over the trials.

For the number of components required to account for 90% of the variance, we found a significant effect of task (F [5] = 88.58; $p < 0.001$). The pairwise comparisons (with Bonferroni's correction) showed that the countermovement jump showed the smallest number of components (mean: 1.16; p-values < 0.050 against all other tasks), and that the high jump showed the largest number of components (mean: 3.22; p-values < 0.050 against all other tasks). All other tasks did not differ in between.

Table 2 shows the Spearman's ρ correlations among the performances of all of the skills. From this, we can observe a high level of association among these skills.

Table 2. Spearman's ρ correlations among all skills.

	CMJ	**SLJ**	**L**	**HJ**	**LJ**
SLJ	0.91				
L	0.36 [ns]	0.41			
HJ	0.78	0.79	0.45		
LJ	0.83	0.86	0.52	0.77	
HT	0.71	0.83	0.38 [ns]	0.73	0.78

CMJ: countermovement jump; SLJ: standing long jump; L: leap; HJ: high jump; LJ: long jump; HT: hurdle transposition. [ns] Nonsignificant correlation.

Provided the largest effect of age on all tasks, we also calculated the Spearman's ρ partial correlations among all tasks, controlling for the effect of age. Table 3 shows the

Spearman's ρ partial correlations among the performances of all of the skills. From this, we see that a large number of the associations observed before can be accounted for by age.

Table 3. Spearman's ρ partial correlations among all skills, controlling for age.

	CMJ	SLJ	L	HJ	LJ
SLJ	0.55				
L	0.17 [ns]	0.28 [ns]			
HJ	0.34 [ns]	0.42	0.34 [ns]		
LJ	0.41	0.58	0.46	0.43	
HT	0.25 [ns]	0.63	0.24 [ns]	0.44	0.52

CMJ: countermovement jump; SLJ: standing long jump; L: leap; HJ: high jump; LJ: long jump; HT: hurdle transposition. [ns] Nonsignificant correlation.

3.2. Proficiency Barrier Assessment

Figure 2 shows the sigmoidal and linear functions fitted to the relationship between countermovement jump and high jump, standing long jump and long jump, and leap and hurdle transposition. As shown in the figures, for all cases, the AICc was smaller for the linear function—despite a similar AICc between linear and sigmoidal functions for the countermovement jump and high jump pair.

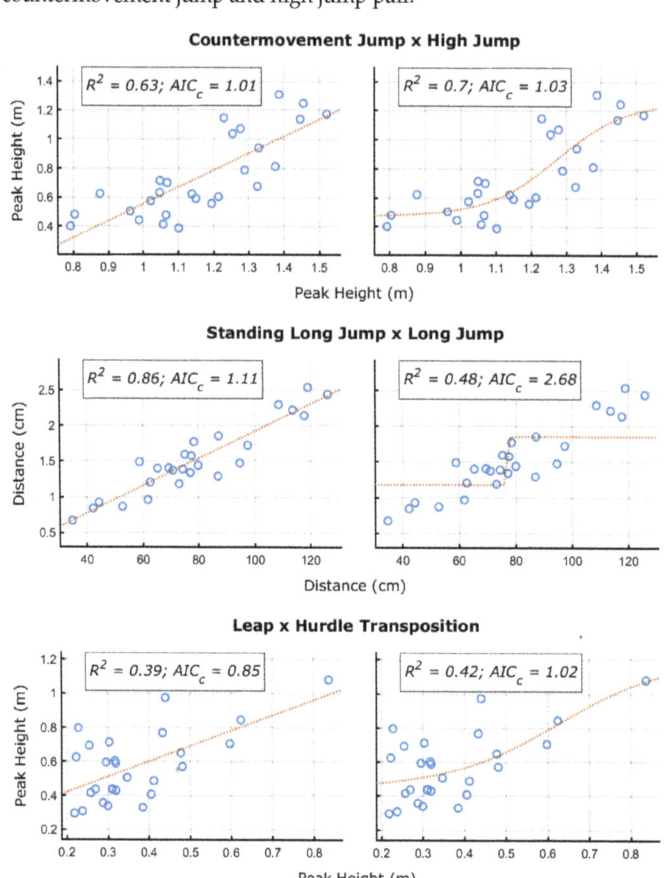

Figure 2. Sport skill performance as a function of the fundamental skill performance. Each circle represents an individual.

From Figure 2, one would then suppose that there is no evidence of a proficiency barrier observing the pairings. Nonetheless, as stated, the issue must be considered also in terms of the movement pattern demonstrated. Figure 3 shows the between-individuals similarity (considering the normalized dot product) in the fundamental movement pattern with individuals sorted in terms of the performance demonstrated of the sport skill.

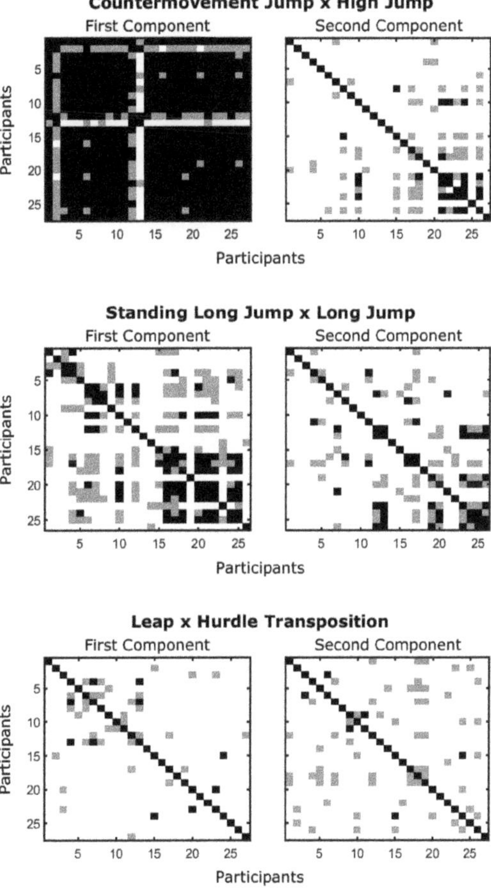

Figure 3. Similarity among individuals (normalized dot product) in their fundamental movement pattern (first and second principal components). Black squares mean normalized dot products higher than 0.9 and gray squares mean normalized dot products higher than 0.8. Individuals are sorted by their performance of the sport skill with higher values (lower right corner) meaning better performances of the sport skill.

As can be observed, for the countermovement jump, the first component is quite similar between almost all participants. Nonetheless, for the second component, only some individuals who showed better performances in the high jump showed large similarity between them. For the standing long jump, both the first and second components showed a cluster of similarity in the movement patterns for individuals with better performances in the long jump. On the contrary, for leap, there was no clear pattern of similarity that emerged.

From these results, we can infer a potential proficiency barrier in two fundamental x sport skills pairings: the countermovement jump and high jump, and the standing long jump and long jump. It is important to note that the performance relationship could demonstrate the sigmoidal relationship only for the countermovement jump and high jump pair and that similarities in the fundamental movement patterns were not necessary for high levels of performance (there are individuals with high performance and no similarity with other high performers). We discuss these issues in Section 4.

3.3. Fundamental Movement Patterns and Performance of Sport Skills

Figure 4 shows the normalized dot product for the fundamental and sport skills pairings as a function of performance of the sport skills and the principal components considered. From the association between the normalized dot product and the performance of the sport skill, the only significant correlation is a weak association between the normalized dot product between countermovement jump and high jump (on the first component) with the performance in the high jump ($\rho = -0.42$; $p = 0.029$). This would imply that those who showed more generalization are the ones with worse results in the sport skill. Note, however, that the normalized dot product values were already low in this case.

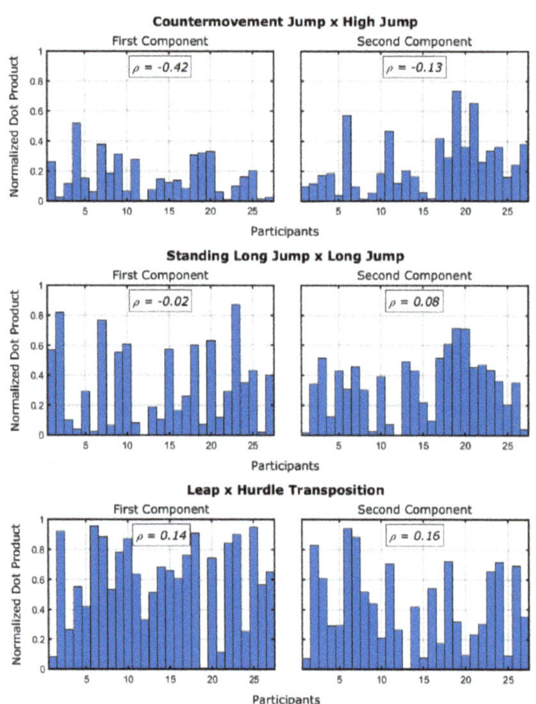

Figure 4. Similarities between the fundamental and sport skills by the same participant (normalized dot product) according to their movement patterns (first and second principal components). Individuals are sorted by their performance of the sport skill, with higher values meaning better performances of the sport skill. The ρ-values represent the Spearman's ρ correlations between the normalized dot products and the performances of the sport skill.

It could be that the pairings showing the proficiency barrier were the ones that, in general, required adaptation. We found an effect of fundamental and sport skills pairing on the similarity of movement patterns (i.e., normalized dot product of the single vector encompassing both the first and second components between the fundamental and sport skills) ($F[2] = 8.67$; $p = 0.013$). The pairwise comparisons (with Bonferroni's correction) showed that the countermovement jump and high jump pairing had lower normalized dot products compared to the leap and hurdle transposition pairing ($p = 0.013$).

Figure 5 shows the difference between the number of components in the fundamental and sport skills pairings as a function of performance of the sport skills. As it is observed, there is no correlation between increase or decrease in the number of components between skills and the performance of the sport skills. As expected from the results in the general association section, the largest difference in number of components occurred for the countermovement jump and high jump pair.

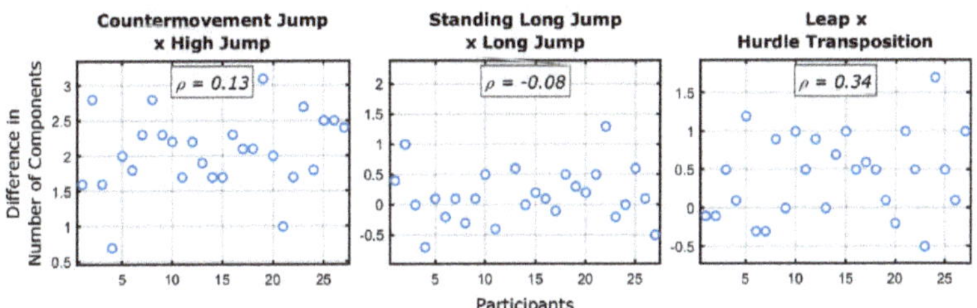

Figure 5. Difference between the fundamental and sport skills (per pairing) in the number of components required to explain at least 90% of the accounted for variance. Individuals were sorted by their performance of the sport skill, with higher values meaning better performances of the sport skill. The ρ values represent the Spearman's ρ correlation between the difference in the number of components and the performance of the sport skill.

4. Discussion

The increased capacity to act in new contexts given previous experiences is one of the cornerstones of human survival through life. Despite numerous claims about the specificity (e.g., [35]) and limited transfer (e.g., [36]) of motor skills, if learning was, indeed, limited to the condition being practiced, one would not have sufficient time to practice and learn all required skills and their variations to survive. Indeed, it is in the motor development and training literature that authors acknowledge the difficulty of reaching high levels of performance in a number of skills and, for this reason, place high importance on early experiences (see [2,4,6,37]). Nonetheless, empirical demonstrations of this importance are still lacking in a vast range of contexts. In fact, little is known about when and how a dependence on previous experiences would be observed.

In the present paper, we investigated how fundamental skills relate to sport-specific skills in track and field. We based this on recent studies on the topic of the proficiency barrier ([9,14–16], see also [17]). These studies provide conflicting findings about the hypothesis that basic components of the so-called fundamental movement skills are necessary for learning more complex or specialized sport skills. Considering the potential different processes that might underlie the relationship between fundamental and sport skills (i.e., generalization and adaptation), our aim was to, first, investigate whether the proficiency barrier would be observed and, second, whether its occurrence was dependent on the required transfer process.

4.1. Proficiency Barrier

From our results, we inferred a proficiency barrier in the relationships between countermovement jump and high jump, and the standing long jump and long jump. In the former pair, both performance (despite slight support for a linear relationship) and movement pattern similarity (between-individuals) supported this inference. In the latter pair, only the movement pattern similarity supported the inference. This might have occurred provided the nonlinear (and redundant) relationship that movement patterns have with performance outcomes (as demonstrated in [15]). This reinforces that if there is a proficiency barrier between fundamental and specific skills, it is demonstrated in the movement kinematics rather than the outcome.

For the leap and hurdle transposition pair, we did not see any sign of a proficiency barrier. This might indicate that there is direct transfer from one to the other. This is a hard argument to hold as the correlation between the two (in terms of performance) was weak and nonsignificant when age was controlled for (however, see the limitations below). Additionally, one might expect that direct transfer may demonstrate a large degree of generalization. At least in terms of the aspects analyzed here (the coordination between joints), we did not find evidence for generalization. One could question whether the leap is fundamental for the hurdle transposition: although the leap requires increasing the step forward, the hurdle transposition requires an attempt to maintain the velocity forward while passing over the hurdle vertically. However, detailed descriptions of all pairs here could lead to similar arguments. As discussed in our limitations section, deciding what is fundamental to what is still an issue.

The two skill pairs supporting the proficiency barrier also showed distinct movement pattern similarity groupings (i.e., black squares in Figure 2). While only the second component showed the expected grouping for the first fundamental/sport skills pair, both the first and second components showed the expected grouping for the second pair (with groupings encompassing a different number of participants and performance levels). The fact that the first principal component of the countermovement jump did not differentiate the levels of performance in the high jump only shows that this movement component is either simple or necessary for the execution of the task. A closer look at the coefficients shows that, for the majority of participants, the first principal component captured the correlated flexion of ankle, knee, and hip—a coupling that they all showed to varying degrees. The second coefficient seems to be a compensatory movement of the knee and hip negatively related to the ankle—something that (1) is not necessary to perform the jumping and (2) seem to be a pattern that requires more practice.

The same logic can be applied to the relationship demonstrated in the first and second components between the standing long jump and long jump: the first component seems to describe an easier to acquire movement component while the second requires long-term practice to be demonstrated. The first component (of the similar group) represents a positive relationship between ankle, knee, hip, and thorax flexions while elbow and shoulder are extending (mostly present at the beginning of the movement). The second component (of the similar group) represents a positive relationship between knee, hip, and shoulder (present toward landing). Both seem to provide a basis for better outcomes in the long jump.

Another important outcome of the present analysis is that, despite the fact that we found common patterns in the fundamental skills that related to the performance of the sport skill for the two pairs, they were not *necessary* for better outcomes. That is, there were some individuals who did not show a similar pattern and had good performances of the sport skill. Thus, contrary to the *strict* proficiency barrier hypothesis [14], which postulates that only those who present proficiency in given components of the fundamental skill will be able to learn the more specific task, there seems to be other paths to learning. This possibility is what O'Keeffe and colleagues [17] showed when testing the transfer from an overarm throw to a dart throw and badminton overhead. They showed that the group who performed only the sport-specific skill (badminton overhead) did improve in the skill, despite having low proficiency in the overarm throw (a fundamental skill).

4.2. Generalization and Adaptation

This leads us to the second big topic of the current study: processes of transfer. From the dynamical systems approach to motor learning and development (see [21,38]), the initial condition of the system (learner) has a large influence on the process of change that the system will go through. From this point of view, learning will always demonstrate transfer effects given that the previous practice will always affect new learning events. The question is the degree of change that the new practice requires from what was previously learned (see [39]). If the new task's demands are in line with the system capacities, there are greater chances of seeing generalization. If the new task's demands require modifications, previous practice might offer a better starting point for exploration and change (which we would refer to as adaptation here). If previous practice is insufficient for dealing with the new task's demands, one would observe a phenomenon similar to the proficiency barrier (see [23] for a similar line of thinking).

Our hypothesis was that generalization would be found when the proficiency barrier is absent. Our results did not fully support our hypothesis. For the countermovement jump and high jump, we have some support: (1) these two skills were the most different in terms of the required number of movement pattern components (indicating the need for adaptation); (2) all participants demonstrated lower similarity values when the fundamental and sport movement patterns were compared (indicating changes in their movement pattern); (3) those with better outcomes in the sport skill were the ones who were able to better change their first movement component; and (4) the clearest signs of a proficiency barrier were demonstrated in this pairing.

However, considering the standing long jump and long jump, individuals varied in their similarity—independent of the sport skill performance. Thus, it might be possible to show either generalization or adaptation and still succeed in the more specific skill. This is a clear example of the potential multiple paths of development. Additionally, the pattern of generalization/adaptation in the leap and hurdle transposition pairing was also quite variable—which reinforces the possibility of individuals succeeding through different paths.

We emphasize that transfer, despite long-term discussions on the issue (e.g., [40–42]), is still far from being understood. We envisage a range of processes (beyond generalization and adaptation) that must be encompassed under the term (e.g., "learning to learn" [43,44]) before definitive understanding of the potential hierarchies that exist in learning and development. In fact, for practitioners, transfer should be of primary concern as interventions (e.g., sports training, rehabilitation, and physical education) are usually a small set of activities planned to provide the largest impact on the maximum range of situations. Without any good ground on the principles of transfer, we see no pathway to design appropriate interventions.

4.3. Limitations

The first limitation of the present study is the arbitrary choice of fundamental and sport skills pairs. Indeed, one could claim that the standing long jump is fundamental for high jump as well (see Table 3 for a potential argument) or that other skills are fundamental rather than the ones chosen here. We have no firm theoretical argument against these possibilities rather than the potential kinematic similarity that these tasks share. In fact, to the best of our knowledge, we see no clear theoretical grounding to suppose that any task is more fundamental than any other in the literature. Even considering Newell's [11] proposed fundamental skills (reaching, standing, and locomotor skills), the question is whether one needs to learn these skills first before attempting to learn skills that are "more complex". Can one learn these skills *while* trying to learn more complex movement patterns? In the initial stages of development, the answer to these questions seems simpler. However, proposed sequences of motor skills to be acquired/practiced in late childhood, sports initiation, and even rehabilitation abound without proper principles to defend them.

The second limitation of the present study is the presence of age as a confounder. As age carries both growth and experiences effects (which can also interact), a superior performance could have occurred because one individual has more strength and, despite bad technique, showed better results than someone who demonstrates the inverse. This type of issue can only be accommodated with larger samples, more tests (to control for different physical abilities), or longitudinal studies. In the present study, as the main focus was on the relationship between movement patterns, we do not see this limitation as a major issue.

The third limitation of the present study is the superficial notion of "components" employed here. Previous studies on the topic utilized assessment based on movement criteria (e.g., [26]). These criteria are descriptions that might involve quantitative aspects (e.g., "the child approaches the ball in a straight path", "the child bounces the ball three times without losing control"), relationships (e.g., "the left arm moves forward while the right arm moves backward"), and timing (e.g., "the lower and upper parts of the trunk face the target at the same time"). Nonetheless, the components extracted here refer only to correlated time-series—relationships—with no reference to quantity or timing. In other terms, our analyses—despite being more detailed and avoiding predefined movement pattern standards—might have been limited to a single dimension of the movement pattern. Thus, for the description of the components above, it was not that the rest of participants did not implement coordinated movements of the mentioned joints, it was just that they do it in ways that are not similar to others. Differences might have occurred in terms of the timing of the motion and the joints that actively participated in the task. Further developments are required to encompass these other dimensions in the type of investigation performed here.

5. Conclusions

As humans learn to perform new motor skills, their potential range of interactions with peers and activities in their contexts increase exponentially. All of this is a result of transfer processes in learning and development. The present study addressed whether the performances of the so-called fundamental skills demonstrated a "necessary" condition for good performances of the sport skills (i.e., the proficiency barrier)—specifically in track and field jump events—and whether such a relationship was dependent on the transfer process occurring. We found that despite the evidence favoring a proficiency barrier, the transfer processes are not related to it. This result seems to point to a multiplicity of paths in motor development.

Author Contributions: Conceptualization, M.M.P.; methodology, M.M.P.; investigation, M.T.S.R. and M.M.P.; resources, M.T.S.R. and F.C.; data curation, M.M.P.; writing—original draft preparation, M.M.P. and M.T.S.R.; writing—review and editing, M.M.P. and M.T.S.R.; supervision, M.M.P. and F.C. All authors have read and agreed to the published version of the manuscript.

Funding: This research received no external funding.

Institutional Review Board Statement: The study was conducted in accordance with the Declaration of Helsinki and approved by the Ethics Committee) of Faculty of Sports—University of Porto (protocol code: 34 2022—approved on 18 January 2023).

Informed Consent Statement: Legal guardians of the participants read and signed an informed consent.

Data Availability Statement: The data and scripts necessary to generate the results of the present study are all available at https://osf.io/bgxa8 (access date: 3 February 2024).

Acknowledgments: We would like to thank João Paulo Vilas-Boas and Pedro Fonseca for the technical support on this manuscript.

Conflicts of Interest: The authors declare no conflicts of interest.

References

1. Seefeldt, V. Developmental Motor Patterns: Implications for Elementary School Physical Education. *Psychol. Mot. Behav. Sport.* **1980**, *36*, 314–323.
2. Clark, J.E.; Metcalfe, J.S. The Mountain of Motor Development: A Metaphor. In *Motor Development: Research and Reviews*; Jane, E., Clark, J.H.H., Eds.; NASPE Pulications: Reston, VA, USA, 2002; Volume 14, pp. 163–190.
3. Johnston, K.; Wattie, N.; Schorer, J.; Baker, J. Talent Identification in Sport: A Systematic Review. *Sports Med.* **2018**, *48*, 97–109. [CrossRef]
4. Côté, J.; Vierimaa, M. The Developmental Model of Sport Participation: 15 Years after Its First Conceptualization. *Sci. Sports* **2014**, *29*, S63–S69. [CrossRef]

5. Goodway, J.D.; Ozmun, J.C.; Gallahue, D.L. *Understanding Motor Development: Infants, Adolescents, Adults*; McGraw-Hill: New York, NY, USA, 2021; ISBN 978-85-8055-180-8.

6. Barnett, L.M.; Stodden, D.; Cohen, K.E.; Smith, J.J.; Lubans, D.R.; Lenoir, M.; Iivonen, S.; Miller, A.D.; Laukkanen, A.; Dudley, D.; et al. Fundamental Movement Skills: An Important Focus. *J. Teach. Phys. Educ.* **2016**, *35*, 219–225. [CrossRef]

7. Wickstrom, R.L. *Fundamental Motor Patterns*; Lea & Febiger: Philladelphia, PA, USA, 1977.

8. Seefeldt, V.; Haubenstricker, J. Patterns, Phases, or Stages: An Analytical Model for the Study of Developmental Movement. In *The Development of Movement Control and Coordination*; Wiley: New York, NY, USA, 1982.

9. Costa, C.L.A.; Cattuzzo, M.T.; Stodden, D.F.; Ugrinowitsch, H. Motor Competence in Fundamental Motor Skills and Sport Skill Learning: Testing the Proficiency Barrier Hypothesis. *Hum. Mov. Sci.* **2021**, *80*, 102877. [CrossRef]

10. Manoel, E.J. Desenvolvimento Motor: Implicações Para a Educação Física Escolar I. *Rev. Paul. Educ. Física* **1994**, *8*, 82–97. [CrossRef]

11. Newell, K.M. What Are Fundamental Motor Skills and What Is Fundamental about Them? *J. Mot. Learn. Dev.* **2020**, *8*, 280–314. [CrossRef]

12. Hulteen, R.M.; Morgan, P.J.; Barnett, L.M.; Stodden, D.F.; Lubans, D.R. Development of Foundational Movement Skills: A Conceptual Model for Physical Activity Across the Lifespan. *Sports Med.* **2018**, *48*, 1533–1540. [CrossRef]

13. Pot, N.; Hilvoorde, I.v.; Afonso, J.; Koekoek, J.; Almond, L. Meaningful Movement Behaviour Involves More than the Learning of Fundamental Movement Skills. *Int. Sports Stud.* **2017**, *39*, 5–20. [CrossRef]

14. dos Santos, F.G.; Pacheco, M.M.; Stodden, D.; Tani, G.; Maia, J.A.R. Testing Seefeldt's Proficiency Barrier: A Longitudinal Study. *IJERPH* **2022**, *19*, 7184. [CrossRef]

15. Pacheco, M.M.; dos Santos, F.G.; Marques, M.T.S.P.; Maia, J.A.R.; Tani, G. Transitional Movement Skill Dependence on Fundamental Movement Skills: Testing Seefeldt's Proficiency Barrier. *Res. Q. Exerc. Sport.* **2021**, *93*, 718–727. [CrossRef]

16. Pacheco, M.M.; Campos, S.D.; Vilela, H.; Freitas, A.S.; dos Santos, C.C.A.; Oliveira, L.M.M.; Godoi Filho, J.R.M. Seefeldt's Proficiency Barrier in the Specialization of a Movement Pattern: Is the Fundamental Movement Skill of Kicking Necessary for Learning the Chipping Technique. In *Motor Development Studies of the Child XVI*; Lagoa, M.J., Coutinho, D., Carvalho, C., Santos, J.O., Viana, J., Silva, G., Eds.; ISMAI: Maia, Portugal, 2023.

17. O'Keeffe, S.L.; Harrison, A.J.; Smyth, P.J. Transfer or Specificity? An Applied Investigation into the Relationship between Fundamental Overarm Throwing and Related Sport Skills. *Phys. Educ. Sport. Pedagog.* **2007**, *12*, 89–102. [CrossRef]

18. Adams, J.A. Historical Review and Appraisal of Research on the Learning, Retention, and Transfer of Human Motor Skills. *Psychol. Bull.* **1987**, *101*, 41–74. [CrossRef]

19. Schmidt, R.A. A Schema Theory of Discrete Motor Skill Learning. *Psychol. Rev.* **1975**, *82*, 225–260. [CrossRef]

20. Braun, D.A.; Mehring, C.; Wolpert, D.M. Structure Learning in Action. *Behav. Brain Res.* **2010**, *206*, 157–165. [CrossRef]

21. Kostrubiec, V.; Zanone, P.-G.; Fuchs, A.; Kelso, J.A.S. Beyond the Blank Slate: Routes to Learning New Coordination Patterns Depend on the Intrinsic Dynamics of the Learner—Experimental Evidence and Theoretical Model. *Front. Hum. Neurosci.* **2012**, *6*, 222. [CrossRef]

22. Thelen, E.; Corbetta, D.; Kamm, K.; Spencer, J.P.; Schneider, K.; Zernicke, R.F. The Transition to Reaching: Mapping Intention and Intrinsic Dynamics. *Child Dev.* **1993**, *64*, 1058–1098. [CrossRef]

23. Tani, G.; Corrêa, U.C.; Basso, L.; Benda, R.N.; Ugrinowitsch, H.; Choshi, K. An Adaptive Process Model of Motor Learning: Insights for the Teaching of Motor Skills. *Nonlinear Dyn. Psychol. Life Sci.* **2014**, *18*, 47–65.

24. Tani, G. Adaptive Process in Motor Learning: The Role of Variability. *Rev. Paul. Educ. Física* **2017**, *55*. [CrossRef]

25. Ross, W.D.; Marfell-Jones, R.J. Cineantropometria. In *Evaluación Fisiológica del Deportista*; Dougall, J.D.M., Wenger, H.A., Green, H.J., Eds.; Paidotribo: Barcelona, Spain, 1995.

26. Ulrich, D. *Test of Gross Motor Development. Examiner's Manual*; Proed: Austin, TX, USA, 2000.

27. Kanko, R.M.; Laende, E.; Selbie, W.S.; Deluzio, K.J. Inter-Session Repeatability of Markerless Motion Capture Gait Kinematics. *J. Biomech.* **2021**, *121*, 110422. [CrossRef]

28. Lahkar, B.K.; Muller, A.; Dumas, R.; Reveret, L.; Robert, T. Accuracy of a Markerless Motion Capture System in Estimating Upper Extremity Kinematics during Boxing. *Front. Sports Act. Living* **2022**, *4*, 939980. [CrossRef]

29. Glazier, P.S.; Davids, K. Constraints on the Complete Optimization of Human Motion. *Sports Med.* **2009**, *39*, 15–28. [CrossRef]

30. Glazier, P.S.; Mehdizadeh, S. In Search of Sports Biomechanics' Holy Grail: Can Athlete-Specific Optimum Sports Techniques Be Identified? *J. Biomech.* **2019**, *94*, 1–4. [CrossRef]

31. Sternad, D. It's Not (Only) the Mean That Matters: Variability, Noise and Exploration in Skill Learning. *Curr. Opin. Behav. Sci.* **2018**, *20*, 183–195. [CrossRef]

32. Bernstein, N.A. *The Co-Ordination and Regulation of Movements*; Pergamon: New York, NY, USA, 1967.

33. Shumway, R.H.; Stoffer, D.S. *Time Series Analysis and Its Applications: With R Examples*; Springer: New York, NY, USA, 2011.

34. Valk, T.A.; Mouton, L.J.; Otten, E.; Bongers, R.M. Fixed Muscle Synergies and Their Potential to Improve the Intuitive Control of Myoelectric Assistive Technology for Upper Extremities. *J. NeuroEng. Rehabil.* **2019**, *16*, 6. [CrossRef]

35. Proteau, L.; Marteniuk, R.G.; Lévesque, L. A Sensorimotor Basis for Motor Learning: Evidence Indicating Specificity of Practice. *Q. J. Exp. Psychol. Sect. A* **1992**, *44*, 557–575. [CrossRef]

36. Schmidt, R.A.; Young, D.E. Transfer of Movement Control in Motor Skill Learning. In *Transfer of Learning: Contemporary Research and Applications*; Cormier, S.M., Hagman, J.D., Eds.; Academic Press: New York, NY, USA, 1987; pp. 47–79.

37. Stodden, D.F.; Langendorfer, S.J.; Goodway, J.D.; Roberton, M.A.; Rudisill, M.E.; Garcia, C.; Garcia, L.E. A Developmental Perspective on the Role of Motor Skill Competence in Physical Activity: An Emergent Relationship. *Quest* **2008**, *60*, 290–306. [CrossRef]

38. Thelen, E.; Smith, L.B. *A Dynamics Systems Approach to the Development of Cognition and Action*; MIT Press: Cambridge, MA, USA, 1994.

39. Pacheco, M.M.; Newell, K.M. Transfer of a Learned Coordination Function: Specific, Individual and Generalizable. *Hum. Mov. Sci.* **2018**, *59*, 66–80. [CrossRef]

40. Thorndike, E.L. The Influence of Improvement in One Mental Function upon the Efficiency of Other Functions. *Psychol. Rev.* **1901**, *8*, 247–261. [CrossRef]

41. Osgood, C.E. The Similarity Paradox in Human Learning: A Resolution. *Psychol. Rev.* **1949**, *56*, 132–143. [CrossRef]

42. Holding, D.H. An Approximate Transfer Surface. *J. Mot. Behav.* **1976**, *8*, 1–9.

43. Harlow, H.F. The Formation of Learning Sets. *Psychol. Rev.* **1949**, *56*, 51–65. [CrossRef] [PubMed]

44. Harlow, H.F.; Harlow, M.K. Learning to Think. *Sci. Am.* **1949**, *181*, 36–39. [PubMed]

Article

Reliability of Xsens IMU-Based Lower Extremity Joint Angles during In-Field Running

Daniel Debertin *, Anna Wargel and Maurice Mohr *

Department of Sport Science, University of Innsbruck, Fürstenweg 185, A-6020 Innsbruck, Austria; anna.wargel@uibk.ac.at
* Correspondence: daniel.debertin@uibk.ac.at (D.D.); maurice.mohr@uibk.ac.at (M.M.)

Abstract: The Xsens Link motion capture suit has become a popular tool in investigating 3D running kinematics based on wearable inertial measurement units outside of the laboratory. In this study, we investigated the reliability of Xsens-based lower extremity joint angles during unconstrained running on stable (asphalt) and unstable (woodchip) surfaces within and between five different testing days in a group of 17 recreational runners (8 female, 9 male). Specifically, we determined the within-day and between-day intraclass correlation coefficients (ICCs) and minimal detectable changes (MDCs) with respect to discrete ankle, knee, and hip joint angles. When comparing runs within the same day, the investigated Xsens-based joint angles generally showed good to excellent reliability (median ICCs > 0.9). Between-day reliability was generally lower than the within-day estimates: Initial hip, knee, and ankle angles in the sagittal plane showed good reliability (median ICCs > 0.88), while ankle and hip angles in the frontal plane showed only poor to moderate reliability (median ICCs 0.38–0.83). The results were largely unaffected by the surface. In conclusion, within-day adaptations in lower-extremity running kinematics can be captured with the Xsens Link system. Our data on between-day reliability suggest caution when trying to capture longitudinal adaptations, specifically for ankle and hip joint angles in the frontal plane.

Keywords: wearable sensors; inertial measurement units; ecological validity; 3D motion analysis; gait analysis; trail running

Citation: Debertin, D.; Wargel, A.; Mohr, M. Reliability of Xsens IMU-Based Lower Extremity Joint Angles during In-Field Running. *Sensors* **2024**, 24, 871. https://doi.org/10.3390/s24030871

Academic Editors: Felipe García-Pinillos, Diego Jaén-Carrillo and Alejandro Pérez-Castilla

Received: 27 December 2023
Revised: 19 January 2024
Accepted: 23 January 2024
Published: 29 January 2024

1. Introduction

The assessment of running physiology and mechanics through wearable sensors has received strong interest from researchers of various disciplines over the last two decades. This trend mirrors the growing understanding that human movement should be studied under real-word conditions to come to valid conclusions about the underlying phenomena that will hold outside of the laboratory [1,2]. The use of wearable inertial measurement units (IMUs) to analyze running kinematics [3], dynamics [4,5], or energetics [6] in real-world settings has become a particular focus within biomechanics and motor control research. Of all available commercial systems offering IMU-based analyses of full-body kinematics, the Xsens Link system (Movella Technologies, Enschede, The Netherlands) has probably found the most frequent application in research settings [7]. The system consists of 17 IMUs that are attached to upper and lower body segments during the movement of interest, while the corresponding software estimates segment positions and relative segment orientations, i.e., joint angles, based on a calibration procedure, a linked-segment skeletal model, and a proprietary sensor fusion algorithm [8].

For example, the Xsens Link system has been applied to study how running kinematics are influenced by unstable surfaces [9], a marathon [10], performance level in trail running [11] or by the cycling-to-running transition in triathletes [12]. Although wearable systems like the Xsens Link offer potentially new insights into running physiology and mechanics, the accurate IMU-based estimation of joint movement is challenging, e.g., due to integration drift in the measured accelerations and angular velocities [13,14], soft-tissue artefacts [15], anatomical calibration errors [16], and/or disagreement between the underlying linked-segment skeletal model and physiological joint articulation [8], to name a few.

Recognizing these challenges, some studies have investigated the validity and reliability of Xsens-based full-body kinematics during walking, running, or other functional tasks [17–23]. In terms of concurrent validity with marker-based optical motion capture, Xsens-based joint angle measurements typically demonstrate good agreement in the sagittal plane (i.e., flexion-extension), but moderate or poor agreement in frontal plane or transverse plane joint angles [17,18,20–22]. Poor agreement in frontal and transverse plane motions can partially be explained by differences in the underlying biomechanical models between Xsens-based and optical motion capture, and those differences could be addressed through respective coordinate system alignments [17,21]. However, the remaining differences in joint angle trajectories are due to technological errors and may be particularly large for ankle joint movements [17].

The reliability of Xsens-based joint angle measurements, specifically for running, is much less well understood. Al-Amri and colleagues [18] investigated the between-day reliability of Xsens-based joint angles during walking, squatting, and jumping. Similar to the pattern of validation studies, between-day reliability was generally high (Intraclass Correlation Coefficients, ICCs > 0.8) with respect to sagittal plane joint angles, while frontal and transverse plane joint angles showed many instances of poor reliability, e.g., ankle eversion during walking (ICC < 0.5). To our knowledge, the only study to investigate the reliability of Xsens-based joint angles during running was a follow-up study by Trott and Al-Amri [19], albeit only published in abstract form. They confirmed good between-day reliability in knee and hip flexion–extension range of motion during treadmill running (ICCs > 0.8), but reported poor between-day reliability of sagittal plane ankle range of motion (ICC < 0.6 for left ankle) [19]. While these two studies provide important insight into the reliability of Xsens-based joint kinematics during walking and running, many open questions remain. First, the between-day reliability of Xsens-based joint angles during running in planes other than the sagittal plane is unknown. This is problematic given that frontal plane joint angles such as hip ab-/adduction and ankle eversion/inversion are frequently discussed in running-related literature, particularly in the field of injury prevention [24]. Second, the reliability of Xsens-based joint angles during over-ground running at a self-selected speed is unknown, which is an important aspect, since IMU-based motion analysis systems are meant to be used outside of the laboratory, where running speed is typically unconstrained. Third, the existing studies determined the between-day reliability of Xsens-based joint angles based on only two testing days, which may not be sufficient to accurately estimate reliability [25]. Finally, and similar to previous validation studies [17,21], it may be valuable to investigate the source of between-day variations in Xsens-based joint angle measurements so that future longitudinal studies, which rely on repeated measurements of running kinematics, can be designed in such a way as to minimize between-day errors.

The specific research objectives of this study were:

I. To determine the between-day, within-day and calibration reliability of discrete hip, knee, and ankle joint angles in the sagittal and frontal planes as quantified by the Xsens Link system during running at a self-selected speed on a stable asphalt surface and an unstable woodchip surface based on more than two measurement sessions each;

II. To investigate potential sources of between-day variations in Xsens-based discrete hip, knee, and ankle joint angles by determining the association of between-day variations in discrete joint angles with between-day variations in running speed and stride frequency, as well as with different running surfaces.

2. Materials and Methods

2.1. Participants and Study Design

A group of 17 recreational runners (8 female, 9 male) volunteered to participate in five running sessions on five separate days for this reliability study. Inclusion criteria were (1) a minimum of one running session per week for at least one year and (2) no disruption of running for more than two weeks in the last six months (e.g., due to lower extremity injury or general overload). This study presents a secondary analysis of data from a previous investigation on the influence of running surface stability on whole-body running kinematics [9]. For this reason, the sample size was guided by an a priori power analysis based on expected effect sizes with respect to the running surface. Nevertheless, we conducted a post-hoc sensitivity analysis for our group of runners based on recommendations regarding how to estimate the sample size for reliability studies [26,27]. Specifically, with a group of $n = 17$ runners and $k = 2$ repeated measurements (i.e., comparison of two runs within the same day or two separate testing days), we were able to estimate ICCs with an expected magnitude of 0.8 (taken from [19] as an estimate for the true ICC) and with the minimum precision set to 0.18. This was an acceptable precision to distinguish poor reliability (ICC < 0.5) from good (ICC > 0.75) and excellent reliability (ICC > 0.9) [28].

The study was conducted in accordance with the Declaration of Helsinki and approved by the local Ethics Committee at the Department of Sport Science, Universität Innsbruck (ethics approval ID 39/2020), and all participants provided written informed consent prior to inclusion in the study. Each participant undertook a series of five testing sessions with inter-session intervals averaging 1.9 ± 1.2 days (range 1–6 days). Each testing session encompassed a 10 min warm-up period with five minutes dedicated to warming up on an asphalt surface and five minutes on a woodchip track surface (see Figure 1 in [9]). Subsequently, kinematic measurements were conducted on each of these surfaces. The sequence in which the surfaces were presented to the participants was balanced–randomized but was held constant across the five testing sessions for each individual. The duration of the warm-up phase served a dual purpose: firstly, to acquaint participants with running while wearing a motion capture suit, and secondly, to establish their preferred running speed, thereby mitigating subsequent alterations in speed during the measurement phase. Participants completed all testing sessions with the same personal running shoes.

2.2. Experimental Protocol

The detailed experimental protocol has been described elsewhere [9]. Briefly, for each testing session, runners wore an Xsens Link suit (Movella Technologies, Enschede, The Netherlands), which includes 17 IMUs distributed on prescribed body segments according to the Xsens Link manual and pre-defined sensor positions for each segment via Velcro stripes and pouches within the suit. The suit sizes varied among participants based on their anthropometry, ensuring a skin-tight fit and comparable relative sensor positions. The Xsens Link suit sampled IMU data at 240 Hz in the "on-body recording mode" such that running IMU data were stored on a data logger within the suit, while data processing was done after the experiment. Following the warm-up, runners completed four runs along a 140 m straight track per surface, leading to a total running distance of 560 m per day and surface. Specifically, runners completed the first 140 m in one direction (run 1), turned around and immediately completed the second 140 m in the other direction (run 2). Following a rest period of 2–3 min (save data, restate instructions, restart data logger), runners completed runs 3 and 4 analogous to runs 1 and 2. To exclude the possible effects of the turn on our reliability estimates, the subsequent analysis only focused on runs 1 and 3 (see Figure 1). Before each new surface, participants performed two separate walking calibration trials according to the manufacturer guidelines. This calibration trial consists of starting in a static standing position (the "N-Pose"), walking forward for a few seconds, turning, and then returning to the initial standing position. During the running trials, participants were instructed to run at a self-selected but constant running speed that they would choose for a 60-minute moderate intensity run. Timing gates in the middle section of the 140 m-long track were used to measure the average running speed (Figure 1).

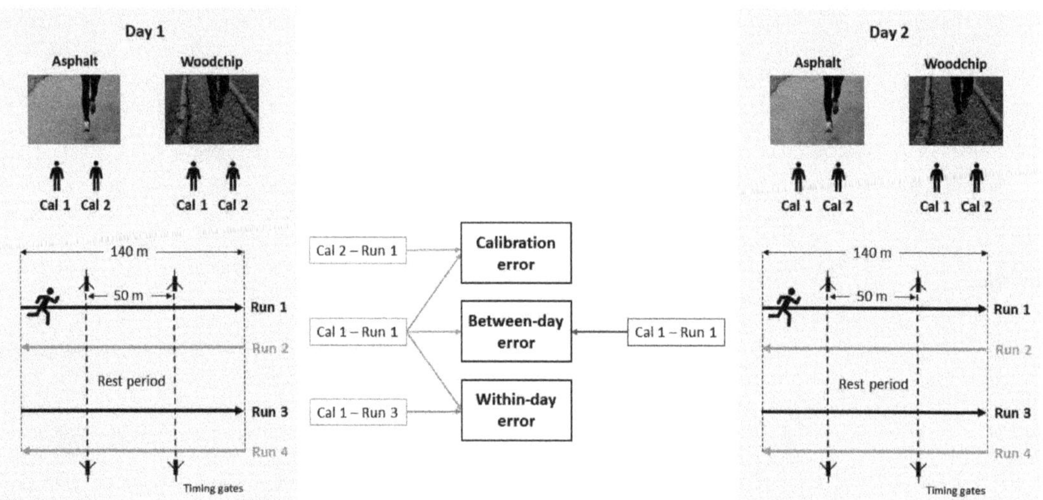

Figure 1. Schematic of experimental protocol and reliability analyses. Note that the calculations of calibration, within-day, and between-day error are only shown for testing day 1 and testing day 2, although the same comparisons were carried out between all five testing days. Runs 2 and 4 (marked in grey) were not included in the analysis. Cal = calibration.

2.3. Data Processing

The Xsens MVN Analyze software (v. 2021.0.1) was used to post-process all running measurements in the "High-definition reprocessing" mode and the "no-level" processing scenario. In this scenario, the position of the pelvis segment is fixed in space and all kinematic quantities are expressed relative to the pelvis, which is the recommended scenario for joint angle analysis in biomechanics [22]. During this step, the Xsens MVN software also determines 3D joint angles for all joints included in the Xsens biomechanical model, which follows the recommendations of the International Society of Biomechanics for defining coordinate systems for the ankle, knee, and hip [29,30]. All further processing steps were conducted in a custom-written Matlab script (The MathWorks Inc. (2023). MATLAB version: 9.14.0 (R2023a), Natick, MA, USA). Individual running gait cycles were identified and segmented based on maxima in the anterior–posterior position of the right foot segment, as described previously [9]. Although this method leads to gait cycles that start slightly before the actual right foot contact, it yields a reliable segmentation approach that should not introduce further error into the analyzed discrete joint angles. Although this method leads to gait cycles that start slightly before the actual right foot contact, we assume its simplicity in yielding a more reliable segmentation approach compared to previous detection methods based on acceleration thresholds. The latter may be prone to false detections in the presence of variable running patterns, speeds, and surfaces, and could therefore bias our reliability comparisons [31]. All joint angle trajectories were then time-normalized, such that each gait cycle consisted of 101 data points with the first and last data point representing subsequent maxima in the right foot anterior–posterior position. We extracted about 40 gait cycles for each run depending on the runners' selected speed while omitting gait cycles during the acceleration and deceleration phases. We then determined one average, time-normalized angle waveform for each investigated joint, calibration–run combination, surface, and testing day for each participant for further analysis. For the purpose of this study, we focused the analysis on initial hip flexion–extension (IHF), initial and peak hip abduction–

adduction (IHA, PHA), initial and peak knee flexion–extension (IKF, PKF), initial and peak ankle dorsiflexion–plantarflexion (IAD, PAD) and initial and peak ankle eversion–inversion (IAI, PAE). Initial values were taken at 0% of gait cycle and peak values during the approximated stance phase (0–40% of gait cycle) (see Figure S1 in Supplementary Materials for an illustration of the variables). We also determined the average running speed (RS) based on the timing gate data and average stride frequency (SF) based on the duration of each of the detected gait cycles.

2.4. Reliability Analysis

To address our first objective (I), we assessed the reliability of each investigated discrete joint angle based on three different comparison types:

1. Between-day comparisons were based on the variation in discrete joint angles between the first runs on each of the five separate days on the same surface (e.g., run 1 on five separate days on the woodchip track, see Figure 1). Specifically, we based reliability estimates on a total of five day-to-day comparisons—Day 1 vs. 2, Day 2 vs. 3, Day 3 vs. 4, Day 4 vs. 5, and Day 5 vs. 2. Day-to-day comparisons were carried out pairwise to enable the comparison of between-day vs. within-day reliability, which also only relied on two measurements, i.e., two runs within a day. While between-day reliability could have been estimated using all five testing days, the resulting confidence intervals of an ICC based on five measurements would have been systematically more narrow compared to ICC estimates based on two measurements, and thus would have hindered a fair comparison of between-day vs. within-day reliability estimates and interpretations. Day 5 was compared to Day 2 to have a fifth comparison pair, equalizing the number of comparisons underlying the median ICCs;

2. Within-day comparisons were based on the variation in discrete joint angles between two separate runs on the same surface and day, and processed with the same calibration trial (e.g., run 1 vs. run 3 on the woodchip track on day 1, see Figure 1). Given the five testing days, we based reliability estimates on a total of five within-day comparisons;

3. Calibration comparison was based on the variation in discrete joint angles between two copies of the same run, but processed with different calibration trials (e.g., run 1 processed with calibration 1 vs. 2 on the woodchip track on day 1, see Figure 1). Given the five testing days, we based reliability estimates on a total of five calibration comparisons.

We initially summarized the absolute between-day differences in each investigated discrete joint angle (between-day error in Figure 1) along with the corresponding absolute within-day differences in those joint angles due to different runs (within-day error in Figure 1), or due to different calibrations (calibration error in Figure 1). For a given day-to-day comparison (e.g., 1-to-2), the within-day differences were determined based on the first of the two days (day 1 for 1-to-2). Then, we determined the average and standard deviation of the absolute between-day, within-day, and calibration-related differences across all five day-to-day comparisons and participants.

The between-day reliability, within-day reliability and calibration reliability of discrete hip, knee, and ankle joint angles were determined according to the intra-class correlation coefficient (ICC) and the minimal detectable change (MDC). The ICC was calculated to represent the degree of absolute agreement between measurements ("case 2" model ICC(A,1)) as defined by McGraw and Wong [32]. The MDC was estimated based on the standard error of measurement (SEM) following the recommendations by Atkinson and Nevill [25]:

$$\text{SEM} = \sqrt{\text{MSE}}, \tag{1}$$

$$\text{MDC} = 1.96 \cdot \text{SEM} \cdot \sqrt{2}, \tag{2}$$

where MSE is the mean squared error of the repeated measures analysis of variance underlying the calculation of the ICC. The reported ICCs and MDCs for between-day, within-day and calibration reliability are the median and range of ICCs across the five testing days (or day-to-day comparisons). ICCs were interpreted according to the guidelines set out by Koo and Li [28] such that cut-off values of ICC < 0.5, $0.5 \leq$ ICC < 0.75, $0.75 \leq$ ICC < 0.9, and ICC $\geq$ 0.9 were categorized and interpreted as showing poor, moderate, good, and excellent reliability. The ICC interpretation could encompass two categories if the minimum ICC fell into a different category than the median ICC across days. In these cases, we interpreted conservatively, e.g., an ICC [range] of 0.95 [0.76, 0.99] was interpreted as good instead of excellent reliability.

To address our second objective (II), we calculated the between-day reliability of running speed and stride frequency according to the aforementioned comparison types. We further curated a "difference matrix" containing the absolute differences in each of the investigated discrete joint angle outcomes for five different day-to-day comparisons (1-to-2, 2-to-3, 3-to-4, 4-to-5, 5-to-2) and two surfaces (asphalt and woodchip), leading to 10 entries for each participant. In parallel, we determined the corresponding absolute between-day differences in running speed and stride frequency. This procedure resulted in a 170 rows × 4 columns error matrix per investigated discrete joint angle (17 participants × 5 day-to-day comparisons × 2 surfaces = 170 rows × 4 columns (difference in joint angle, speed, stride frequency, and the surface)). Then, we analyzed whether the running surface and/or between-day differences in speed and stride frequency were significant predictors of between-day differences in Xsens-based joint angles using one linear mixed model for each investigated joint angle variable. The linear mixed model computations were conducted in jamovi software (v. 2.3.21) [33] using the *GAML*j module with its default settings [34] (version 2.6.6). Within each investigated model, running speed, stride frequency and running surface were defined as fixed effects ($\overline{b}$), while the intercept within each runner cluster (aj) was defined as a random effect reflected by the model Equation (3):

$$\hat{y}_{cj} = \overline{a} + \left(a_j - \overline{a}\right) + \overline{b}_{Speed} \cdot x_{speed,\ ij} + \overline{b}_{SF} \cdot x_{SF,\ cj} + \overline{b}_{Surface,cj} \cdot x_{Surface,\ cj} + \epsilon_{cj} \tag{3}$$

where $\hat{y}_{cj}$ is the predicted absolute between-day difference in a given joint angle for a given runner-cluster j and a given day-to-day comparison c, $\overline{a}$ and a_j are the average and runner-dependent intercepts, $\overline{b}$ is the average coefficient across runners, and ϵ is an error term representing unexplained variance.

3. Results

3.1. Runner Characteristics

The group of investigated runners had an average age ($\pm$SD) of 26 ± 2 years. Female runners ($n = 8$) were 1.68 ± 0.07 cm tall with a body mass of 60 ± 7 kg. Male runners ($n = 9$) were 1.82 ± 0.03 cm tall with a body mass of 73 ± 7 kg. On average, runners in this sample completed 2.6 ± 1.4 running sessions per week with a session duration of 58 ± 19 min and a weekly running distance of 29 ± 23 km. Such runners can be classified as recreational runners [35].

3.2. Between-Day, Within-Day and Calibration Reliability of Xsens-Based Lower Extremity Joint Angles

The average magnitude of between-day variations in discrete joint angles was generally larger—about twice as high—compared to the corresponding magnitude of within-day variations due to either separate runs or different calibration files (BD vs. WD and BD vs. Cal in Figure 2). This was true for all analyzed joints and both surfaces. No clear trend emerged when comparing the magnitude of within-day variations in discrete joint angles between separate runs (WD), or from the same run but based on different calibration files (Cal). The one exception was the frontal plane ankle angle (IAI, PAE in Figure 2), for which within-day differences due to calibration were more pronounced compared to the differences between runs. Similarly, no clear trend emerged when comparing the magnitude of joint angle variations between running surfaces. Again, the frontal plane ankle angles were the exception, with more pronounced calibration-related differences for runs on the woodchip track compared to runs on asphalt.

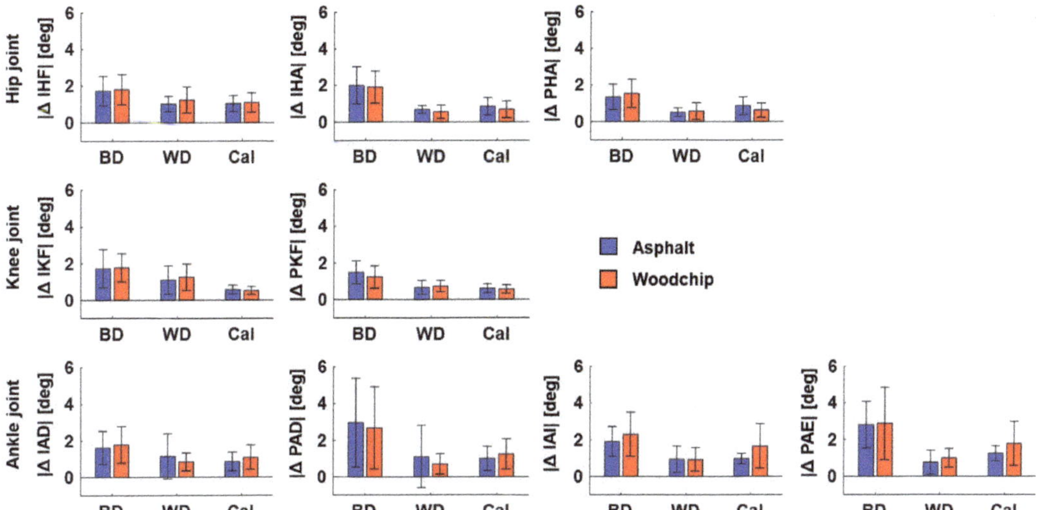

Figure 2. Mean absolute differences (±SD) in Xsens-based lower extremity joint angles between runs on different days (BD), different runs within the same day but processed with the same calibration file (WD) and the same run within the same day but processed with different calibration files (Cal). Blue and red bars show results for runs on asphalt and the woodchip track, respectively. IHF = initial hip flexion, IHA = initial hip ab-/adduction, PHA = peak hip adduction, IKF = initial knee flexion, PKF = peak knee flexion, IAD = initial ankle dorsiflexion, PAD = peak ankle dorsiflexion, IAI = initial ankle inversion, PAE = peak ankle eversion.

The ICC medians and ranges (min and max ICCs) based on five comparisons for between-day, within-day and calibration reliability are summarized in Table 1, split by variable category (joint angles for the hip, knee and ankle as well as spatio-temporal parameters), split by running surface (asphalt and woodchip), and color-coded according to their reliability category (poor, moderate, good, excellent). With regard to the hip joint, hip flexion (IHF) showed good between-day reliability, while hip add-/abduction showed poor (IHA) to moderate (PHA) between-day reliability for both surfaces. Within-day reliability and calibration reliability were excellent for hip flexion (IHF) and predominantly good for hip add-/abduction (IHA and PHA). Discrete knee joint angles in the sagittal plane (IKF and PKF) revealed good to excellent between-day reliability, as well as excellent within-day and calibration reliability for both surfaces. With respect to the ankle, between-day reliability was moderate to good for initial ankle angles (IAD and IAI), whereas peak ankle angles (PAD and PAE) showed poor between-day reliability. For within-day reliability, all ankle angles showed excellent median ICCs (>0.9); however, some minimum ICCs were poor (PAD on asphalt) to moderate (IAD on asphalt, PAD on woodchip), leading to low reliability categories for these angles. Similarly, all ankle angles showed good to excellent calibration reliability with the exception of PAD and IAI on the woodchip track (moderate reliability).

Table 1. ICCs for between-day, within-day and calibration reliability of Xsens-based lower extremity joint angles, as well as running speed and stride frequency during running on an asphalt and woodchip surface. Orange, yellow, light green, and dark green coloring indicate poor, moderate, good, and excellent reliability according to Koo and Li [28].

Group	Variable	Grand Mean [1] (deg)		Between-Day Reliability ICC Median [Range]		Within-Day Reliability ICC Median [Range]		Calibration Reliability ICC Median [Range]	
		Asphalt	Woodchip	Asphalt	Woodchip	Asphalt	Woodchip	Asphalt	Woodchip
Hip	Initial hip flexion (IHF)	34.77	37.02	0.92 [0.89, 0.95]	0.96 [0.83, 0.96]	0.98 [0.95, 0.98]	0.98 [0.91, 0.99]	0.97 [0.96, 0.99]	0.97 [0.96, 0.98]
	Initial hip ab-/adduction (IHA)	−0.15	0.38	0.52 [0.44, 0.68]	0.49 [0.29, 0.62]	0.96 [0.87, 0.98]	0.94 [0.86, 0.99]	0.91 [0.83, 0.94]	0.93 [0.77, 0.99]
	Peak hip adduction (PHA)	−9.42	−8.00	0.83 [0.70, 0.85]	0.74 [0.63, 0.77]	0.97 [0.92, 0.99]	0.96 [0.74, 0.99]	0.89 [0.87, 0.95]	0.96 [0.87, 0.97]
Knee	Initial knee flexion (IKF)	7.53	12.18	0.92 [0.90, 0.95]	0.92 [0.89, 0.96]	0.97 [0.96, 0.98]	0.97 [0.95, 0.99]	0.99 [0.99, 0.99]	0.99 [0.99, 1.00]
	Peak knee flexion (PKF)	42.53	41.69	0.82 [0.75, 0.88]	0.89 [0.85, 0.93]	0.96 [0.95, 0.97]	0.96 [0.95, 0.98]	0.97 [0.96, 0.97]	0.98 [0.97, 0.99]
Ankle	Initial ankle dorsiflexion (IAD)	6.83	5.56	0.90 [0.82, 0.94]	0.88 [0.87, 0.95]	0.97 [0.52, 0.99]	0.98 [0.90, 0.99]	0.97 [0.96, 0.98]	0.96 [0.91, 0.99]
	Peak ankle dorsiflexion (PAD)	21.68	21.95	0.38 [0.23, 0.52]	0.47 [−0.02, 0.49]	0.93 [0.30, 0.99]	0.97 [0.61, 0.99]	0.88 [0.82, 0.97]	0.91 [0.57, 0.93]
	Initial ankle inversion (IAI)	−7.47	−7.81	0.81 [0.77, 0.90]	0.72 [0.64, 0.85]	0.96 [0.92, 0.98]	0.95 [0.87, 0.98]	0.95 [0.93, 0.97]	0.84 [0.73, 0.93]
	Peak ankle eversion (PAE)	11.83	11.10	0.55 [0.53, 0.72]	0.65 [0.46, 0.72]	0.97 [0.94, 0.98]	0.97 [0.92, 0.98]	0.94 [0.89, 0.96]	0.84 [0.77, 0.94]
Spatio-tem-poral	Running speed (RS)	3.47	3.38	0.93 [0.84, 0.95]	0.93 [0.76, 0.96]	0.94 [0.84, 0.97]	0.97 [0.90, 0.98]	*not applicable*	
	Stride frequency (SF)	1.40	1.39	0.88 [0.81, 0.88]	0.73 [0.55, 0.87]	0.93 [0.90, 0.95]	0.89 [0.73, 0.95]		

[1] Grand mean represents the average variable across the first run on each testing day and across all participants.

A closer look into the median MDCs (Figure 3) confirms the lower reliability, i.e., larger MDCs, of comparisons of runs between days vs. within days. For within-day comparisons, the MDCs range from a minimum of 1.30° for peak hip adduction on asphalt to a maximum of 2.95° for initial knee flexion on the woodchip track. For between-day comparisons, MDCs range from a minimum of 3.17° for peak knee flexion on the woodchip track to a maximum of 8.22° for peak ankle dorsiflexion on asphalt. There were only two variables where the median MDC was smaller than or equal to the average change of the respective joint angle between the two running surfaces: (1) Within-day comparisons of peak hip adduction (median MDC of 1.30–1.38° with an average surface difference of 1.42°) and (2) within-day and between-day comparisons of initial knee flexion (median MDC of 2.38–2.95° (within-day) and median MDC of 4.60–4.73° (between-day) with an average surface difference of 4.65°).

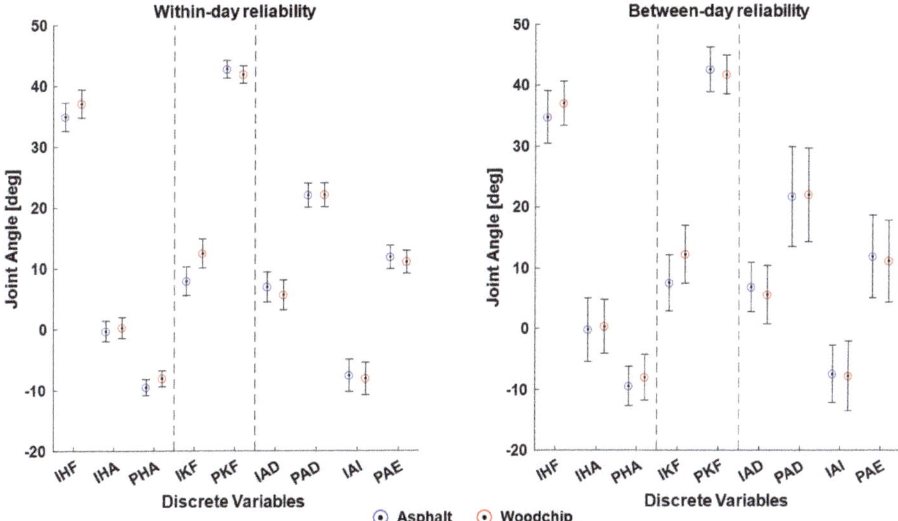

Figure 3. Within-day and between-day reliability of Xsens-based lower extremity joint angles during running. Circles (blue = asphalt, red = woodchip) indicate the grand mean joint angles across runners and testing sessions. Error bars represent the mean angle plus-minus one median minimal detectable change (based on five reliability estimates). IHF = initial hip flexion, IHA = initial hip ab-/adduction, PHA = peak hip adduction, IKF = initial knee flexion, PKF = peak knee flexion, IAD = initial ankle dorsiflexion, PAD = peak ankle dorsiflexion, IAI = initial ankle inversion, PAE = peak ankle eversion.

3.3. Potential Sources of Between-Day Variations in Xsens-Based Lower Extremity Joint Angles

In addition to the measurement technology, we investigated variations in the participants' running styles (expressed through running speed and stride frequency) and the running surface as potential explanatory factors for the magnitude of between-day variations in joint angle outcomes. Table 1 indicates that running speed showed good between-day reliability. The mixed model analyses further showed that between-day differences in running speed were a significant predictor for between-day differences in initial hip flexion, and explained 26% of the respective variance (Table 2). On average, an absolute between-day difference in running speed of 1 m/s was associated with an absolute between-day difference in initial hip flexion of 4.23°. Stride frequency showed good reliability on asphalt but only moderate reliability on the woodchip track (Table 1). According to the mixed model analyses, 6% (peak knee flexion) and 12% (peak ankle dorsiflexion and eversion) of the absolute between-day differences can be explained through differences in stride frequency (Table 2). There was no significant association between the running surface and any of the analyzed between-day joint angle differences.

Table 2. Model summary for the association of absolute between-day differences in Xsens-based lower extremity joint angles with the running surface and absolute between-day differences in running speed and stride frequency. Green coloring: These models contained at least one significant coefficient ($p < 0.05$) of between-day differences in the respective joint angle. Grey coloring: These models did not contain a significant coefficient ($p < 0.05$) for the differences between the individual days in the respective joint angle.

Group	Variable	R-Squared Marginal	Between-Day Difference in Running Speed (b_{speed}) (deg per m/s)	Between-Day Difference in Stride Frequency (b_{SF}) (deg per 1/s)	Surface (Woodship—Asphalt, $b_{Surface}$) (deg)
Hip	Initial hip flexion	0.26	4.23 (3.06, 5.40)	2.55 (−6.43, 11.54)	−0.02 (−0.41, 0.36)
	Initial hip ab-/adduction	0.01	−0.69 (−2.18, 0.80)	−1.05 (−13.51, 11.40)	−0.08 (−0.53, 0.37)
	Peak hip abduction	0.02	−0.48 (−1.56, 0.60)	4.89 (−4.20, 13.99)	0.18 (−0.15, 0.49)
Knee	Initial knee flexion	0.01	0.15 (−1.25, 1.54)	6.01 (−5.71, 17.73)	0.01 (−0.40, 0.43)
	Peak knee flexion	0.06	0.04 (−0.99, 1.07)	11.32 (2.71, 19.94)	−0.31 (−0.62, −0.00)
Ankle	Initial ankle dorsiflexion	0.02	1.03 (−0.46, 2.53)	0.92 (−11.58, 13.41)	0.14 (−0.30, 0.59)
	Peak ankle dorsiflexion	0.12	−0.57 (−3.12, 1.99)	52.10 (30.00, 74.19)	−0.54 (−1.28, 0.19)
	Initial ankle inversion	0.03	−0.60 (−2.25, 1.04)	10.04 (−3.58, 23.66)	0.35 (−0.14, 0.85)
	Peak ankle eversion	0.12	−0.34 (−2.66, 1.99)	41.01 (21.74, 60.28)	−0.12 (−0.82, 0.58)

4. Discussion

The objective of this study was to investigate the between-day and within-day reliability of Xsens-based lower extremity joint angles during running at a self-selected speed in the field while considering a range of potential sources for between-day variations in joint angle outcomes, including system calibration, variation in running speed and stride frequency, and the running surface.

4.1. Within-Day Reliability of Xsens-Based Lower Extremity Joint Angles

When comparing two subsequent runs on the same day, initial and peak hip and knee joint angles generally showed good to excellent reliability with high ICCs and relatively low MDCs. For peak hip adduction and initial knee flexion, the MDCs were smaller than the average change in these joint angles when running on asphalt vs. the less stable woodchip track. A similar trend was observed for the initial hip flexion angle. Taken together, these results can be interpreted to show that the IMU-based Xsens Link system has sufficient reliability and is well suited to detect within-day adaptations in sagittal knee and sagittal/frontal hip joint movement in response to different running surfaces or comparable interventions, such as variable footwear. Importantly, this statement only applies to study designs that do not require a re-calibration between experimental conditions given the added measurement error of the calibration procedure.

The median within-day ICCs for ankle angles also pointed towards good to excellent within-day reliability, but showed some instances of only moderate or poor reliability in the sagittal plane (e.g., minimum ICC of 0.30 for peak ankle dorsiflexion in Table 1). These instances of lower reliability cannot be explained by calibration issues given that the same calibration trial was used to process the two contrasted runs. Therefore, we assume that instances of low within-day reliability in ankle angles may result from undesired relative movements between the foot IMU sensor and the running shoe throughout a running trial, and thus an invalid calibration of the sensor-to-segment orientation. Future investigators should ensure a reliable method of fastening the foot IMU sensor and avoiding undesired sensor movement for any surface or running speed setting, e.g., by securing the sensor with additional tape instead of the Velcro-based attachment provided by the manufacturer.

4.2. Between-Day and Calibration Reliability of Xsens-Based Lower Extremity Joint Angles

The between-day reliability was clearly lower than the within-day reliability for all investigated joint angles, as demonstrated by the lower ICCs, higher MDCs, and higher absolute between-run differences. Nevertheless, good to excellent between-day reliability was observed for initial hip flexion, initial and peak knee flexion, and ankle dorsiflexion. In parallel, these angles showed excellent calibration reliability. For initial knee flexion in particular, the between-day MDC was still smaller than the average change in knee flexion between the asphalt and woodchip surfaces. This suggests that Xsens-based running analyses may be reliable enough to monitor longitudinal changes in knee flexion angles and potentially initial hip and ankle sagittal angles over time, provided those changes have a magnitude of at least $4°$.

In contrast, between-day reliability was lower for initial hip add-/abduction and ankle inversion (moderate reliability), and was poor for peak hip adduction, peak ankle dorsiflexion, and peak ankle eversion. This finding agrees with previous reliability studies of Xsens-based motion capture showing low reliability for frontal plane joint angles during gait in general [18], with a particularly high measurement error for the ankle joint [17,19]. Consistent with this finding is the observation that peak ankle dorsiflexion and initial ankle inversion were the only variables to show only moderate calibration reliability. Robert-Lachaine and colleagues [36] estimated the reliability of the N-Pose that is used as part of the Xsens calibration procedure and that defines the neutral segment orientations [22]. They showed that the orientation of the ankle's axes of rotation can vary by up to $3°$ following repeated calibrations, which could contribute to poor between-day reliability. However, for peak ankle eversion and peak hip adduction, we observed poor between-day reliability despite good calibration reliability. Two possible explanations for this disconnect are: First, our study design only allowed us to estimate within-day calibration reliability, i.e., based on two calibrations on the same day. It is possible that runners' calibration movements deviated significantly more between calibrations on different days. Second, between-day differences in IMU placement on the thigh, shank, and shoe may lead to a randomly varying influence of soft-tissue and shoe deformation on peak frontal plane hip and ankle angles during the stance phase of running. This would be an additional source of between-day errors independent of the calibration [15].

The moderate to poor between-day reliability and high MDCs for ankle and hip angles in the frontal plane are of particular concern given that (1) these angles are often discussed in the context of running injuries [24], and (2) these angles have a small range of motion. Therefore, for the example of peak ankle eversion, the MDC of 6.7° already covers more than 30% of the ankle inversion–eversion range of motion during the gait cycle (cf. Supplementary Figure S1). In summary, it appears difficult to reliably track hip and ankle angles in the frontal plane with the IMU-based Xsens system across different days. Based on the results of this study, Xsens-based measurements can currently not be recommended for monitoring longitudinal changes in frontal plane hip and ankle angles during running. One idea to potentially improve the between-day reliability in frontal plane joint angles would be to monitor the exact N-Pose used by runners on the first measurement day (e.g., based on goniometric measurements, foot prints, or photographs), and then help runners to attain an almost identical N-Pose on following measurement days [36].

4.3. Potential Sources of Between-Day Variations in Xsens-Based Lower Extremity Joint Angles

The absolute between-day differences of the investigated Xsens-based lower extremity joint angles provide an overview of their absolute reliability in the original unit of measurement (see Figure 2). This analysis has revealed that discrepancies in discrete joint angles between measurements on two different days are about twice as large as discrepancies between measurements on the same day, confirming the trends observed in the ICCs and MDCs. Further, the analysis showed that within-day variations of joint angles have a similar magnitude for comparisons of (1) two separate runs processed using the same calibration trial and (2) the same run processed with two different calibration trials. The first comparison reveals variations in joint angles due to (1a) runner-internal processes, such as the ability to reproduce the same movement pattern under comparable external conditions, and (1b) measurement errors at the level of the IMU sensors, the Xsens sensor fusion algorithm, or changes in the sensor-to-segment alignment following undesired sensor movements (see Section 4.1). The second comparison reveals variations in joint angles due to (2a) the runner-internal ability to perform repeated calibration movements in a similar way, and/or (2b) measurement errors at the level of the Xsens calibration algorithm.

In conjunction, the observations in Figure 2 suggest that—when running at a self-selected speed outdoors—about half of the between-day differences in Xsens-based joint angles may be explained by day-to-day variability internal to the runner, while the other half may result from technical errors within the measurement system, including sensor placement, calibration, and IMU measurement errors. Additional error sources that were not accounted for in this study include environmental factors such as temperature or precipitation [37]. The fact that the movement pattern of runners varies from day to day is not new; e.g., Benson and colleagues showed that at least five self-paced outdoor running sessions on different days are necessary to fully characterize a runner's movement pattern [38]. Our study adds that when the running movement is quantified on different days using the Xsens Link suit, the between-day variations will include an additional technology-based source of variation that is of similar magnitude to the internal source of variation.

4.4. Comparison with the Reliability of Other Assessments of 3D Running Kinematics

Besides the gold standard of marker-based optical motion capture (OMC), IMU-based methods and markerless (computer vision-based) methods have become viable alternatives for the assessment of 3D running kinematics over the last two decades. Therefore, it was of interest to contrast the reliability estimates between the IMU-based Xsens system presented here with previous reliability studies investigating the gold standard or emerging markerless motion analysis techniques. Table 3 shows comparisons of between-day ICCs of the current study with those from two previous studies reporting the between-day reliability of discrete lower-extremity joint angles during running at a self-selected speed based on OMC [39,40], and with one study based on markerless motion analysis [41]. In general, markerless motion analysis showed the same (three out of nine variables) or superior (six out of nine variables) between-day reliability category compared to the Xsens-based approach of the current study. This trend towards the improved reliability of markerless motion analysis likely stems from the minimal dependence of this technology on user input with respect to sensor placement or subject-specific calibration procedures [42]. When comparing the Xsens-based and OMC-based approaches, superior reliability depended on the specific variable and the marker model used in the OMC-based approach. All three technologies showed good to excellent between-day reliability (ICCs > 0.87) for initial ankle dorsiflexion, indicating that all three motion analysis approaches are suited to monitoring day-to-day changes in the foot strike angle. Knee and hip flexion outcomes also showed good between-day reliability for all three technologies, with the exception of the OMC approach, when using the Plug-in-Gait marker model, resulting in only poor to moderate reliability. The partially inferior reliability of the OMC approach combined with the Plug-in-Gait marker model may result from the strong dependence of this technique on reliable marker placement [39]. While an in-depth analysis of reliability differences between systems was out of the scope of this manuscript, Table 3 can be used as guidance when trying to select the most reliable measurement system for a given discrete joint angle during running.

Table 3. ICCs for between-day reliability of Xsens-based lower extremity joint angles during running compared to previously reported ICC between-day reliability estimates of marker-based optical motion capture (OMC) and markerless motion analysis. Orange, yellow, light green, and dark green coloring indicate poor, moderate, good, and excellent reliability according to Koo and Li [28].

Category	Variable	Current Study IMU-Based (Xsens) ICC Median [Range]		Okahisa et al. (2023) [39] OMC (Vicon) ICC		Stoneham et al. (2019) [40] OMC (Vicon) ICC [95% CI]	Moran et al. (2023) Markerless [41] (Theia/Visual3D) ICC [95% CI]
		Asphalt	Woodchip	Lab Floor PiG *	Lab Floor CGM2 *	Lab Floor PiG *	Lab Floor
Hip	Initial hip flexion	0.92 [0.89, 0.95]	0.96 [0.83, 0.96]	0.80	0.87	0.89 [0.72, 0.96]	0.86 [0.80, 0.91]
	Initial hip ab-/adduction	0.52 [0.44, 0.68]	0.49 [0.29, 0.62]	0.67	0.75	0.31 [0.54, 0.93]	0.75 [0.64, 0.83]
	Peak hip abduction	0.83 [0.70, 0.85]	0.74 [0.63, 0.77]	0.45	0.84	0.69 [0.31, 0.88]	0.91 [0.87, 0.94]
Knee	Initial knee flexion	0.92 [0.90, 0.95]	0.92 [0.89, 0.96]	0.74	0.81	0.76 [0.44, 0.91]	0.88 [0.83, 0.92]
	Peak knee flexion	0.82 [0.75, 0.88]	0.89 [0.85, 0.93]	0.74	0.89	0.73 [0.49, 0.92]	0.90 [0.86, 0.94]
Ankle	Initial ankle dorsiflexion	0.90 [0.82, 0.94]	0.88 [0.87, 0.95]	0.87	0.93	0.95 [0.86, 0.98]	0.94 [0.90, 0.96]
	Peak ankle dorsiflexion	0.38 [0.23, 0.52]	0.47 [−0.02, 0.49]	0.68	0.86	0.85 [0.63, 0.95]	0.76 [0.66, 0.84]
	Initial ankle inversion	0.81 [0.77, 0.90]	0.72 [0.64, 0.85]	*not available*		0.69 [0.31, 0.88]	0.72 [0.61, 0.81]
	Peak ankle eversion	0.55 [0.53, 0.72]	0.65 [0.46, 0.72]	*not available*		0.74 [0.41, 0.90]	0.83 [0.75, 0.88]

* PiG = Plug-in-Gait model; CGM2 = Conventional Gait Model 2.

4.5. Limitations

We did not conduct a systematic between-rater comparison, and thus our results should technically not be generalized to other raters. However, a previous between-rater reliability study on squatting, jumping and walking movements concluded that Xsens-based joint angles are not influenced by user expertise [18]. This is not surprising because the only real influence of the user on the measurement procedure results from the placement of the IMU sensors, which is predetermined and guided by Velcro straps inside the Xsens Link suit and video tutorials of the manufacturer. Therefore, we assume that the influence of the rater is negligible for Xsens-based joint angle measurements relative to the influence of re-calibrating the suit, and further, that the reported within- and between-day reliability estimates can be used for the study design and sample size estimations of future running-related motion analysis studies.

5. Conclusions

During running at a self-selected speed outdoors, the IMU-based Xsens Link suit derives estimates of lower-extremity joint angles in the sagittal and frontal planes that generally show good to excellent reliability between repeated runs on the same day. Provided the sensors do not require a re-calibration between experimental conditions, the Xsens Link suit is well suited to capturing within-day adaptations in the movement pattern of the lower extremities in response to different running surfaces or similar interventions. For repeated measurements on different days, the Xsens Link suit retained good to excellent between-day reliability for sagittal plane hip, knee, and ankle angles just before foot contact, and can be used to reliably monitor longitudinal changes in these angles if these changes exceed $4°$. Hip and ankle frontal plane angles showed only poor to moderate between-day reliability, likely due to a higher calibration error, and thus, the Xsens Link suit can currently not be recommended for use in monitoring day-to-day changes in those variables. Potential ways to improve between-day reliability include (1) controlling running speed (specifically for initial hip flexion), (2) controlling stride frequency (specifically for peak knee and ankle angles), and (3) standardizing the subject-specific calibration pose (N-Pose) between testing days (specifically for frontal plane angles).

Supplementary Materials: The following supporting information can be downloaded at: https://www.mdpi.com/article/10.3390/s24030871/s1, Figure S1: Xsens-based joint angles as a function of time during the running gait cycle for five testing days of one exemplary runner.

Author Contributions: Conceptualization, M.M.; methodology, M.M.; software, M.M.; formal analysis, D.D. and M.M.; data curation, D.D., A.W. and M.M.; writing—original draft preparation, D.D., A.W. and M.M.; writing—review and editing, D.D., A.W. and M.M.; visualization, D.D., A.W. and M.M.; project administration, M.M. All authors have read and agreed to the published version of the manuscript.

Funding: This research received no external funding.

Institutional Review Board Statement: The study was conducted in accordance with the Declaration of Helsinki and approved by the local Ethics Committee at the Department of Sport Science, Universität Innsbruck (ID 39/2020), and all participants provided written informed consent prior to inclusion in the study.

Informed Consent Statement: Informed consent was obtained from all subjects involved in the study.

Data Availability Statement: All raw data underlying this study have been uploaded to the Mendeley Data Repository: Mohr, Maurice; Debertin, Daniel; Wargel, Anna; Peer, Lukas; De Michiel, Alessia; van Andel, Steven; Federolf, Peter (2023), "Whole-body kinematic adaptations to running on an unstable, irregular, and compliant surface: Data, code, and Supplemental Files", Mendeley Data, V4, https://doi.org/10.17632/j4bck84rpr.5 (accessed on 15 January 2024).

Conflicts of Interest: The authors declare no conflicts of interest.

References

1. Mohr, M.; Federolf, P.; Pepping, G.-J.; Stein, T.; van Andel, S.; Weir, G. Editorial: Human movement and motor control in the natural environment. *Front. Bioeng. Biotechnol.* **2023**, *11*, 1210173. [CrossRef] [PubMed]
2. Dorschky, E.; Camomilla, V.; Davis, J.; Federolf, P.; Reenalda, J.; Koelewijn, A.D. Perspective on 'in the wild' movement analysis using machine learning. *Hum. Mov. Sci.* **2023**, *87*, 103042. [CrossRef] [PubMed]
3. Benson, L.C.; Clermont, C.A.; Bošnjak, E.; Ferber, R. The use of wearable devices for walking and running gait analysis outside of the lab: A systematic review. *Gait Posture* **2018**, *63*, 124–138. [CrossRef] [PubMed]
4. Dorschky, E.; Nitschke, M.; Seifer, A.-K.; van den Bogert, A.J.; Eskofier, B.M. Estimation of gait kinematics and kinetics from inertial sensor data using optimal control of musculoskeletal models. *J. Biomech.* **2019**, *95*, 109278. [CrossRef] [PubMed]
5. Jaén-Carrillo, D.; Roche-Seruendo, L.E.; Cartón-Llorente, A.; Ramírez-Campillo, R.; García-Pinillos, F. Mechanical Power in Endurance Running: A Scoping Review on Sensors for Power Output Estimation during Running. *Sensors* **2020**, *20*, 6482. [CrossRef]
6. Slade, P.; Kochenderfer, M.J.; Delp, S.L.; Collins, S.H. Sensing leg movement enhances wearable monitoring of energy expenditure. *Nat. Commun.* **2021**, *12*, 4312. [CrossRef]
7. Zeng, Z.; Liu, Y.; Hu, X.; Tang, M.; Wang, L. Validity and Reliability of Inertial Measurement Units on Lower Extremity Kinematics During Running: A Systematic Review and Meta-Analysis. *Sports Med. Open* **2022**, *8*, 86. [CrossRef]
8. Roetenberg, D.; Luinge, H.; Slycke, P. XSens MVN: Full 6DOF Human Motion Tracking Using Miniature Inertial Sensors. *XSens Technol.* **2013**, *1*, 1–9.
9. Mohr, M.; Peer, L.; De Michiel, A.; van Andel, S.; Federolf, P. Whole-body kinematic adaptations to running on an unstable, irregular, and compliant surface. *Sports Biomech.* **2023**, 1–15. [CrossRef]
10. Zandbergen, M.; Buurke, J.; Veltink, P.; Reenalda, J. Quantifying and correcting for speed and stride frequency effects on running mechanics in fatiguing outdoor running. *Front. Sports Act. Living* **2023**, *5*, 1085513. [CrossRef]
11. Genitrini, M.; Fritz, J.; Stöggl, T.; Schwameder, H. Performance Level Affects Full Body Kinematics and Spatiotemporal Parameters in Trail Running-A Field Study. *Sports* **2023**, *11*, 188. [CrossRef]
12. Fraeulin, L.; Maurer-Grubinger, C.; Holzgreve, F.; Groneberg, D.A.; Ohlendorf, D. Comparison of Joint Kinematics in Transition Running and Isolated Running in Elite Triathletes in Overground Conditions. *Sensors* **2021**, *21*, 4869. [CrossRef]
13. Roetenberg, D.; Slycke, P.J.; Veltink, P.H. Ambulatory position and orientation tracking fusing magnetic and inertial sensing. *IEEE Trans. Biomed. Eng.* **2007**, *54*, 883–890. [CrossRef] [PubMed]
14. Picerno, P. 25 years of lower limb joint kinematics by using inertial and magnetic sensors: A review of methodological approaches. *Gait Posture* **2017**, *51*, 239–246. [CrossRef] [PubMed]
15. Cereatti, A.; Bonci, T.; Akbarshahi, M.; Aminian, K.; Barré, A.; Begon, M.; Benoit, D.L.; Charbonnier, C.; Maso, F.D.; Fantozzi, S.; et al. Standardization proposal of soft tissue artefact description for data sharing in human motion measurements. *J. Biomech.* **2017**, *62*, 5–13. [CrossRef]
16. Hughes, G.T.G.; Camomilla, V.; Vanwanseele, B.; Harrison, A.J.; Fong, D.T.P.; Bradshaw, E.J. Novel technology in sports biomechanics: Some words of caution. *Sports Biomech.* **2021**, 1–9. [CrossRef]
17. Robert-Lachaine, X.; Mecheri, H.; Larue, C.; Plamondon, A. Validation of inertial measurement units with an optoelectronic system for whole-body motion analysis. *Med. Biol. Eng. Comput.* **2017**, *55*, 609–619. [CrossRef] [PubMed]
18. Al-Amri, M.; Nicholas, K.; Button, K.; Sparkes, V.; Sheeran, L.; Davies, J.L. Inertial Measurement Units for Clinical Movement Analysis: Reliability and Concurrent Validity. *Sensors* **2018**, *18*, 719. [CrossRef] [PubMed]
19. Trott, E.; Al-Amri, M. The reliability of inertial measurement units in estimating lower limb joint angles during treadmill running. *Gait Posture* **2022**, *97*, S182–S183. [CrossRef]
20. Zhang, J.-T.; Novak, A.C.; Brouwer, B.; Li, Q. Concurrent validation of Xsens MVN measurement of lower limb joint angular kinematics. *Physiol. Meas.* **2013**, *34*, N63–N69. [CrossRef] [PubMed]
21. Mavor, M.P.; Ross, G.B.; Clouthier, A.L.; Karakolis, T.; Graham, R.B. Validation of an IMU Suit for Military-Based Tasks. *Sensors* **2020**, *20*, 4280. [CrossRef]
22. Schepers, M.; Giuberti, M.; Bellusci, G. Xsens MVN: Consistent tracking of human motion using inertial sensing. *Xsens Technol.* **2018**, *8*, 1–8.
23. Nijmeijer, E.M.; Heuvelmans, P.; Bolt, R.; Gokeler, A.; Otten, E.; Benjaminse, A. Concurrent validation of the Xsens IMU system of lower-body kinematics in jump-landing and change-of-direction tasks. *J. Biomech.* **2023**, *154*, 111637. [CrossRef]
24. Willwacher, S.; Kurz, M.; Robbin, J.; Thelen, M.; Hamill, J.; Kelly, L.; Mai, P. Running-Related Biomechanical Risk Factors for Overuse Injuries in Distance Runners: A Systematic Review Considering Injury Specificity and the Potentials for Future Research. *Sports Med.* **2022**, *52*, 1863–1877. [CrossRef]
25. Atkinson, G.; Nevill, A.M. Statistical methods for assessing measurement error (reliability) in variables relevant to sports medicine. *Sports Med.* **1998**, *26*, 217–238. [CrossRef] [PubMed]
26. Borg, D.N.; Bach, A.J.E.; O'Brien, J.L.; Sainani, K.L. Calculating sample size for reliability studies. *PM&R* **2022**, *14*, 1018–1025.
27. Bonett, D.G. Sample size requirements for estimating intraclass correlations with desired precision. *Stat. Med.* **2002**, *21*, 1331–1335. [CrossRef] [PubMed]
28. Koo, T.K.; Li, M.Y. A Guideline of Selecting and Reporting Intraclass Correlation Coefficients for Reliability Research. *J. Chiropr. Med.* **2016**, *15*, 155–163. [CrossRef] [PubMed]

29. Wu, G.; Siegler, S.; Allard, P.; Kirtley, C.; Leardini, A.; Rosenbaum, D.; Whittle, M.; D'Lima, D.D.; Cristofolini, L.; Witte, H.; et al. ISB recommendation on definitions of joint coordinate system of various joints for the reporting of human joint motion—Part I: Ankle, hip, and spine. *J. Biomech.* **2002**, *35*, 543–548. [CrossRef]

30. Wu, G.; Cavanagh, P.R. ISB recommendations for standardization in the reporting of kinematic data. *J. Biomech.* **1995**, *28*, 1257–1262. [CrossRef] [PubMed]

31. Benson, L.C.; Clermont, C.A.; Watari, R.; Exley, T.; Ferber, R. Automated accelerometer-based gait event detection during multiple running conditions. *Sensors* **2019**, *19*, 1483. [CrossRef]

32. McGraw, K.O.; Wong, S.P. Forming inferences about some intraclass correlation coefficients. *Psychol. Methods* **1996**, *1*, 30. [CrossRef]

33. The Jamovi Project (2023). Jamovi (Version 2.3.21) [Computer Software]. Available online: https://www.jamovi.org (accessed on 20 December 2023).

34. Gallucci, M. GAMLj Models. 1 January 2023. Available online: https://gamlj.github.io/book/booklet.html#booklet (accessed on 2 November 2023).

35. Honert, E.C.; Mohr, M.; Lam, W.K.; Nigg, S. Shoe feature recommendations for different running levels: A Delphi study. *PLoS ONE* **2020**, *15*, e0236047. [CrossRef]

36. Robert-Lachaine, X.; Mecheri, H.; Larue, C.; Plamondon, A. Accuracy and repeatability of single-pose calibration of inertial measurement units for whole-body motion analysis. *Gait Posture* **2017**, *54*, 80–86. [CrossRef] [PubMed]

37. Ahamed, N.U.; Kobsar, D.; Benson, L.; Clermont, C.; Kohrs, R.; Osis, S.T.; Ferber, R. Using wearable sensors to classify subject-specific running biomechanical gait patterns based on changes in environmental weather conditions. *PLoS ONE* **2018**, *13*, e0203839. [CrossRef] [PubMed]

38. Benson, L.C.; Ahamed, N.U.; Kobsar, D.; Ferber, R. New considerations for collecting biomechanical data using wearable sensors: Number of level runs to define a stable running pattern with a single IMU. *J. Biomech.* **2019**, *85*, 187–192. [CrossRef]

39. Okahisa, T.; Matsuura, T.; Tomonari, K.; Komatsu, K.; Yokoyama, K.; Iwase, J.; Yamada, M.; Sairyo, K. Between-day reliability and minimum detectable change of the Conventional Gait Model 2 and Plug-in Gait Model during running. *Gait Posture* **2023**, *100*, 171–178. [CrossRef]

40. Stoneham, R.; Barry, G.; Saxby, L.; Wilkinson, M. Measurement error of 3D kinematic and kinetic measures during overground endurance running in recreational runners between two test sessions separated by 48 h. *Physiol. Meas.* **2019**, *40*, 024002. [CrossRef]

41. Moran, M.F.; Rogler, I.C.; Wager, J.C. Inter-Session Repeatability of Marker-Less Motion Capture of Treadmill Running Gait. *Appl. Sci.* **2023**, *13*, 1702. [CrossRef]

42. Kanko, R.M.; Laende, E.; Selbie, W.S.; Deluzio, K.J. Inter-session repeatability of markerless motion capture gait kinematics. *J. Biomech.* **2021**, *121*, 110422. [CrossRef] [PubMed]

Article

Effects of High-Load Bench Press Training with Different Blood Flow Restriction Pressurization Strategies on the Degree of Muscle Activation in the Upper Limbs of Bodybuilders

Kexin He [1,†], Yao Sun [2,†], Shuang Xiao [1], Xiuli Zhang [3], Zhihao Du [1,4,*] and Yanping Zhang [2,*]

1   School of P.E. and Sports, Beijing Normal University, Beijing 100875, China; 11112018073@bnu.edu.cn (K.H.); 18074334809@163.com (S.X.)
2   College of Physical Education, China University of Mining and Technology, Xuzhou 221116, China; sunyao5188@163.com
3   College of Physical Education (Main Campus), Zhengzhou University, Zhengzhou 450052, China; zxl_zzu_edu@126.com
4   School of Sports Science, Jishou University, Jishou 416000, China
*   Correspondence: dzh_cumt_edu@163.com (Z.D.); kdzhangyanping@126.com (Y.Z.)
†   These authors contributed equally to this work.

Citation: He, K.; Sun, Y.; Xiao, S.; Zhang, X.; Du, Z.; Zhang, Y. Effects of High-Load Bench Press Training with Different Blood Flow Restriction Pressurization Strategies on the Degree of Muscle Activation in the Upper Limbs of Bodybuilders. *Sensors* 2024, 24, 605. https://doi.org/10.3390/s24020605

Academic Editor: David Niederseer

Received: 11 December 2023
Revised: 29 December 2023
Accepted: 5 January 2024
Published: 17 January 2024

Abstract: Background: The aim of this study was to investigate the effects of different pressurization modes during high-load bench press training on muscle activation and subjective fatigue in bodybuilders. **Methods**: Ten bodybuilders participated in a randomized, self-controlled crossover experimental design, performing bench press training under three different pressurization modes: T1 (low pressure, high resistance), T2 (high pressure, high resistance), and C (non-pressurized conventional). Surface EMG signals were recorded from the pectoralis major, deltoid, and triceps muscles using a Delsys Trigno wireless surface EMG during bench presses. Subjective fatigue was assessed immediately after the training session. **Results**: (1) Pectoralis major muscle: The muscle activation degree of the T1 group was significantly higher than that of the blank control group during the bench press ($p < 0.05$). The muscle activation degree of the T2 group was significantly higher than that of the C group during the bench press ($p < 0.05$). In addition, the muscle activation degree of the T2 group was significantly higher than that of the T1 group during the first group bench press ($p < 0.05$). (2) Deltoid muscle: The muscle activation degree of the T2 group during the third group bench press was significantly lower than the index values of the first two groups ($p < 0.05$). The muscle activation degree in the experimental group was significantly higher than that in the C group ($p < 0.05$). The degree of muscle activation in the T2 group was significantly higher than that in the T1 group during the first bench press ($p < 0.05$). (3) Triceps: The muscle activation degree of the T1 group was significantly higher than the index value of the third group during the second group bench press ($p < 0.05$), while the muscle activation degree of the T2 group was significantly lower than the index value of the first two groups during the third group bench press ($p < 0.05$). The degree of muscle activation in all experimental groups was significantly higher than that in group C ($p < 0.05$). (5) RPE index values in all groups were significantly increased ($p < 0.05$). The RPE value of the T1 group was significantly higher than that of the C group after bench press ($p < 0.05$). The RPE value of the T1 group was significantly higher than that of the C group after bench press ($p < 0.05$). In the third group, the RPE value of the T1 group was significantly higher than that of the C and T2 groups ($p = 0.002$) ($p < 0.05$). **Conclusions**: The activation of the pectoralis major, triceps brachii, and deltoid muscles is significantly increased by high-intensity bench press training with either continuous or intermittent pressurization. However, continuous pressurization results in a higher level of perceived fatigue. The training mode involving high pressure and high resistance without pressurization during sets but with 180 mmHg occlusion pressure and pressurization during rest intervals yields the most pronounced overall effect on muscle activation.

Keywords: blood flow restriction; bench press; upper-extremity muscle groups; muscle activation level

1. Introduction

BFRT, also known as blood flow restriction training, is an exercise training method that uses a binding cuff at the near end of the limbs of the participants to completely block venous blood return, while reducing but maintaining arterial blood flow in the muscles, resulting in blood accumulation in the limbs, and triggering a stronger stress physiological response than traditional training [1]. BFRT is widely used in competitive sports training, mass fitness, and sports injury rehabilitation [2].

At present, international experimental studies on blood flow restriction show that this training method is a mainstream research topic, and the research consensus is that low-intensity resistance training combined with different pressure interventions can be superior to non-pressure training in strength performance. When pressure reaches a certain threshold, low-intensity blood flow restriction can cause strength gains and muscle hypertrophy similar to traditional high-intensity resistance training. Domestic and international exercise experiments on pressurized resistance have resulted in a series of studies on the effects of different pressurized resistance modes on muscular strength and endurance. In these studies, common compression resistance training mode combinations include low-intensity exercise (20%1 RM~30%1 RM) combined with low, medium, and high occlusion pressures (100 mmHg~300 mmHg) [3], moderate-intensity exercise (30%1 RM to 50%1 RM) combined with low-to-medium occlusion pressures (100 mmHg to 200 mmHg) [4], and high-intensity exercise (70%1 RM) combined with low occlusion pressure (100 mmHg~150 mmHg) [5–7]. However, little research has been conducted on the pressurized resistance mode of high-intensity exercise combined with medium and high pressures, mainly because sustained high occlusion pressures can lead to a significant increase in the risk of potential safety issues such as cardiovascular injuries, muscle injuries, and other injury risks [8], which to some extent can lead to the occurrence of sports injuries.

From the current stage of experimental studies on pressurized resistance at home and abroad, the means of pressurization include continuous pressurization and intermittent pressurization. Continuous pressurization consists of maintaining continuous pressurization throughout the entire exercise period, including rest between sets, while intermittent pressurization consists of two modes: pressurization during the exercise period but depressurization during the intermittent period or depressurization during the exercise period and pressurization during the intermittent period. Two recent experimental studies [9,10] were conducted by combining moderate-to-high intensity exercise with a high degree of occlusion pressure by arranging the participants to have no pressurization during exercise and medium-to-high pressurization during the rest intervals between sets and conducting an experimental study on acute resistance training of the lower extremities. This kind of pressurization intervention can prevent the safety problems caused by the continuous restriction of blood flow due to the continuous high pressure as well as address the paucity of research regarding a high-pressure high-resistance model in pressurized resistance training. It should be noted that the above two experimental studies focused on the effects of a high-pressure high-resistance training model of pressurized resistance training on the muscle strength and muscle dimensions of the lower extremities, but the neuromuscular properties of this model have not been examined in the upper-extremity muscles. Previous studies have clearly indicated that pressurized resistance training can bring about greater muscle activation than traditional resistance training and that pressurized stimulation has a better effect on neuromuscular properties [11,12]. Therefore, it is reasonable to investigate the acute changes in muscle activation in the upper extremities during high-pressure high-resistance bench press training by means of changing the mode of pressurization.

2. Materials and Methods

2.1. Participants

This study calculated the number of participants through G*Power software 3.1 version, and finally selected 10 participants. The participants in this experiment were 10 bodybuilders from the College of Physical Education and Sports, Beijing Normal University, and the inclusion criteria for the participants were as follows: (1) more than three years of resistance training and proficiency in the bench press exercise; (2) a 1 RM in a bench press of at least 1.2 times their body mass; and (3) no athletic injuries in the last six months. The basic information of the participants is shown in Table 1. The purpose of this experiment, the method of implementation, and the possible associated risks were fully explained to the participants before the experiment, and consent was obtained from the faculty coach and the athletes. Before the test, the participants were informed of the training movements involved in the experiment and the interventions that differed from normal training, and each subject was allowed to wear the pressurization equipment during training for familiarization one week before the experiment. All participants signed informed consent forms prior to the experiment, which strictly followed the Helsinki Declaration [13] and passed the review of the Ethics Committee of Zhengzhou University (ZZUIRB2022-JCYXY0016). The experimental site was the National Fitness Center of Physical Education College of Beijing Normal University.

Table 1. Basic information of the participants.

Age	Height (cm)	Body Mass (kg)	Shoulder Width (cm)	Bench Press 1 RM (kg)
23.67± 1.73	174.22± 4.06	79.17 ± 8.28	44.50 ± 4.28	106.33 ± 10.48

Inclusion criteria: (1) bodybuilders aged 20–25 years; (2) no major diseases or chronic diseases; (3) proficient in standard bench press training movements; (4) bench press 1 RM level of at least 1.2 times body mass.

Exclusion criteria: (1) muscle strain, tendon inflammation, fracture, and other sports injuries that occurred in the past six months; (2) rich experience in resistance training; the total training years must reach three years or more; (3) use of specialized illegal drugs that promote muscle growth; (4) Reliance on protective equipment such as power belts, wrist guards, and other equipment to complete the lying push.

2.2. Methods

2.2.1. Experimental Design and Intervention Program

The 72 h before the formal experiment was the preparation stage. First, the participants were recruited, and then the basic information of the participants was recorded. The parameters included age, height, body mass, and bench press 1 RM for each subject. See Figure 1 for details

On the day of the formal experimental test, participants first warmed up and then performed a maximum voluntary isometric contraction (MVIC) test and pressurized bench press surface electromyography test on the target muscles of the upper limb. Each subject performed three different modes of bench press training: C (no-pressurization mode): bench press training at an exercise intensity of 70%1 RM; T1 (low-pressurization, high-resistance mode): bench press training at an exercise intensity of 70%1 RM with continuous pressurization and an occlusion pressure of 100 mmHg; and T2 (high pressurization, high resistance): bench press training at an exercise intensity of 70%1 RM with intermittent pressurization denoted by a depressurization phase during the exercise period and a pressurization phase with an occlusion pressure of 180 mmHg during the intervals between the sets. The participants performed all three modes of training protocols as 3 continuous sets of 8 repetitions each, with a 2 min interset rest interval for bench press training. In this study, the right target muscle groups of the pectoralis major, deltoid, and triceps brachii were selected for the MVIC test (the order of muscle testing was randomized and parallel), and the changes in electromyography were recorded by using a 3-channel Delsys Trigno wireless surface electromyography signal acquisition system at the same time as the MVIC test. Five minutes after the completion of the MVIC test, the participants underwent pressurized bench press training, and Theratools BFR pressurized bench press equipment was used. Theratools BFR pressurization equipment was used for pressurized intervention under the following conditions: the bundle pressure was 30 mmHg; the location of the pressurized cuff was near the proximal end of the arm; the surface EMG signal acquisition was synchronized with the pressurized bench press training; and the sequence of pressurized bench press conditions in different modes was randomized and counterbalanced. The time interval between the two different modes was 72 h, not only to prevent muscle damage caused by the pressurized resistance training from affecting the participants' test status [14] but also to reduce the mutual interference effect between the different modes.

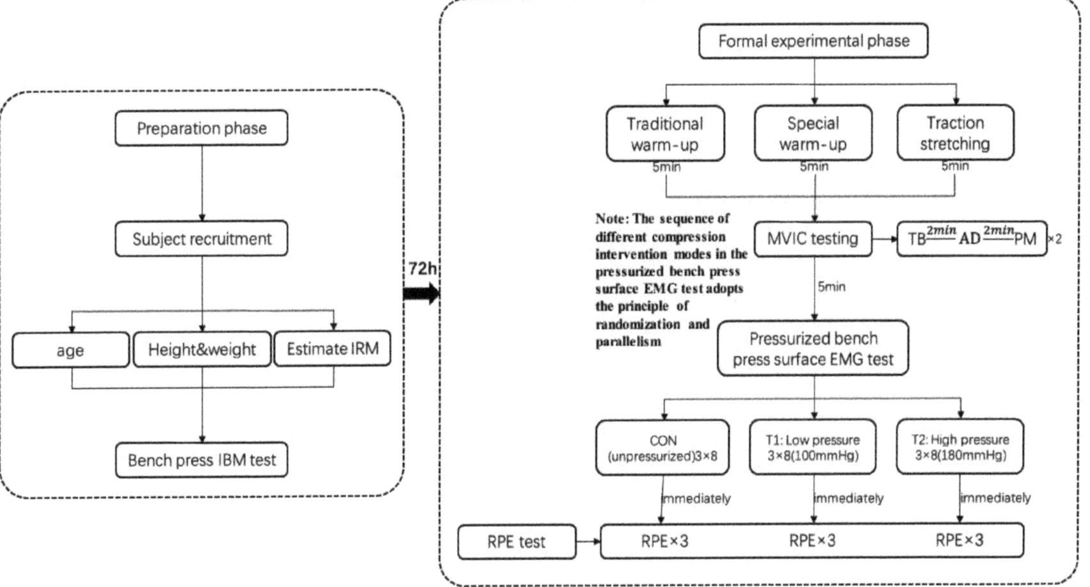

Figure 1. Flowchart of the experiment.

2.2.2. Testing of Experimental Indicators

(1)　Bench Press 1 RM Test

Seventy-two hours prior to the experimental phase, participants performed a 1 RM bench press test, and their personal data were collected, including 1. height, 2. body mass, 3. age, and 4. shoulder width. Participants were sequenced through a parallel and randomized lottery for the experimental protocol. After data collection was completed, a bench press 1 RM test was performed using a standardized warm-up procedure for each test using a bicycle ergometer with an upper-body component for approximately 5 min with a resistance of 100 W and a pedaling frequency between 70 and 80 rpm. Next, a warm-up was performed for upper-body muscle groups, including mobility exercises for the shoulders and chest. After the general warm-up was completed, a specialized warm-up was performed. Participants performed 15, 10, and 5 repetitions of the bench press using 20%, 40%, and 60% of the estimated 1 RM. After warming up, the 1 RM test is as follows: (1) increase body mass by 4–9 kg from 80% of the estimated 1 RM per successive attempt, conservatively repeat 3–5 times, then rest for 2 min. (2) Continue to increase the body mass by 4–9 kg, then complete 2–3 repetitions, rest, then increase by 4–9 kg while trying to lift the 1 RM. (3) If one attempt is successful, continue to increase the body mass by 4–9 kg, but if the attempt fails, reduce the body mass by 2–4 kg and measure 1 RM over the course of 3–5 attempts. This process was repeated until failure. The 1 RM for all participants was determined in 5 trials.

In the 1 RM test and the formal experiment, participants were asked to bench press with a constant rhythm and trajectory (using a metronome to control the subject's rhythm, the duration of the centrifugal phase (point A → point B) was 2 s, and the centripetal phase (point B → point C) was 1 s to try to ensure consistency of the movement time). Participants were instructed to keep their head, shoulders, and hips in contact with the bench, lower the barbell until it touches the chest during its descent, and fully extend the elbows at the end of the concentric phase to achieve a valid repetition in the bench press movement.

Participants' hands were placed on the barbell in a fixed position equal to 150% of the individual's peak shoulder width. Participants were not allowed to use weightlifting belts, wrist wraps, elbow sleeves, or other assistive equipment during the exercise period, and all repetitions were directly supervised by an experienced physical trainer.

(2) Maximum random isometric contraction (MVIC) test for target muscle groups of the upper limb

On the day of the official test, after the same warm-up as in the bench press 1 RM test, the MVIC test was first performed using a 3-channel Delsys Trigno wireless surface EMG signal acquisition system to measure and analyze the surface EMG signals of the muscles. Based on the biomechanical characteristics of the bench press training maneuver, three muscles of the upper limb were included, pectoralis major, deltoid, and triceps brachii, and the electrodes were located on the muscle belly of each of these three muscles with reference to anatomical characteristics. Before placing the gel-coated self-adhesive electrodes, the hair covering the muscle was shaved, and the skin was cleaned of dirt and sweat with alcohol, which reduced the skin's resistance while still being able to ensure effective attachment of the electrode sensors. According to Konrad's program [15], the integrated EMG values were collected for each muscle under the MVIC test. The test method was as follows (2 MVIC tests were performed for each muscle):

① Pectoralis muscle (Pectorals): The elbow joint was at 90 degrees, push-ups were performed, the tester pressed down on the subject's shoulders and gradually exerted force, the subject pushed upward as hard as he could to push off the ground, and the tester put up resistance to him and insisted on the action for 3–5 s to collect the electromyographic data of the pectoralis muscle. ② Anterior Deltoid: The subject was in a sitting position, the arm was flexed forward to the epigastric region, the tester pressed the subject's wrist, the subject's arm was flexed upward with maximum force, the tester exerted downward force against it, and the electromyographic data of the anterior deltoid were collected by maintaining the action for 3–5 s. ③ Triceps brachii: The subject was in a sitting position, the arm was naturally hanging down, the upper and lower arm were at 90 degrees, the tester grasped the wrist of the subject, the subject performed a lower-arm extension movement downward with maximum force, and the tester provided upward resistance, maintaining the movement for 3–5 s to collect the electromyographic data of the triceps brachii muscle.

(3) Pressurized bench press training surface EMG test

After completing the MVIC test, the participants took a rest for 5 min. During this interval, the tester implemented pressure intervention on the participants as a follow-up test. In this experiment, Theratools BFR pressure equipment was adopted. The equipment consists of a pressure pump and a binding cuff. The binding method is Velcro tape, and the pressurized part is the tropanes of the deltoid muscle of the athlete's arm. The binding pressure is 30 mmHg, the blocking pressure during exercise is 100 mmHg at low pressure and 180 mmHg at high pressure, respectively, and the pressurized method is continuous pressure at low pressure; that is, pressure is applied throughout the bench press. The high pressure adopts intermittent pressure: intermittent pressure in each group of exercises, and pressure is reduced during exercise. The sequence of EMG testing on the surface of the bench press in different pressurization modes was randomized and counterbalanced, and the time interval between the two different pressurization modes was 72 h, with a 2 min interval between each set. The pressurized bench press surface EMG test was performed after all preparations were made. The maximum heart rate and immediate postexercise blood pressure were selected as indirect indicators to judge the training intensity and training safety of the different conditions for the participants, and the greater the maximum heart rate and immediate postexercise blood pressure were, the greater the intensity of the pressurized resistance exercise intervention was considered to be [3].

(4)   Subjective fatigue (RPE) test

Prior to the start of the experiment, the standardized meanings of the RPE values were explained to the participants, and after each set of bench press training in the different pressurization modes, the participants were tested for subjective fatigue, and the grades were recorded. The test was performed using Zourdos' novel scale [16], and the levels ranged from 1 to 10, in which the first four levels were scored using subjective exertion: levels 1–2 indicated no exertion at all, and levels 3–4 indicated slight exertion. Levels 5–10 were scored using the number of repetitions in reserve (RIR): levels 5–6 indicated that the subject could have performed 4–6 more repetitions at the conclusion of that pressurized bench press set, level 7 indicated that 3 more repetitions were possible, level 8 indicated that 2 more repetitions were possible, level 9 indicated that 1 more repetition was possible, and level 10 indicated exhaustion.

2.3. Statistical Methods

The EMG data analyzed above were analyzed using Excel 2010 and SPSS 17.0 statistical software, and all the data were expressed in the form of mean ± standard deviation (M ± SD). A repeated-measure two-way ANOVA (number of exercise conditions × mode of pressurization) was used to statistically analyze the RMS standard values of upper-limb muscle groups during high-intensity bench press training in different modes of pressurization, and LSD was used for multiple comparisons of means when the interaction was significant. Finally, subjective fatigue after training in each condition was statistically analyzed using paired-sample t tests. The significance level of the test of difference for all the above analyses was taken as $p < 0.05$.

3. Results

3.1. Multiple Comparison Analysis

A two-way (model of pressurization × exercise condition) repeated-measure ANOVA revealed that the mode of pressurization had a significant effect on the change in activation values of the pectoralis major, deltoid, and triceps brachii ($p < 0.05$); the exercise condition had a significant effect on the change in activation values of the deltoid and triceps brachii ($p < 0.05$); and there was no significant effect of the mode of pressurization and exercise condition on the change in activation values of all muscles ($p > 0.05$). The EMG raw signal is detailed in Figure 2.

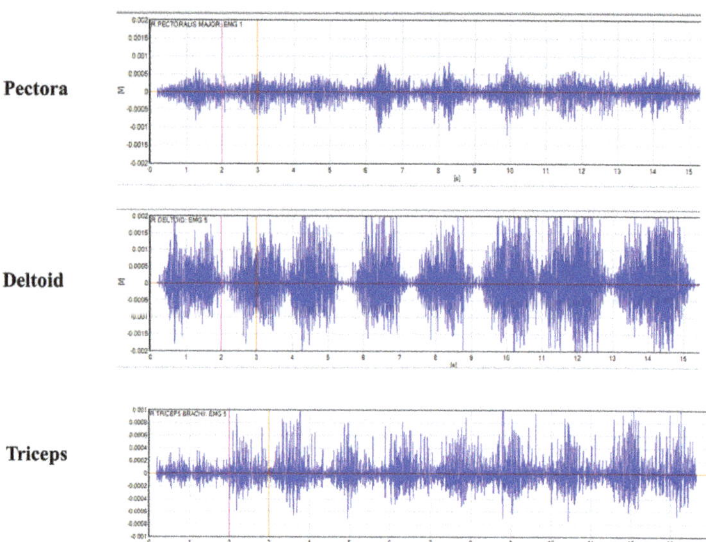

Figure 2. Raw EMG signals of the target muscle during bench press performed by one subject.

3.2. Changes in %MVIC Values of Target Muscles in Each Set of Bench Press Training in Different Pressurization Modes

The results showed (Tables 2 and 3) that pectoralis major muscle activation values were significantly higher in T1 than in C when performing the first ($p = 0.027$), second ($p = 0.005$), and third sets of bench presses ($p = 0.011$) ($p < 0.05$); muscle activation values were significantly higher in T2 than in C when performing the first ($p = 0.000$) and second ($p = 0.003$) sets of bench presses ($p < 0.05$); and muscle activation values were also significantly greater in T2 than in T1 when performing the first ($p = 0.016$) set of bench presses ($p < 0.05$).

Table 2. Results of Levene's homogeneity test.

Pressurized Mode	Pectoralis Major Muscle (Across the Top of the Chest)		Deltoid Muscle (Over the Shoulder)		Triceps Brachii (Back of the Upper Arm)	
	F	P	F	P	F	P
pectoralis major muscle (across the top of the chest)	0.077	0.926	0.177	0.839	0.768	0.475
deltoid muscle (over the shoulder)	0.631	0.541	0.049	0.952	0.391	0.68
triceps brachii (back of the upper arm)	0.036	0.964	0.021	0.979	0.149	0.862

Based on the results of Levene's homogeneity test, it was found that $p > 0.05$ for all groups, i.e., the sample variance was equal for all groups, thus allowing for a two-factor repeated-measures ANOVA.

Table 3. Effect of pressurization mode, exercise condition, and their interaction on %MVIC values of target muscles.

	Pectoralis Major Muscle (Across the Top of the Chest)		Deltoid Muscle (Over the Shoulder)		Triceps Brachii (Back of the Upper Arm)	
	F	P	F	P	F	P
Pressurized mode	15.931	0.000 *	57.209	0.000 *	41.198	0.000 *
sports condition	0.881	0.421	3.648	0.046 *	3.251	0.047 *
interaction	1.971	0.134	1.511	0.214	0.722	0.551

Note: "*" represents a significant difference; significant difference is $p < 0.05$.

Deltoid muscle activation in T2 was significantly lower than that in T1 ($p = 0.032$) and T2 ($p = 0.000$) during the third set of bench presses ($p < 0.05$); muscle activation in T1 was significantly higher than that in C during the first ($p = 0.000$), second ($p = 0.014$), and third ($p = 0.015$) sets of bench presses ($p < 0.05$); and muscle activation in T2 was significantly higher than that in C during the first ($p < 0.000$), second ($p = 0.000$), and third sets ($p < 0.000$) of bench presses ($p < 0.05$). The value of muscle activation in T2 was also significantly greater than that of T1 ($p < 0.05$) when performing the second set of bench presses ($p = 0.000$).

Triceps brachii muscle activation was significantly greater in T1 ($p = 0.014$) during the second set of bench presses and in T2 during the third set of bench presses ($p < 0.05$) but significantly less than in Condition I ($p = 0.005$) and T2 ($p = 0.002$) during the third set of bench presses ($p = 0.000$). Muscle activation values were significantly higher than those of C ($p < 0.05$), but during the third set of bench presses ($p = 0.000$), muscle activation was significantly lower than that of C ($p < 0.005$). T2 muscle activation values were significantly higher than those of C when performing the first ($p = 0.000$) and second ($p = 0.000$) sets of bench presses ($p < 0.05$), but the degree of muscle activation during the third set of bench presses ($p = 0.000$) was significantly less than that of C ($p < 0.05$); see Table 4 for details.

Table 4. List of changes in %MVIC values for each muscle in each set of bench press training in different pressurization modes.

Muscle Name	Pressurized Mode	3 Sets of Bench Press Training (8 + 8 + 8)		
		Set I	Set II	Set III
pectoralis major muscle (across the top of the chest)	Sustained low pressure (T1)	43.73 ± 19.97 *Δ	46.30 ± 20.01 *	47.33 ± 21.05 *
	Intermittent high pressure (T2)	61.21 ± 23.27 *	53.46 ± 15.56 *	44.79 ± 18.61
	Unpressurized (C)	34.33 ± 16.64	34.15 ± 19.98	36.28 ± 19.19
deltoid muscle (over the shoulder)	Sustained low pressure (T1)	64.02 ± 15.09 *	58.39 ± 13.66 *#	58.56 ± 12.27 *
	Intermittent high pressure (T2)	73.85 ± 16.31 *§	75.40 ± 15.89 *§	65.85 ± 14.99 *
	Unpressurized (C)	48.83 ± 12.56	48.63 ± 12.88	48.85 ± 15.08
triceps brachii (back of the upper arm)	Sustained low pressure (T1)	46.42 ± 22.25 *	56.85 ± 16.45 *§	32.27 ± 19.75 *
	Intermittent high pressure (T2)	54.64 ± 13.79 *§	57.62 ± 13.46 *§	33.41 ± 16.79 *
	Unpressurized (C)	40.32 ± 17.43	47.46 ± 12.98	36.49 ± 18.37

Note: "*" indicates that there is a significant difference in the change in muscle activation level between the pressurized experimental condition and Condition C; "Δ" indicates that there is a significant difference in the change in muscle activation level between Condition T2 and Condition T1 in the first set of bench presses; "#" indicates that there is a significant difference in the change in muscle activation values between Condition T2 and Condition T1 in the second set of bench presses; "§" indicates that there is a significant difference in the change in muscle activation level values between the three sets of bench presses within each experimental condition, with a significant difference of $p < 0.05$.

3.3. Subjective Fatigue Results after High-Load Bench Press Training in Different Pressurization Modes

The results showed that with the increase in the number of bench press sets (Table 5), both the pressurized experimental condition and the nonpressurized C RPE index value increased significantly ($p < 0.05$); after the first bench press set, the T1 RPE value was significantly greater than the index value of the C ($p = 0.030$); after the second bench press set, the T1 ($p = 0.016$) RPE value was significantly greater than the index value of the C; after the third bench press set, the T1 RPE value was significantly greater than the index values of C ($p = 0.022$) and T2 ($p = 0.002$) ($p < 0.05$), and there was no significant difference between T2 and C ($p = 0.121$) ($p > 0.05$).

Table 5. List of subjective fatigue RPE scores of participants after bench press training in different pressurization modes ($n = 10$).

Subjective Fatigue	Pressurized Mode	Training Group		
		Set I	Set II	Set III
RPE value	Sustained low pressure (T1)	6.22 ± 0.67 *§	7.11 ± 1.17 *§#	8.67 ± 1.22 *
	Intermittent high pressure (T2)	5.89 ± 1.05 §	6.80 ± 1.12 §#	8.00 ± 1.41 Δ
	Unpressurized (C)	5.00 ± 1.00 §	5.89 ± 0.60 §#	7.11 + 1 17

Note: "*" indicates that there is a significant difference in the RPE value comparing the pressurized experimental condition with Condition C; "Δ" indicates that there is a significant difference in the RPE value between Conditions T1 and T2; "§" indicates that there is a significant difference in the RPE value comparing the first set of bench presses with the second and third sets within each condition. Within each condition, there were significant differences in RPE values between the first set of bench presses and the second and third sets; "#" indicates that there were significant differences in RPE values between the second and third sets of bench presses within each condition ($p < 0.05$).

4. Discussion

The main purpose of this study was to analyze the effects of high-load bench press training with different pressure modes on upper-limb muscle activation characteristics and the subjective fatigue of bodybuilders.

(1)   Analysis of changes in the pectoralis major activation level

This study found that the degree of pectoral major muscle activation in bodybuilders during bench press training was significantly higher than that in group C. The results of this study support the study of Che Tongtong [12], who arranged 10 basketball players to perform four groups (30–15–15–15) of low-intensity pressure (160 mmHg) bench press training, and also observed that the activation degree of the pectoralis major muscle in the pressure group was significantly greater than that in the non-pressure group. A large number of previous studies have shown that during the contraction process of skeletal muscle with blood flow restriction [7,17–22], there will be intramuscular hypoxia due to the limitation of blood flow, as well as the accumulation of lactate, hormones, and other metabolites (overload). A series of stress responses through the stimulation of afferent nerve centers III and IV can be fed back and cause a large amount of muscle fiber recruitment, which may lead to the inhibition of α-motor neurons. Thus, the body may lead to inhibition of α-motor neurons so that the organism may recruit more muscle fibers to maintain force and prevent conduction failure, which is manifested by a significant increase in muscle activation. In addition, a recent review article [23] suggests that the application of pressurized stimuli during resistance training leads to an increase in the proportion of anaerobic metabolic energy supply, resulting in rapid fatigue of type-I muscle fibers, which induces an increase in the recruitment of high-threshold type-II muscle fibers, which further increases muscle activation.

Che et al. [12] found that in four groups of low-intensity pressure training, the activation degree of pectoralis major was significantly greater than that of the non-pressure group only during the bench press training of the third and fourth groups. However, the difference in this study is that the activation degree of pectoralis major in the three groups of bench press training was significantly greater in the pressure experimental group than in the C group, and the activation degree of pectoralis major in the first two groups of bench press was significantly higher in the T2 group than in the C group. However, there was no significant change between the third group of squats. The possible reasons can be attributed to differences in exercise intensity and specialization. First of all, previous studies only adopted 30%1 RM, which may be due to the low metabolic pressure in the first two groups and failed to induce the recruitment of muscle fibers with a higher threshold due to the low-intensity training in the first two groups. In this study, the exercise intensity of 70%1 RM was used for training, so a large mechanical load pressure was generated during the bench press in the first group, resulting in a gradual increase in metabolic pressure. Therefore, high-intensity resistance training can recruit high-threshold muscle fibers more quickly, so a significant increase in pectoralis major muscle activation was observed earlier in this study (in the first set of bench presses), and this study also confirmed that exercise intensity has a significant effect on neuromuscular adaptation induced by pressure training [24]. However, there was no significant change between the conditions during the third set of bench presses. The possible reasons can be attributed to the differences in exercise intensity and specialization: first, the previous study only used a 30%1 RM load, while the present study involved training with 70%1 RM as the exercise intensity, and the mechanical load was significantly greater than that of the former; therefore, different metabolic pressures were generated so that due to the lower metabolic pressures of low-intensity training in the first two sets, higher-threshold motor unit recruitment did not occur, whereas the high-intensity resistance training was able to more quickly recruit those motor units. Second, the participants in the present study were bodybuilders, whereas those in the previous study were basketball players. Theoretically, the bodybuilders underwent more resistance training and may have had greater neuromuscular recruitment of the pectoralis major muscle during the bench press. Relatively speaking, the daily training of bodybuilding athletes mainly focuses on resistance training. Bench press is also the core action of daily upper-limb training, and its frequency is significantly higher than that of other sports. Therefore, bodybuilders as participants in this bench press test showed a better performance of recruitment ability of the pectoral major muscle [25].

Notably, the present study showed that the degree of pectoralis major activation in T2 was significantly greater than that in T1 during the first set of bench presses, indicating that the use of a blood flow restriction pressurization strategy with pressure relief during the training period and high pressurization during the interset interval (180 mmHg) was more effective in promoting the degree of pectoralis major activation than continuous low pressurization (100 mmHg) during bench press training at an exercise intensity of 70%1 RM. In this regard, The above research results are consistent with the dose–effect relationship of pressure resistance training proposed by Lei et al. [23]; after reviewing the training effects of pressurized resistance training, Lei et al. [23] pointed out that there was an "inverted U-shaped" quantitative relationship between the amount of pressurization and muscle performance in pressurized resistance training, i.e., within the appropriate range of pressurization, with an increase in occlusion pressure, the stronger the occlusion effect, the greater the stimulation, and a greater stimulation will be more effective than continuous low pressure (100 mmHg). When the pressure reaches a certain high value, the stress and adaptation of the organism reach an optimal level, and the training effect is optimized; however, when the pressure is higher than the critical value, the metabolic load stimulus is too strong to exceed the limit of the organism's adaptation, resulting in a decrease in the effect of training, which is related to the fatigue of the exercise. The results of the present study support the quantitative effect relationship proposed in the above study [23]. In addition, there was no significant difference between the two experimental conditions in the third set of bench presses, but the low-pressure T1 was significantly higher than C, but there was no significant difference in the high-pressure T2 condition. According to the above dose–effect relationship [23], it can also be inferred that the metabolic pressure caused by intermittent high-pressure training may exceed the limit of the participants, suggesting that this may be related to the gradual accumulation of greater exercise fatigue caused by high occlusion pressure.

(2)   Analysis of changes in deltoid activation level

The results of the present study showed that deltoid activation was significantly higher in the three sets of bench press training in both the pressurized experimental conditions, T1 and T2, than in C. In addition, when the first set of bench presses was performed, T2 had significantly greater deltoid activation than T1. The results also support a previous study by the Che team [12], in which 10 basketball players performed five sets (30–15–15–15–15) of low-intensity stress (160 mmHg) bench press training and found that deltoid muscle activation was significantly greater under stress conditions than under nonstress conditions. However, Che's study [12] found that deltoid muscle activation increased significantly only in the fourth group of bench press, while in this study, deltoid muscle activation was similar to or even superior to that of the pectoralis major muscle, including the third group of bench press under T2, and deltoid muscle activation under stressful T1 and T2 was significantly higher than that under stress-free C in the three groups of high-intensity training. The stronger effect of blood flow restriction on deltoid activation can be attributed to the site of the pressurized band. After conducting low-intensity pressure resistance training experiments, Loenneke et al. [26] pointed out that pressure stimulation during exercise training significantly reduced venous backflow, so that after subsequent compression band decompression, a pressure gradient conducive to peripheral blood flow into muscle fibers was quickly formed. Under the pressure gradient, extracellular fluid flooded into the cells and tissues of the exercising muscle group. At the same time, due to the stimulation of exercise training and stress, the accumulation of metabolic stress products will produce stronger stimulation of peripheral and central chemoreceptors. At this time, the metabolic stress products accumulated due to exercise training and pressurization stimulation will produce stronger stimulation to peripheral and central chemoreceptors, which will ultimately promote the synthesis of hormone proteins for myocyte growth and development. This may be the main reason for the more pronounced changes in muscle activation in the deltoid than in the pectoralis major muscle [27].

One thing that differentiates the change in the activation level of the pectoralis major is that the activation level of the deltoid muscle during the third set of bench presses performed during T2 was significantly less than the activation level during the first two sets of bench presses. This may be because as exercise continues, the intermittent high-pressure mode occurs due to increased metabolic stress and accumulation of metabolic waste products as a result of the greater occlusion pressure, which causes exercise fatigue and ultimately reduces the level of muscle activation in the participants.

(3)   Analysis of changes in triceps activation

The results of this study showed that the triceps activation degree of the T1 group was significantly lower than that of the second group during the third set of bench presses, and the triceps activation degree of the T2 group was significantly lower than that of the first two groups during the third set of bench press; that is, with the progress of each group of bench press, the triceps activation degree of the pressure group was continuously reduced, while the reduction degree of the T2 group was greater than T1. The reason for this phenomenon may be that, on the one hand, the binding part of the upper-limb compression training is the upper arm; that is, the whole muscle of the triceps is bound by the compression band. In the case of continuous compression, although the blocking pressure is small (100 mmHg), the muscle will produce certain discomfort during the bench press with the progress of training, and the training effect will be reduced. On the other hand, pressure training requires participants to exercise under the dual stimulation of "pressure" and "resistance". Therefore, with the continuous progress of exercise, this training mode leads to factors such as muscle and nerve disorder, continuous consumption of energy substances, and a decrease in energy metabolism and energy supply rate, resulting in the exercise-induced fatigue of participants and ultimately reducing the activation degree of triceps. It should be mentioned that according to the overall change in muscle activation degree, the third group of bench press training will lead to a greater decrease in the activation degree of the triceps muscle than that of the pectoralis major and deltoid muscles. The possible reason may be that the compression band binding part of the upper arm near the proximal heart during press bench press training basically covers the triceps muscle. The occluded pressure and metabolic pressure on the triceps reach the peak, so fatigue occurs most significantly.

The results of this experiment also showed that even though blood flow restriction resulted in a significant decrease in triceps activation as the bench press continued, the activation of the triceps was still significantly higher with blood flow restriction compared to nonpressurized traditional resistance training, and the changes in this muscle were basically the same as those in the pectoralis major and deltoid muscles, which supports the results of a previous study [12]. The results of the present experiment confirm that blood flow restriction is able to promote better neuromuscular properties of the triceps brachii muscle than no blood flow restriction.

4.1. Analysis of Subjective Fatigue Test Results

The research [11] confirms that RPE is a method to monitor subjective fatigue during exercise and to effectively monitor and regulate the intensity of loads in sports. The results of this study showed that RPE values increased significantly in all conditions as each set of bench presses was performed. It is not difficult to understand that, with or without pressurization intervention, continuous resistance training produces a certain amount of metabolic stress, which results in the body experiencing neuromuscular disruption, depletion of energy substances, a decrease in energy metabolism, an increase in the metabolic stress response, and an accumulation of metabolic wastes, which can result in performance decrements as well as physiological fatigue of the body; therefore, the RPE values of the experimental and C groups are increasingly large.

The present study also found that the T1 had significantly higher RPE values than the C in all three bench press sets, and the findings support previous research from Vieiea's research team [28]. The design of a comparative test revealed that blood flow restriction produced a stronger subjective feeling of fatigue compared to traditional high-intensity resistance training. Schwiete et al. [29] also indicated that perceived pain was significantly higher in the pressurized resistance training experimental condition than in the nonpressurized C. Domestic scholars, such as Che Tong et al. [11] also observed that the RPE value of the blood flow restriction condition was significantly higher than that of the nonpressurized condition after putting basketball players through low-impact compression squats. However, in the third set, the T1 index value after bench press was significantly higher than that of the T2 condition, while there was no significant difference between T2 and C. This suggests that it may be more difficult for participants to use the continuous low-pressure mode of pressurization. Despite the large amount of pressurization, the intermittent high-pressure mode of pressurization with pressure relief during the exercise period can give participants a certain amount of time to rest during the pressure relief period, which can provide adequate time to restore the energy supply and remove metabolic waste so that subjective fatigue is lower. Therefore, the subjective fatigue of the participants was lower when using the intermittent pressurization mode, and this method is more favorable for bodybuilders to perform pressurized high-resistance training.

4.2. Practical Application

The purpose of this study is to compare the changes in the muscle activation degree of muscle groups in high-intensity bench press training under different pressure modes, explore the adaptability of the high-pressure high-resistance training model, and expand the application field of compression resistance training so as to provide a theoretical basis for bodybuilders to enhance neuromuscular adaptation and improve muscle mass (muscle hypertrophy/muscle recruitment).

4.3. Advantages of the Study

Intermittent pressure intervention was used in this study, and surface-integrated myography was used as an outcome index in high-intensity resistance training. This method can not only prevent safety problems such as cardiovascular injury and muscle injury due to the blood flow limitation caused by continuous high pressure but also fill the research gap in neuromuscular adaptation of a high-pressure and high-resistance training model.

4.4. Study Limitations

First, this study only monitored participants' heart rate and blood pressure to control load intensity to prevent exercise risk. However, the implementation of pressure intervention in high resistance training also involves risk factors such as blood pressure, so hemodynamic indicators need to be monitored to improve the applicability of different pressure resistance training models. Secondly, this study mainly used surface-integrated myography and RPE as the main outcome indicators, but based on the characteristics of pressure training, changes in blood composition are also another important indicator. Since pressure training tends to lead to more metabolic stress, changes in the expression of metabolic stress products such as lactic acid, nitric oxide, and interleukin in blood components have not been thoroughly studied.

4.5. Future Research Directions

Further research could explore the persistence and potential long-term adaptation of the acute effects found in this study by conducting long-term compression resistance exercise interventions. Long-term trials will be designed to more fully assess the potential effects of exercise training on physical function, physiological parameters, and athletic performance. In addition, long-term experiments can provide more time windows to observe and analyze possible individual differences and adaptive changes, focusing on the design and execution of long-term experiments to gain insight into the long-term effects of exercise training and its potential mechanisms. It also provides a more reliable scientific basis for improving training programs and optimizing sports performance.

5. Conclusions

(1) In high-intensity pressurized bench press training, both continuous and intermittent pressurization significantly increased the activation of the pectoralis major, triceps brachii, and deltoid muscles, but the continuous pressurization method caused a stronger subjective fatigue sensation. (2) In terms of the overall training effect, the high-pressure, high-resistance training mode of no pressurization during the training period and pressurization during the intervals between sets with an occlusion pressure of 180 mmHg had the best enhancement effect.

Author Contributions: Conceptualization, K.H. and Y.S.; methodology, S.X. and X.Z.; software, S.X. and K.H.; validation, K.H., Y.S. and S.X.; formal analysis, Z.D.; investigation, S.X.; resources, K.H.; data curation, Z.D.; writing—original draft preparation, K.H.; writing—review and editing, Y.Z.; visualization, Y.Z.; supervision, Z.D.; project administration, X.Z. and Y.Z.; funding acquisition, X.Z. All authors have read and agreed to the published version of the manuscript.

Funding: This study was supported by the National Social Science Foundation of China (Grant No. 21BTY035).

Institutional Review Board Statement: This study was conducted in accordance with the Helsinki Declaration and approved by the Ethics Committee of Zhengzhou University (No. ZZUIRB2022-JCYXY0016 and approved on 22 September 2022).

Informed Consent Statement: Participants provided written informed consent to participate in the study. The potential risks and benefits of participating in this study were explained in advance to each participant. All participants provided signed notices prior to participation.

Data Availability Statement: All data are published on the figshare platform: 10.6084/m9 figshare. 24412597.

Acknowledgments: The authors thank the Laboratory of Sports Biomechanics at Beijing Normal University for its support in data collection. At the same time, I would like to express my gratitude to Lei Guobin, a student from Wuhan Sport University, for his unwavering assistance throughout the experimental process and paper writing as my assistant!

Conflicts of Interest: The authors declare no conflict of interest.

References

1. Loenneke, J.P.; Thiebaud, R.S.; Abe, T.; Bemben, M.G. Blood flow restriction pressure recommendations: The hormesis hypothesis. *Med. Hypotheses* **2014**, *82*, 623–626. [CrossRef]
2. Liu, S.; Wang, S.; Ji, W. Progress in the application and limitation of stress training. *Chin. J. Rehabil. Med.* **2022**, *37*, 6.
3. Yu, L.; Wang, Z.; Gao, J.; Zhu, X. Effects of short-term pressure strength training on body composition and cardiovascular function in adult males. *J. Beijing Sport Univ.* **2022**, *43*, 132–139.
4. Suga, T.; Okita, K.; Morita, N.; Yokota, T.; Hirabayashi, K.; Horiuchi, M.; Takada, S.; Takahashi, T.; Omokawa, M.; Kinugawa, S.; et al. Intramuscular metabolism during low-intensity resistance exercise with blood flow restriction. *J. Appl. Physiol. (1985)* **2009**, *106*, 1119–1124. [CrossRef] [PubMed]
5. Lu, J.; Liu, S.; Sun, P.; Li, W.; Lian, Z. Effects of different pressure blood flow restriction combined with low intensity resistance training on lower limb muscle and cardiopulmonary function of college students. *Chin. J. Appl. Physiol.* **2020**, *36*, 595–599.
6. Christopher, A.F.; Jeremy, P.L.; Lindy, M.R.; Robert, S.T.; Michael, G.B. Methodological considerations for blood flow restricted resistance exercise. *J. Trainology* **2012**, *1*, 14–22.
7. Yuan, W.T. Research on the effects of lower extremity pressurization combined with different resistance training on human function. *Genom. Appl. Biol.* **2018**, *37*, 8.
8. Wei, J.; Li, B.; Feng, L.; Li, Y. Methodological factors and potential safety issues of blood flow restriction training. *China Sports Sci. Technol.* **2019**, *55*, 3–12.
9. Torma, F.; Gombos, Z.; Fridvalszki, M.; Langmar, G.; Tarcza, Z.; Merkely, B.; Naito, H.; Ichinoseki-Sekine, N.; Takeda, M.; Murlasits, Z.; et al. Blood flow restriction in human skeletal muscle during rest periods after high-load resistance training down-regulates miR-206 and induces Pax7. *J. Sport Health Sci.* **2021**, *10*, 470–477. [CrossRef]
10. Davids, C.J.; Næss, T.C.; Moen, M.; Cumming, K.T.; Horwath, O.; Psilander, N.; Ekblom, B.; Coombes, J.S.; Peake, J.; Raastad, T.; et al. Acute cellular and molecular responses and chronic adaptations to low-load blood flow restriction and high-load resistance exercise in trained individuals. *J. Appl. Physiol. (1985)* **2021**, *131*, 1731–1749. [CrossRef]
11. Che, T.; Yang, T.; Liang, Y.; Li, Z. Effects of lower extremity low-intensity compression half-squat training on muscle activation and subjective fatigue of core muscle group. *Sports Sci.* **2021**, *41*, 59–66.
12. Che, T.; Li, Z.; Zhao, Z.; Wei, W.; Sun, K.; Chen, C. Effects of low-intensity bench press training on muscle activation and subjective fatigue. *J. Chengdu Univ. Phys. Educ.* **2022**, *48*, 123–130.
13. World Medical Association. World Medical Association Declaration of Helsinki: Ethical principles for medical research involving human participants. *JAMA* **2013**, *310*, 2191–2194. [CrossRef]
14. Sonkodi, B.; Kopa, Z.; Nyirády, P. Post Orgasmic Illness Syndrome (POIS) and Delayed Onset Muscle Soreness (DOMS): Do They Have Anything in Common? *Cells* **2021**, *10*, 14–28. [CrossRef]
15. Konrad, P. The abc of emg. In *A Pactical Introduction to Kinesiological Electromyography*; Noraxon U.S.A. Inc.: Scottsdale, AZ, USA, 2005; Volume 1, pp. 30–35.
16. Zourdos, M.C.; Goldsmith, J.A.; Helms, E.R.; Trepeck, C.; Halle, J.L.; Mendez, K.M.; Cooke, D.M.; Haischer, M.H.; Sousa, C.A.; Klemp, A.; et al. Proximity to Failure and Total Repetitions Performed in a Set Influences Accuracy of Intraset Repetitions in Reserve-Based Rating of Perceived Exertion. *J. Strength Cond. Res.* **2021**, *35*, S158–S165. [CrossRef] [PubMed]
17. Oliveira, J.; Campos, Y.; Leitão, L.; Arriel, R.; Novaes, J.; Vianna, J. Does Acute Blood Flow Restriction with Pneumatic and Non-Pneumatic Non-Elastic Cuffs Promote Similar Responses in Blood Lactate, Growth Hormone, and Peptide Hormone? *J. Hum. Kinet.* **2020**, *74*, 85–97. [CrossRef]
18. Zheng, B.; Zhang, Z. Study on the effects of different pressurized resistance training modes on human physiological and biomechanical characteristics after exercise. *J. Southwest Norm. Univ. Nat. Sci. Ed.* **2021**, *46*, 9.
19. Zhao, Z.; Cheng, J.; Wei, W.; Sun, K.; Wang, M. Effects of pressure training and traditional muscle building training on some hormones and bioactive factors of elite male handball players. *China Sports Sci. Technol.* **2019**, *55*, 20–29.
20. Yinghao, L.; Jing, Y.; Yongqi, W.; Jianming, Z.; Zeng, G.; Yiting, T.; Shuoqi, L. Effects of a blood flow restriction exercise under different pressures on testosterone, growth hormone, and insulin-like growth factor levels. *J. Int. Med. Res.* **2021**, *49*, 3000605211039564. [CrossRef]
21. Li, Z.; Wei, W.; Zhao, Z.; Sun, K.; Gao, W.; Xiao, Z. Effects of different degree of blood flow restriction on serum growth hormone and testosterone secretion during low-intensity resistance exercise. *China Sports Sci. Technol.* **2020**, *56*, 38–43.
22. Bemben, D.A.; Sherk, V.D.; Buchanan, S.R.; Kim, S.; Sherk, K.; Bemben, M.G. Acute and Chronic Bone Marker and Endocrine Responses to Resistance Exercise With and Without Blood Flow Restriction in Young Men. *Front. Physiol.* **2022**, *13*, 837631. [CrossRef]
23. Lei, S.; Zhang, M.; Ma, C.; Gao, W.; Xia, X.; Dong, K. Muscle effect, dose-effect relationship and physiological mechanism of pressure resistance training. *Chin. J. Tissue Eng.* **2023**, *27*, 4254–4264.
24. Schwiete, C.; Franz, A.; Roth, C.; Behringer, M. Effects of Resting vs. Continuous Blood-Flow Restriction-Training on Strength, Fatigue Resistance, Muscle Thickness, and Perceived Discomfort. *Front. Physiol.* **2021**, *12*, 663665. [CrossRef] [PubMed]
25. Chernozub, A.; Manolachi, V.; Tsos, A.; Potop, V.; Korobeynikov, G.; Manolachi, V.; Sherstiuk, L.; Zhao, J.; Mihaila, I. Adaptive changes in bodybuilders in conditions of different energy supply modes and intensity of training load regimes using machine and free weight exercises. *PeerJ* **2023**, *11*, e14878. [CrossRef] [PubMed]

26. Loenneke, J.P.; Fahs, C.A.; Rossow, L.M.; Abe, T.; Bemben, M.G. The anabolic benefits of venous blood flow restriction training may be induced by muscle cell swelling. *Med. Hypotheses* **2012**, *78*, 151–154. [CrossRef]
27. Lambert, B.; Hedt, C.; Daum, J.; Taft, C.; Chaliki, K.; Epner, E.; McCulloch, P. Blood Flow Restriction Training for the Shoulder: A Case for Proximal Benefit. *Am. J. Sports Med.* **2021**, *49*, 2716–2728. [CrossRef]
28. Vieira, A.; Gadelha, A.B.; Ferreira-Junior, J.B.; Vieira, C.A.; Soares Ede, M.; Cadore, E.L.; Wagner, D.R.; Bottaro, M. Session rating of perceived exertion following resistance exercise with blood flow restriction. *Clin. Physiol. Funct. Imaging* **2015**, *35*, 323–327. [CrossRef]
29. Suga, T.; Okita, K.; Morita, N.; Yokota, T.; Hirabayashi, K.; Horiuchi, M.; Takada, S.; Omokawa, M.; Kinugawa, S.; Tsutsui, H. Dose effect on intramuscular metabolic stress during low-intensity resistance exercise with blood flow restriction. *J. Appl. Physiol. (1985)* **2010**, *108*, 1563–1567. [CrossRef]

Article

Improving Balance and Movement Control in Fencing Using IoT and Real-Time Sensorial Feedback [†]

Valentin-Adrian Niță [1],[*] **and Petra Magyar** [2]

[1] Department of Communications, University Politehnica of Timișoara, 300006 Timișoara, Romania

[2] Department of Physical and Sports Education, West University of Timisoara, 300223 Timișoara, Romania; petra.magyar94@e-uvt.ro

[*] Correspondence: valentin.nita@upt.ro

[†] This paper is an extended version of our conference paper: Nita, V.A.; Magyar, P. Smart IoT device for measuring body angular velocity and centralized assessing of balance and control in fencing. In Proceedings of the 2023 International Symposium on Signals, Circuits and Systems (ISSCS), Iasi, Romania, 13–14 July 2023; pp. 1–4. https://doi.org/10.1109/ISSCS58449.2023.10190900.

Abstract: Fencing, a sport emphasizing the equilibrium and movement control of participants, forms the focal point of inquiry in the current study. The research endeavors to assess the efficacy of a novel system designed for real-time monitoring of fencers' balance and movement control, augmented by modules incorporating visual feedback and haptic feedback, to ascertain its potential for performance enhancement. Over a span of five weeks, three distinct groups, each comprising ten fencers, underwent specific training: a control group, a cohort utilizing the system with a visual real-time feedback module, and a cohort using the system with a haptic real-time feedback module. Positive outcomes were observed across all three groups, a typical occurrence following a 5-week training regimen. However, noteworthy advancements were particularly discerned in the second group, reaching approximately 15%. In contrast, the improvements in the remaining two groups were below 5%. Statistical analyses employing the Wilcoxon signed-rank test for repeated measures were applied to assess the significance of the results. Significance was solely ascertained for the second group, underscoring the efficacy of the system integrated with visual real-time feedback in yielding statistically noteworthy performance enhancements.

Keywords: sensorial feedback; real-time monitor; IoT; balance and movement control in fencing

Citation: Niță, V.-A.; Magyar, P. Improving Balance and Movement Control in Fencing Using IoT and Real-Time Sensorial Feedback. *Sensors* **2023**, *23*, 9801. https://doi.org/10.3390/s23249801

Academic Editors: Felipe García-Pinillos, Alejandro Pérez-Castilla and Diego Jaén-Carrillo

Received: 27 October 2023
Revised: 27 November 2023
Accepted: 11 December 2023
Published: 13 December 2023

1. Introduction

The sports landscape is transforming through technology integration, particularly with the increasing prevalence of wearable devices utilizing inertial measurement units (IMUs). The extant body of research substantiates the efficacy of these devices in enhancing the training experience [1–8]. Traditionally positioned on the hip [9], alternative placements such as the wrist [10,11], thigh [12], knee [13], or even the back [14] are viable options contingent upon the specific demands of a given sport. Notably, across a spectrum of sports like tennis [15], football [16], basketball [17], handball [18], hockey [19], and martial arts [20], insufficient attention has been directed towards fencing in the existing research literature.

Fencing, characterized by its occurrence on a specialized surface measuring 14 m in length and 1.5 m in width, akin to a chessboard for two participants, places significance on the positional dynamics between adversaries. Irrespective of the opponent's physical attributes, celerity, or strength, effective counteraction hinges upon the judicious manipulation of distance and positioning. To optimize such elements, fencers must cultivate speed, agility, force, endurance, and an acute sense of balance and movement control through specific training. While extensive literature exists for the enhancement of speed and agility in general [21–24] and within the domain of fencing [25], the assessment and improvement of hand control [26,27], there is a conspicuous absence of tools or devices for measuring balance and movement control.

Usually, IMUs are used for assessing injury risks, as in [9–12] and [14], while the proposed system is used to evaluate balance and movement control performance. It is essential to mention that the proposed system considers the particular limitations of the movements made in fencing.

Current coaching practices rely heavily on visual assessment by experienced coaches, which is particularly challenging when dealing with a large cohort of athletes—perhaps 100 fencers in training and 10 to 20 in competition. Unfortunately, the traditional approach necessitates a sequential evaluation, limiting the feasibility of continuous assessment. In addressing this gap, a pioneering Internet of Things (IoT) system has been developed and successfully tested for automated real-time measurement of balance and movement control [28]. This innovative system empowers coaches to conduct comprehensive and instantaneous evaluations of balance and movement control for all their fencers, presenting a transformative solution to the existing challenges in the field. In his book, *This is Fencing!*, Ziemowit Wojciechowski (one of the world's most renowned and sought-after foil coaches with a long and illustrious record of success) speaks about the importance of performance analysis, which can be "qualitatively based on observations or quantitatively based on factual or statistical data" [29].

Figure 1 illustrates a simplified representation of the motion dynamics in a fencing game. As depicted, the primary motion predominantly occurs along the X-axis. In this context, it is crucial to note that the torso's angular velocity along the Z-axis is generally close to zero, with exceptions occurring during specific actions like counterattacks and close encounters. Additionally, any rotation around the X-axis by the fencer leads to undesirable side imbalance and is a behavior that should be avoided in all circumstances.

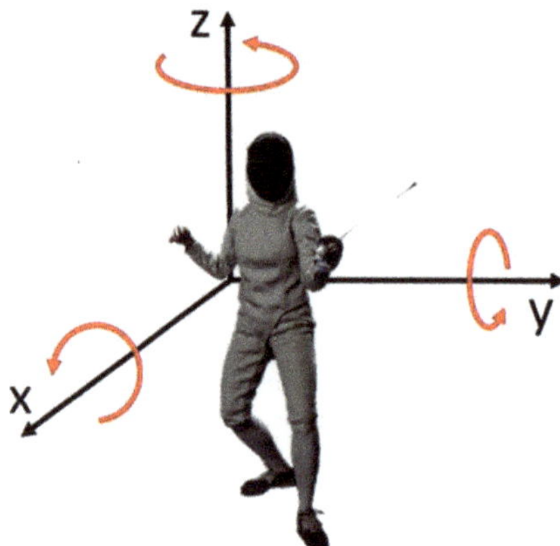

Figure 1. Movement on the fencing piste [28].

The sole permissible rotation, albeit limited in magnitude, occurs along the Y-axis. This is due to the unique leg movement involved in fencing: when moving forward, the front leg is raised and advanced, followed by the back leg, resulting in a slight rearward tilt. Conversely, the back foot is repositioned before the front leg, causing a subtle forward tilt when moving backward. Because of this specific footwork, a fencer's torso should ideally maintain a nearly constant zero angular velocity around the X and Z axes. Moreover, if the movement is executed with proper balance control, the angular velocity around the Y-axis should be kept to a minimum. A professional fencer's movement should closely resemble a train on tracks, with smooth back-and-forth motion and minimal tilting, ideally exhibiting angular velocities of 0 along all three axes (X, Y, and Z).

The angular velocity of the torso can be monitored in time using a gyroscope worn on the back of the fencer.

This current study examines an expanded iteration of the system introduced in prior work [28], aiming to enhance its functionalities beyond measuring and monitoring balance and movement control. The extended system is designed to actively improve fencers' capacities by incorporating real-time haptic and visual feedback mechanisms. The experimental design involves three distinct groups of fencers: a control group comprising 10 participants who will undergo training without the utilization of the proposed system, another group of 10 fencers who will engage with the system incorporating visual feedback, and a third group of 10 fencers who will interact with the system comprising haptic feedback. This intervention will span five weeks.

2. Materials and Methods

Figure 2 delineates the primary components integrated into the proposed system. At the core of this system is the "Gyroscope sensor", a pivotal element employed for real-time monitoring of fencers' balance. This is achieved by analyzing angular velocity along the fencer's torso's X, Y, and Z axes. The "Balance and movement control monitor" is an Android application used by fencers and coaches for comprehensive performance tracking over temporal intervals. The "Haptic feedback" module is embodied by a specialized smartwatch designed explicitly for fencing [30]. This device emits vibrations if the gyroscope sensor detects imbalances surpassing a predefined threshold. This threshold is adjustable per the fencers' anticipated performance levels, categorized based on their proficiency levels—ranging from beginner to professional. Concurrently, the "Visual feedback" mechanism manifests as a device featuring a colored LED signaling system. The LED emits a light green hue when the fencer's movements align with acceptable balance standards corresponding to their skill levels. Conversely, an orange signal is activated if the fencer's performance falls below the designated reference level, calibrated in accordance with their experience.

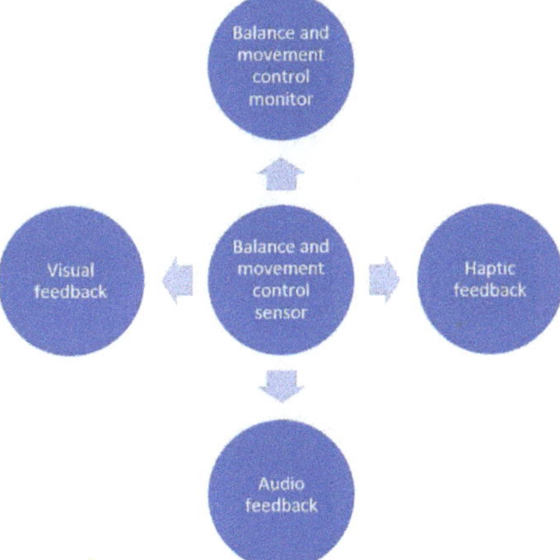

Figure 2. Main system blocks are used for enhancing balance and movement control in fencing.

The "Audio feedback" component, still in the developmental phase, incorporates the utilization of headphones. It is designed to emit a distinct auditory cue when fencers exhibit suboptimal balance and movement control levels. It is imperative to note that this article focuses on the analysis of the "Visual feedback" and "Haptic features", while the "Audio feedback" feature remains under active development.

2.1. Balance and Movement Control Sensor

The fundamental constituent within the envisaged sensor is a gyroscope, a device instrumental in the measurement and preservation of orientation and angular velocity. Consisting of a rotating wheel or rotor affixed to gimbals, this apparatus facilitates unimpeded rotation in all directions. Upon the initiation of rotor motion, its axis of rotation remains steadfast, unaffected by the device's movements.

The gyroscope's functioning is grounded in the conservation of angular momentum. As the rotor spins, it possesses a fixed amount of angular momentum, resisting alterations in its orientation. Consequently, if the device is rotated, the rotor maintains its original orientation, causing the gimbals to revolve around it.

Figure 3 depicts the hardware components of the balance sensor, including an 1100 mAh battery, a Wemos Lolin 32 Lite (an Arduino-type board) equipped with Wi-Fi capabilities, and an MPU6500 sensor with a 3-axis gyroscope (offering programmable full-scale ranges of ±250, ±500, ±1000, and ±2000 degrees per second, and minimal noise at 0.01 degrees per second per square root of Hertz), an accelerometer, and a digital motion processor. The accelerometer features user-programmable full-scale ranges of ±2 g, ±4 g, ±8 g, and ±16 g. Initial sensitivity calibration for both sensors minimizes production-line calibration needs. This device is designed to operate in temperatures ranging from $-40\,°C$ to $80\,°C$ and with voltages between 1.71 V and 3.6 V. With Wi-Fi continuously active, it typically consumes an average of 150 mA, providing about 7 h of autonomy with a 1100 mAh battery. This translates to approximately three fencing training sessions before requiring recharging. The total estimated cost of the components is roughly $15, with an additional $5 for a 3D-printed enclosure (as shown in Figure 4) and a strap for wearing the sensor on the back, positioned between the shoulders.

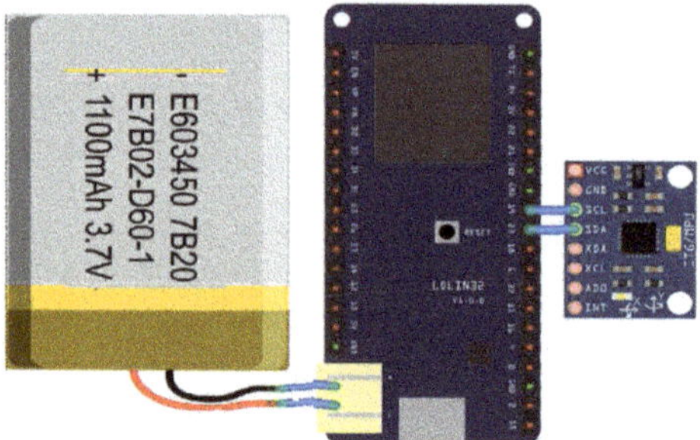

Figure 3. The balance sensor hardware [28].

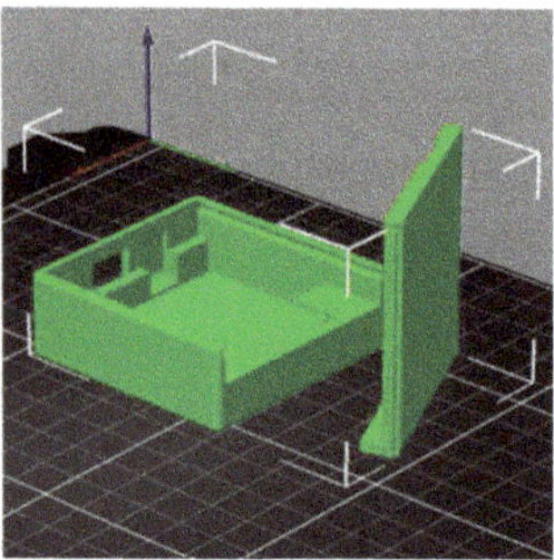

Figure 4. 3D model of the sensor enclosure [28].

Recognizing the niche nature of fencing as a sport and the limited demand for such a device, the enclosure (Figure 3) has been designed for 3D printing per order. It has dimensions of 60 mm by 60 mm and a height of 20 mm, featuring a screwless design for easy manual assembly. The design is based on the model available at [31]. The total 3D printing time is approximately 3 h, requiring around 30 g of PLA filament. The final enclosure weighs less than 100 g and can be comfortably worn with a back strap without hindering a fencer's performance on the fencing piste.

The equilibrium sensor depicted in Figure 3 establishes connectivity with the internet through Wi-Fi, facilitating data storage in a cloud-based database. This configuration allows fencers to monitor their performance metrics about balance contemporaneously. Authorization to access this data is contingent upon individual fencer preauthorization, and further permissions can be granted to share this information with their respective coaches. This collaborative feature furnishes coaches with a comprehensive overview of the progress exhibited by all athletes under their tutelage.

Data access is facilitated for both fencers and coaches, subject to explicit permission granted by the fencers, and is executed through a server infrastructure. Data processing occurs on the server, while the smartphone application retains a transient copy of the data. This temporary copy is automatically deleted if fencers revoke access privileges for a specific coach.

Two discrete cohorts of fencers were meticulously selected to establish benchmarks utilizing the data from the intelligent balance sensor. The initial group, encompassing novice practitioners, comprised ten individuals with fewer than twelve months of fencing experience. Conversely, the second group, comprised of proficient fencers, consisted of ten individuals with a notable 4 to 5 years of fencing practice. The observation of these fencers transpired during their traversal along a 14 m fencing strip, both at 50% of their maximum speed and at full acceleration, denoted as 100% speed, as detailed in a prior study [28].

To conduct a performance evaluation between novice and experienced fencers, a meticulous measurement of their angular velocity along the X, Y, and Z axes was undertaken, employing a time resolution of 100 milliseconds. The resultant data were aggregated for each fencer by applying the mean absolute deviation (MAD).

MAD, as described in [32], is a statistical metric for assessing the average deviation between individual data points and the mean of the dataset. Its calculation involves determining the absolute difference between each data point and the dataset's mean, summing these fundamental differences, and dividing the total by the number of data points in the set.

The mean absolute deviation (MAD) is expressed in the same units as the original dataset, which measures how dispersed the data are from their mean. A higher MAD value in our datasets indicates that a fencer exhibits unsteady or imbalanced movement, while a lower MAD signifies a fencer with advanced fencing skills. We chose MAD as a preferable alternative to the standard deviation, another commonly used measure of data spread. Unlike standard deviation, MAD is less affected by outliers, making it particularly valuable when extreme values, such as those caused by a contra attack resulting in high angular velocity on the Z-axis, might distort standard deviation calculations.

Each fencer's movement control and balance performance were quantified using a trio of values: MAD of angular velocity on the X, Y, and Z axes. Subsequently, an unsupervised machine learning algorithm, K-means, was employed to partition the results of the two groups and evaluate whether these results could effectively differentiate between novice and advanced fencers.

K-means clustering, a well-established unsupervised machine learning algorithm [33], categorizes and segments data into clusters based on their similarities. This algorithm divides data points into k clusters with a centroid or central point. The algorithm initiates with random centroids and assigns data points to the nearest centroid. Centroids are then updated to reflect the mean of the given data points. This process repeats until convergence, achieved when centroids no longer change significantly or when a maximum number of iterations is reached.

Figure 5 displays the results at 50% speed, while Figure 6 presents the results at 100% speed. These figures are two-dimensional, considering only the MAD of angular velocity on the X and Y axes, though clustering also employs the Z-axis data. The automatic separation into two clusters remains consistent in both scenarios, demonstrating the reliability of the monitored data for distinguishing between experienced and inexperienced fencers. The centroids obtained can serve as reference points to identify fencers making progress in their movement control and those who may benefit from additional preparation.

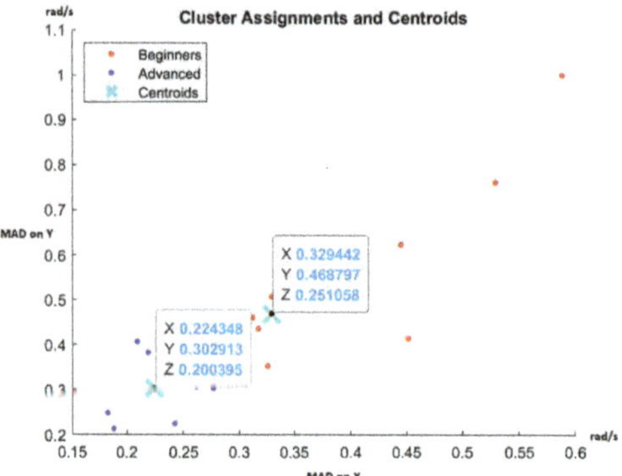

Figure 5. Cluster assignments using the K-means algorithm at 50% speed [28].

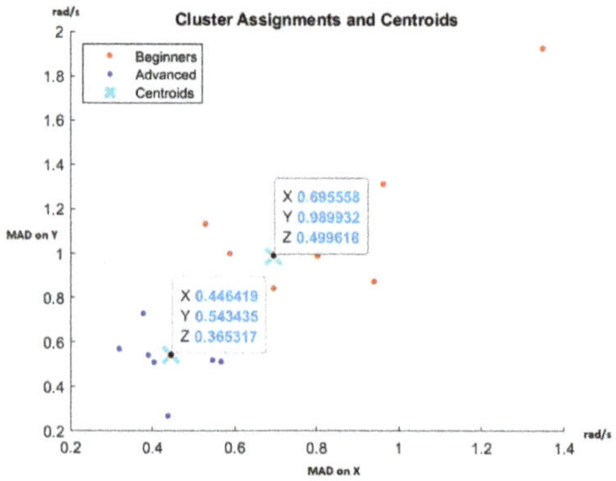

Figure 6. Cluster assignments using the K-means algorithm at 100% speed [28].

Furthermore, these data enable coaches to spot outliers, such as highly talented fencers who could be primed for high-performance training or those with consistently poor results, who may be better suited for recreational fencing rather than pursuing international medals. It is important to note that these proposed indicators rely solely on balance and movement control and should not be the sole performance metrics. They can be complemented by indicators of speed, reaction time, and precision abilities to enhance the selection of fencers with potential for high performance [28].

2.2. Haptic Feedback Module

The haptic module is in the form of a smartwatch developed specifically for fencing [30].

In Figure 7, the schematic representation illustrates the configuration of the haptic module integrated into the proposed system. This module is designed to furnish real-time feedback to fencers in the event of heightened imbalance, as expounded upon in greater detail in [30], which provides an exhaustive delineation of its features. For the immediate context, two primary functions are harnessed: the vibration motor, constituting the haptic feedback mechanism directed towards the fencers, and the Wi-Fi capabilities, essential for the reception of real-time commands from the balance sensor. At the core of this module resides an Arduino-based board, specifically the NodeMCU, as depicted in Figure 7. This board encapsulates a Wi-Fi ESP ESP8266 microchip, confined within a compact two by three cm rectangle, incurring an approximate cost of five euros.

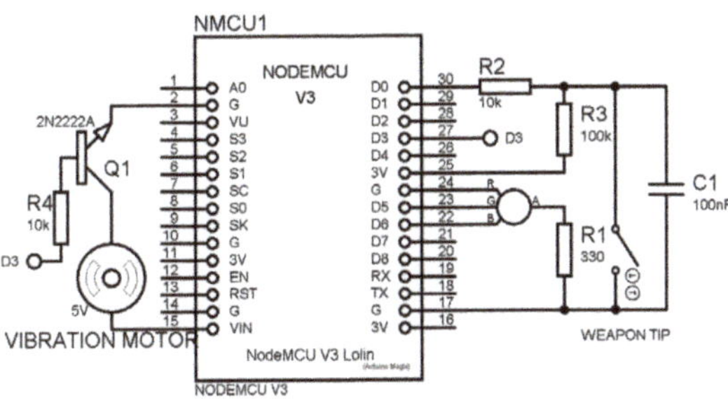

Figure 7. Haptic module schematic design [30].

Figure 8 illustrates the schematic representation of the enclosure design for the haptic module.

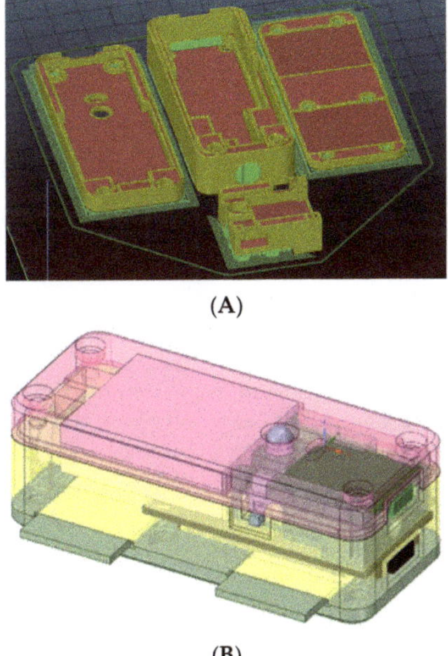

(A)

(B)

Figure 8. Haptic module 3D enclosure design. (**A**) Unfolded 3D case design based on four separated parts. (**B**) 3D model designed for 3D printing of the low-cost semi-scoring machine [30].

2.3. Visual Feedback Module

The visual feedback module facilitates interaction with the balance sensor by utilizing the identical NodeMCU employed in the haptic feedback module. This NodeMCU is intricately linked to an 8.6 cm WS2812 RGB LED ring. The WS2812 delineates a smart LED light source family characterized by integrating the control circuit and the RGB chip within a compact 5050-packaged unit. Figure 9 depicts the configuration of the visual feedback module. The LED light is strategically positioned at one extremity of the fencing strip, ensuring its perpetual visibility to the fencer undergoing training. In this scenario, fencers are necessitated to maintain continuous attention on the LED, mirroring the visual vigilance imperative in a fencing bout where constant visual analysis of the opponent is required. The LED emits a light green hue when the fencer's movements align with acceptable balance standards corresponding to their skill levels. Conversely, an orange signal is activated if the fencer's performance falls below the designated reference level, calibrated in accordance with their experience.

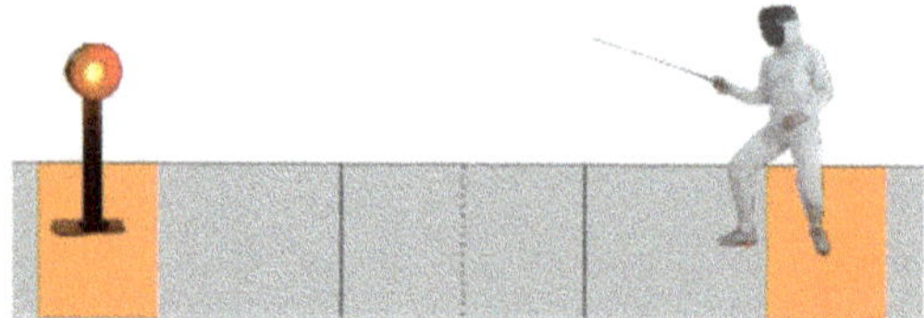

Figure 9. Visual feedback module setup.

2.4. Population and Sample

This investigation engaged juvenile athletes from the ACS Floreta Fencing Club in Timișoara aged 11 to 14. Explicit written authorization from the parents or legal guardians of the athletes was secured to facilitate their involvement in the research. To augment the authenticity and scholarly import of the study, a meticulous selection procedure was implemented, adhering to precisely delineated inclusion and exclusion criteria. These criteria were systematically classified into two distinct groups to ensure transparency, as outlined below:

1. Inclusion criteria:

    - Subjects must be between 11 and 14 years old at the time of selection.
    - They should have 4–5 years of experience in fencing.
    - They must have the written consent of their parents/legal guardians for their participation in the study.

2. Exclusion criteria:

    - Unmotivated absence from training sessions (not more than 4 times/5 weeks) and tests.
    - Excused absence from more than four training sessions during the study. This kind of absence can be encountered in the context of competitions, training camps, school exams, or the occurrence of some illnesses.

The athletes were divided into three distinct groups: the control group (CG), the experimental group utilizing the visual feedback module (VG), and the experimental group employing the haptic feedback module (HG). Homogeneity across the groups was maintained concerning age, gender, and training proficiency, and the allocation process was executed through randomization, employing a lot-drawing method for both female and male participants.

Each group adhered to a regimen of four training sessions per week, wherein 30 min per session was designated explicitly for targeted exercises aimed at improving movement control. Notably, all groups engaged in identical exercises, with the sole divergence being that the VG and HG groups had access to the system incorporating the feedback module.

The evaluative benchmark comprises a structured 4-2-2-4 scenario using three strategically positioned poles at varying distances. These poles are situated at the commencement line, 2 m from the starting line, and 4 m from the starting line, respectively. Fencers are tasked with traversing the prescribed course, involving movement from the start point to the 4 m pole and back, followed by a sequence of movements from the starting point to the 2 m pole and back repeated twice. Subsequently, they navigate once more from the start point to the 4 m pole and back. Cumulatively, this entails forward and backward movements of 12 m each, encompassing seven alterations in movement direction. Comprehensive assessments were conducted on all fencers at the initiation of the study and subsequently repeated after a 5-week interval from the commencement of the investigation.

The angular velocity during movement is compared with reference thresholds. Determination of reference thresholds emanates from the cluster's centroid associated with advanced fencers, as depicted in Figure 6, and is delineated in Table 1. A performance falling below the reference threshold indicates commendable results, while a performance surpassing the threshold is deemed undesirable. To quantify improvements in balance and movement control throughout the study, the temporal aspect has been encapsulated by measuring the time percentage. This metric represents the proportion of time relative to the entirety of the benchmark test, during which fencers perform above the reference thresholds. To mitigate the potential impact of measurement errors throughout the comprehensive evaluation, an over-threshold performance is defined when at least two thresholds are exceeded or when all three thresholds are concurrently surpassed.

Table 1. Reference thresholds for measuring balance and movement control in fencing.

X-Axis Threshold [rad/s]	Y-Axis Threshold [rad/s]	X-Axis Threshold [rad/s]
0.447	0.543	0.365

The comparative analysis between the initial and final test results entailed a relative average comparison within each group. Additionally, the Wilcoxon signed-rank test for repeated measures was employed to ascertain the statistical significance of improvements within each group. Furthermore, the inter-group improvements were subject to scrutiny through the Kruskal-Wallis [34] test, applied to the difference vectors derived from the initial and final measurements. Pairwise comparisons between groups, specifically CG vs. VG, CG vs. HG, and VG vs. HG, were conducted using the Mann-Whitney U test [35], with Bonferroni-corrected p-values [36]. Notably, non-parametric tests were selected for these analyses due to the limited sample size. The objective was to monitor the statistical significance of the outcomes, considering the inherent constraints associated with the modest number of samples.

3. Results

3.1. Average Absolute Evaluation of the Efficacy of the Proposed Training Feedback Modules

In the Supplementary Materials, we can see the change in the performance of the evaluated groups in terms of total movement time and unbalanced time in Tables S1–S6. Furthermore, Tables 2 and 3 serve as central repositories for critical outcomes, facilitating a comparative analysis of data from both absolute and relative perspectives. Across all three groups, marginal enhancements are discerned concerning the total movement time, indicative of a negligible evolution in the speed of the fencers over the 5-week testing interval. Upon scrutiny of the balance time, nominal progress is observed within the control group, a marginal improvement in the cohort subjected to the haptic feedback module, and a notable 18% amelioration in unbalanced time for the group utilizing the visual feedback module for training. This improvement is measured in the context of unbalance detected on all three axes and a 15.5% improvement when considering unbalance on two of the three. However, despite an overall improvement in the average, individualized examination of results, as delineated in the Supplementary Materials, reveals instances where certain fencers exhibit deterioration in both movement and balance control. Figures 10–12 provide a visual means for the comparative assessment of overall average results for the Control Group (CG), the Visual Feedback Group (VG), and the Haptic Feedback Group (HG).

Table 2. Overall average comparison, Week 0 vs. Week 5.

Group	Week 0 Average Total Time [s]	Week 5 Average Total Time [s]	Week 0 Average Unbalance Time [%]—2/3	Week 5 Average Unbalance Time [%]—2/3	Week 0 Average Unbalance Time [%]—3/3	Week 5 Average Unbalance Time [%] 3/3
CG	9.952	9.928	34.366	32.775	17.152	16.905
VG	10.661	10.951	33.005	27.891	17.141	13.867
HG	10.933	10.948	31.028	29.816	16.58	15.766

Table 3. Relative comparison, Week 0 vs. Week 5.

Group	Week 0 vs. Week 5 Total Time [%]	Week 0 vs. Week 5 Unbalance—2/3 [%]	Week 0 vs. Week 5 Unbalance—2/3 [%]
CG	0.24174	4.629576	1.440065
VG	−2.64816	15.49462	19.1004
HG	−0.137011	3.906149	4.90953

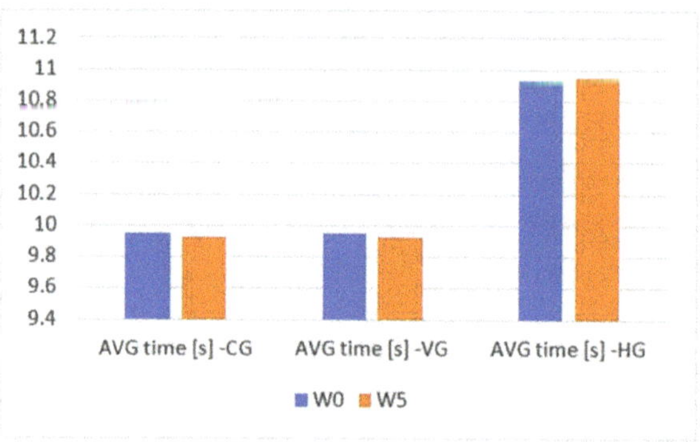

Figure 10. Average movement time comparison W0 vs. W5 for CG, VG, HG.

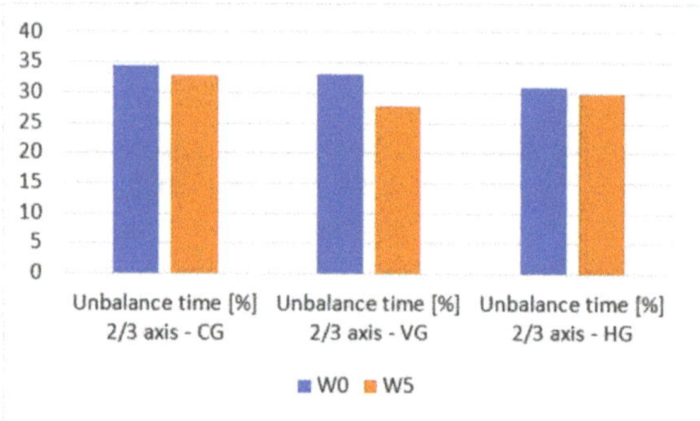

Figure 11. Unbalance time [%] 2/3 axis comparison W0 vs. W5 for CG, VG, HG.

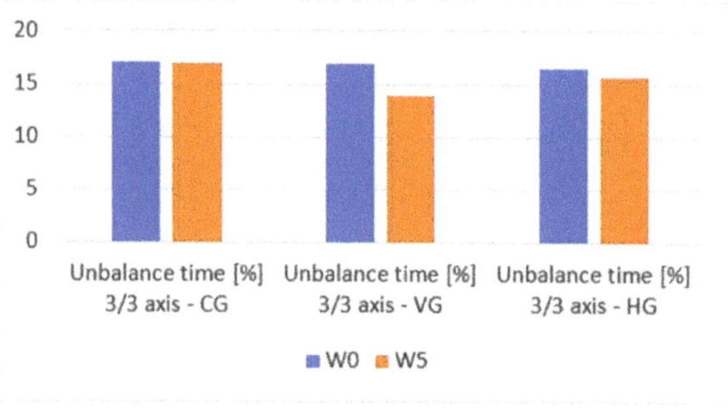

Figure 12. Unbalance time [%] 3/3 axis comparison W0 vs. W5 for CG, VG, HG.

3.2. Relative Evaluation of the Efficacy of the Proposed Training Feedback Modules Using the Wilcoxon Signed-Rank Test for Repeated Measures

When evaluating group performances based solely on the overall absolute average, a potential limitation exists wherein a minority of subjects may demonstrate substantial improvements. At the same time, the majority may experience marginal enhancements or, in some instances, a decline in performance. Relying solely on the average might suggest significant improvement attributable to a particular method, yet random cases could influence such observations. To ascertain the genuine impact of the proposed modules on enhancing balance and movement control in fencing, a secondary assessment of the results is conducted using the Wilcoxon signed-rank test for repeated measures [37]. The detailed outcomes of this test, comparing unbalance time on 2 out of 3 axes and 3 out of 3 axes for CG, VG, and HG, are presented in Tables S7–S12 in the Supplementary Materials.

Table 4 shows the overall results of the Wilcoxon test applied to all three test groups. Two overall results are of fundamental importance: the Wilcoxon test result and the critical value. If the test result is under the critical value, the improvements in balance and movement control are statistically significant; if it is not, they are not statistically significant. On the first line, we can see the results for the control group; in both situations, unbalanced time on two out of three axes and unbalanced time on three out of three axes, we can see that the Wilcoxon test points out that the improvements are not statistically significant. Even when we analyze Table 3, minor improvements are obtained from an average point of view. The same thing results for the group that has been training using the proposed system with haptic feedback. Real improvements were obtained from the group that was trained using the system with a visual feedback system.

Table 4. Wilcoxon signed-rank test for repeated measures results, Week 0 vs. Week 5.

Group/ Critical Value 8 $p = 0.05$	Wilcoxon Test Result W0 vs. W5 Unbalance—2/3 [%]	Wilcoxon Test Result W0 vs. W5 Unbalance—2/3 [%]
CG	15	22
VG	4	6
HG	14	21

3.3. Vector Difference Comparison of the Final and Initial Testing between the Three Groups Using the Kruskal–Wallis and Mann–Whitney U Tests, with Bonferroni-Corrected p-Values

An examination of the outcomes depicted in Figures 10–12 reveals discernible improvements in the performance of all groups following five weeks of training, specifically regarding movement control and balance. However, a meticulous inspection of Table 4 discerns that statistically significant improvements, at the 5% significance level, are exclusively evident for the group that availed the visual feedback module. In this subsection, we aim to test two hypotheses:

1. Each of the three groups has been drawn from a population with an identical distribution.
2. Following the five-week training period, the enhancements realized by one group demonstrate statistical significance when compared to the other groups within the study.

Examining the first hypothesis will involve the comparison of performance pairs, namely CG vs. VG, CG vs. HG, and VG vs. HG, applying the Mann–Whitney U test to the initial measurements.

The validation of the second hypothesis will be conducted through the application of the Kruskal–Wallis test to the vector of differences encompassing all groups, derived from the initial and final measurements and also by applying the Mann–Whitney U test with Bonferroni-corrected p-values.

Upon scrutiny of each group's performance through the Mann–Whitney U test, as elucidated in Tables 5 and 6, it is ascertained that the first hypothesis is true. According to the Mann–Whitney U test outcomes, the groups substantiate the same population at the commencement of the experimental period.

Table 5. Mann–Whitney U test applied to unbalance time Week 0—2/3 axis.

	CG	VG
VG	0.7913	-
HG	0.5708	0.6232

Table 6. Mann–Whitney U test applied to unbalance time Week 0—3/3 axis.

	CG	VG
VG	1	-
HG	0.9097	0.9097

In Figures 13 and 14, the vectors of differences between Week 0 and Week 5 are juxtaposed for all three groups, specifically focusing on unbalance time (%) detected on 2/3 axes and 3/3 axes. The comparative analysis reveals more pronounced differences when scrutinizing Group VG against Groups CG and HG, albeit the statistical significance level remains slightly below 20%. It is imperative to acknowledge the relatively modest strength of the second hypothesis, underscoring the necessity for caution, given the constrained 5-week testing interval, the limited sample size, and the fact that improvements have been detected for all groups. These constraints substantiate the recourse to non-parametric statistics. Future investigations warrant extending the study to encompass fencers from diverse geographical locations. Fencing is particularly niche in Romania, where the 30 individuals constituting the control and test groups represent approximately 20% of all fencers in the country with comparable experience.

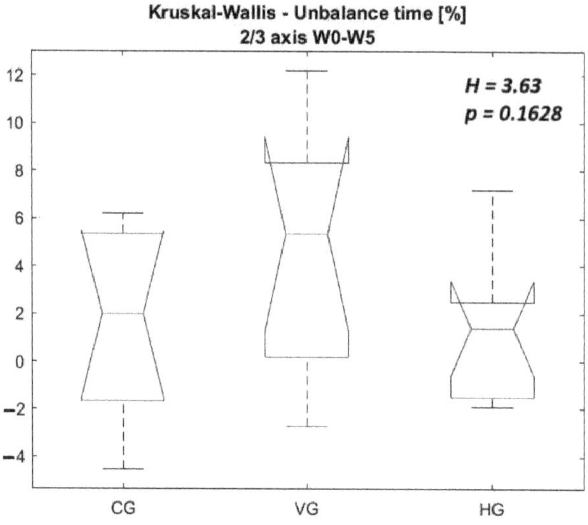

Figure 13. Kruskal–Wallis to the vector of differences between Week 0 and Week 5 for all groups—unbalance time [%] 2/3 axis.

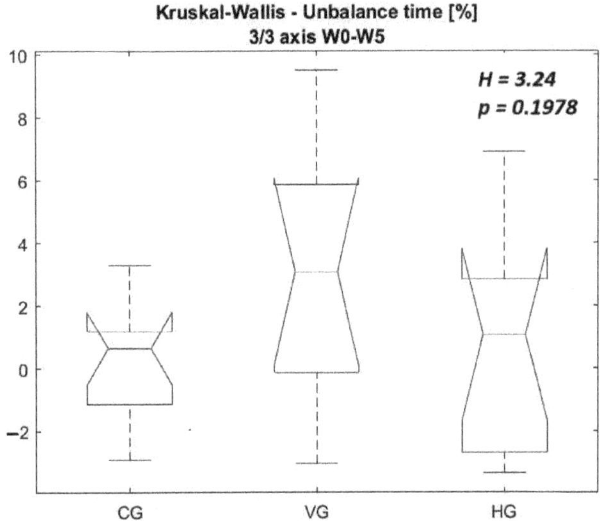

Figure 14. Kruskal–Wallis to the vector of differences between Week 0 and Week 5 for all groups—unbalance time [%] 3/3 axis.

A comprehensive examination of the outcomes in Tables 7 and 8 shows that CG and HG exhibit comparable performances. However, discernible distinctions emerge in the performance of VG, albeit with a statistically low significance level when juxtaposed with the other groups.

Table 7. Mann–Whitney U test, with Bonferroni-corrected p-values applied to vector difference of unbalance time Week 0—2/3 axis.

	CG	VG
VG	0.3239	-
HG	1	0.1513

Table 8. Mann–Whitney U test, with Bonferroni-corrected p-values applied to vector difference of unbalance time Week 0—3/3 axis.

	CG	VG
VG	0.2082	-
HG	1	0.3972

4. Discussions

The present study introduces an augmented system featuring feedback modules, initially devised for the real-time monitoring of balance and movement control based on an MPU6500 accelerometer and a gyroscope sensor. In [38], an extensive analysis of the MPU6500 sensor's reliability and performance is presented, revealing its widespread utilization as a low-cost and dependable sensor in various applications, including ground and aerial robotics. Notably, the gyroscope's measurement error is demonstrated to be comparable to 10^{-4} [rad/s], which is one thousand times less than the designated reference threshold.

The extended system is tested to ascertain its efficacy in enhancing performance through real-time feedback modules, transcending its initial performance-monitoring role as outlined in [28]. Two distinct setups were employed to train two groups of fencers, while a control group served as a reference. One setup incorporated a visual feedback module, while the other used a haptic one.

All groups were randomly constituted and encompassed fencers boasting 4–5 years of experience. The visual feedback module incorporates an RGB LED light, dynamically signaling to fencers in real-time with a green hue when their movement aligns with a predefined reference value, as obtained from [28], and an orange indication when their movements surpass the reference value. In the second configuration, a smartwatch imparts haptic feedback through vibrations when fencers exceed the reference value.

The three groups underwent assessment initially at the commencement of the experiment and subsequently after a 5-week interval. In the control group and the groups employing the haptic feedback module, marginal enhancements were noted; however, statistical significance was not achieved according to the Wilcoxon test. The configuration manifesting tangible improvements is the monitoring system with visual feedback, demonstrating an average improvement exceeding 15%. Importantly, these results are substantiated by the Wilcoxon test, attesting to their statistical significance. In addition, Section 3.3 examines the vectors depicting the disparities between final measurements and initial measurements by applying the Kruskal–Wallis test and Mann–Whitney U test, with p-values adjusted using the Bonferroni correction. The outcomes of these supplementary analyses suggest that the likelihood of uniform improvements across all three groups, from a statistical perspective, is marginally below 20%. While this figure exceeds the conventional 5% threshold, it is imperative to consider the study's limited training window of 5 weeks and the relatively small sample size of 10 subjects per group. Nonetheless, it is essential to underline that the initial test results, before the 5-week training window, were compared using the Mann–Whitney U test. The outcome was that the groups substantiated the same population at the commencement of the experimental period.

Additionally, as elucidated in the introduction, coaches routinely conduct analogous assessments of balance and movement control in their fencers through subjective observations [29]. The system's objective evaluations were corroborated by the subjective assessments of ACS Floreta club coaches for 28 out of 30 fencers (93.33%) across both test groups and the control group, encompassing two sets of tests.

In future endeavors, developing and testing a new module centered on audio feedback is imperative. Notably, fencers are accustomed to deriving cues from their opponents' visual and auditory signals to act or react efficiently. Simultaneously, comprehensive research endeavors are warranted, encompassing both novice fencers and those at the high-performance level.

The findings of this study carry significant implications for advancing technologies and training methodologies in the realm of sports by using IMUs. By showcasing the potential of the Internet of Things (IoT) and real-time sensorial feedback using IMUs, the study underscores the capacity of these innovations to augment performance in fencing. Typically, IMUs are employed to assess injury risks rather than enhance performance. This demonstration of efficacy may catalyze further exploration and innovation within sports technology and training methodologies.

Strengths and Limitations

The main strengths of this study were that now, for the first time, there is a tool that can be used to improve balance and movement control in fencing based on real-time feedback, which demonstrated real impact for one setup that uses visual feedback clues. This tool works based on objectively quantifying a fencer's balance and movement control performance by analyzing the torso angular velocity. However, there are some limitations to our analysis that should be noted. First, the number of fencers in the experimental groups and control group was only 10 per group. Notably, the entire population of 30 persons represents around 20% of all the fencers in Romania, which met the selection criteria from Section 2.4. Because of this, non-parametric statistics have been used to analyze the results' statistical significance. Second, the test window was only five weeks long, limiting the expected improvements. The size of the test window has been selected to minimize absenteeism in the training of the test subject due to competitions, training camps, school exams, or the occurrence of some illnesses. It is important to note that the fencers have been selected from the same club to ensure that all subjects undergo similar training sessions.

5. Conclusions

The current investigation yields several noteworthy conclusions. Firstly, it substantiates the efficacy of the real-time monitoring system with visual feedback in enhancing balance and movement control among fencers. The recorded average improvement surpassing 15% proved statistically significant, as confirmed by the Wilcoxon signed-rank test. This outcome aligns with previous research highlighting the effectiveness of wearable devices employing inertial measurement units (IMUs) across diverse sports for optimizing training experiences.

Secondly, the study underscores the superiority of the visual feedback module over the haptic feedback module in augmenting performance. This observation suggests that for fencers, visual feedback may possess greater salience and informativeness in the context of refining balance and movement control. Nevertheless, additional research is imperative to validate this finding and explore potential synergies from integrating visual and haptic feedback within the real-time monitoring system.

Thirdly, it is crucial to acknowledge certain limitations in the study that may influence the interpretation of the results. The relatively modest sample size and the confined 5-week testing period are notable constraints. Furthermore, the study exclusively focused on fencing, warranting future research endeavors to replicate and broaden the generalizability of the findings.

In conclusion, exploring balance and movement control improvement in fencing through integrating the Internet of Things (IoT) and real-time sensorial feedback presents encouraging outcomes for performance enhancement. The real-time monitoring system, particularly with visual feedback, emerges as a potent tool for refining balance and movement control in fencers.

Supplementary Materials: The following supporting information can be downloaded at https://www.mdpi.com/article/10.3390/s23249801/s1: Table S1: Week 0 measurements of the control group; Table S2: Week 0 measurements of the test group with visual feedback; Table S3: Week 0 measurements of the test group with haptic feedback; Table S4: Week 5 measurements of the control group; Table S5: Week 5 measurements of the test group with visual feedback; Table S6: Week 5 measurements of the test group with haptic feedback; Table S7: The Wilcoxon test was applied to the control group based on unbalanced time on 2 out of 3 axes; Table S8: The Wilcoxon test was applied to the control group based on unbalanced time on 3 out of 3 axes; Table S9: The Wilcoxon test was applied to the group training with visual feedback based on unbalanced time on 2 out of 3 axes; Table S10: The Wilcoxon test was applied to the group training with visual feedback based on unbalanced time on 3 out of 3 axes; Table S11: The Wilcoxon test was applied to the group training with visual feedback based on unbalanced time on 2 out of 3 axes; Table S12: The Wilcoxon test was applied to the group training with visual feedback based on unbalanced time on 3 out of 3 axes.

Author Contributions: V.-A.N. contributed to the study design and concept, V.-A.N. and P.M. contributed to the acquisition of data, V.-A.N. performed the data processing and statistical analysis, P.M. conducted the fencing training during the 5-week testing window, V.-A.N. and P.M. drafted and reviewed the manuscript. All authors have read and agreed to the published version of the manuscript.

Funding: This research received no external funding.

Institutional Review Board Statement: The study was conducted according to the Ethical Regulations and Guidelines of the Politehnica University of Timisoara, Romania. https://www.upt.ro/Informatii_etica-si-deontologie_164_ro.html (accessed on 8 August 2022).

Informed Consent Statement: GDPR data protection information was provided and respected, and informed consent was obtained from all subjects involved in the study.

Data Availability Statement: The data presented in this study are available on request from the corresponding author.

Conflicts of Interest: The authors declare no conflict of interest.

References

1. Vezocnik, M.; Juric, M.B. Average Step Length Estimation Models' Evaluation Using Inertial Sensors: A Review. *IEEE Sens. J.* **2019**, *19*, 396–403. [CrossRef]
2. Pérez-Chirinos Buxadé, C.; Fernández-Valdés, B.; Morral-Yepes, M.; Tuyà Viñas, S.; Padullés Riu, J.M.; Moras Feliu, G. Validity of a Magnet-Based Timing System Using the Magnetometer Built into an IMU. *Sensors* **2021**, *21*, 5773. [CrossRef] [PubMed]
3. Demrozi, F.; Pravadelli, G.; Bihorac, A.; Rashidi, P. Human activity recognition using inertial, physiological and environmental sensors: A comprehensive survey. *IEEE Access* **2020**, *8*, 210016–210836. [CrossRef] [PubMed]
4. Yuan, Q.; Chen, I.-M. Human Velocity and Dynamic Behavior Tracking Method for Inertial capture System. *Sens. Actuators* **2012**, *183*, 123. [CrossRef]
5. Rajkumar, A.; Vulpi, F.; Bethi, S.R.; Wazir, H.K.; Raghavan, P.; Kapila, V. Wearable Inertial Sensors for Range of Motion Assessment. *IEEE Sens. J.* **2020**, *20*, 3777–3787. [CrossRef]
6. Yuan, Q.; Chen, I.-M. Localization and Velocity Tracking of Human via 3 IMU Sensors. *Sens. Actuators A Phys.* **2014**, *212*, 25–33. [CrossRef]
7. Wu, M.; Fan, M.; Hu, Y.; Wang, R.; Wang, Y.; Li, Y.; Wu, S.; Xia, G. A real-time tennis level evaluation and strokes classification system based on the Internet of Things. *Internet Things* **2022**, *17*, 100494. [CrossRef]
8. López-Nava, I.H.; Muñoz-Meléndez, A. Wearable Inertial Sensors for Human Motion Analysis: A Review. *IEEE Sens. J.* **2016**, *16*, 7821–7834. [CrossRef]
9. Ekelund, U.; Tarp, J.; Steene-Johannessen, J.; Hansen, B.H.; Jefferis, B.; Fagerland, M.W.; Whincup, P.; Diaz, K.M.; Hooker, S.P.; Chernofsky, A.; et al. Dose-response associations between accelerometry measured physical activity and sedentary time and all-cause mortality: A systematic review and harmonised meta-analysis. *BMJ* **2019**, *366*, l4570. [CrossRef]
10. Leroux, A.; Xu, S.; Kundu, P.; Muschelli, J.; Smirnova, E.; Chatterjee, N.; Crainiceanu, C. Quantifying the Predictive Performance of Objectively Measured Physical Activity on Mortality in the UK Biobank. *J. Gerontol. A Biol. Sci. Med. Sci.* **2021**, *76*, 1486–1494. [CrossRef]
11. Strain, T.; Wijndaele, K.; Dempsey, P.C.; Sharp, S.J.; Pearce, M.; Jeon, J.; Lindsay, T.; Wareham, N.; Brage, S. Wearable-device-measured physical activity and future health risk. *Nat. Med.* **2020**, *26*, 1385–1391. [CrossRef]
12. Klenk, J.; Dallmeier, D.; Denkinger, M.D.; Rapp, K.; Koenig, W.; Rothenbacher, D.; ActiFE Study Group. Objectively Measured Walking Duration and Sedentary Behaviour and Four-Year Mortality in Older People. *PLoS ONE* **2016**, *11*, e0153779. [CrossRef]
13. McGrath, T.; Stirling, L. Body-Worn IMU-Based Human Hip and Knee Kinematics Estimation during Treadmill Walking. *Sensors* **2022**, *22*, 2544. [CrossRef]
14. Moon, K.S.; Gombatto, S.P.; Phan, K.; Ozturk, Y. Extraction of Lumbar Spine Motion Using a 3-IMU Wearable Cluster. *Sensors* **2023**, *23*, 182. [CrossRef]
15. Williams, B.K.; Sanders, R.H.; Ryu, J.H.; Bourdon, P.C.; Graham-Smith, P.; Sinclair, P.J. Static and dynamic accuracy of a magnetic-inertial measurement unit used to provide racket swing kinematics. *Sport Biomech.* **2019**, *18*, 202–214. [CrossRef]
16. Blair, S.; Duthie, G.; Robertson, S.; Hopkins, W.; Ball, K. Concurrent validation of an inertial measurement system to quantify kicking biomechanics in four football codes. *J. Biomech.* **2019**, *73*, 24–32. [CrossRef]
17. Straeten, M.; Rajai, P.; Ahamed, M.J. Method and Implementation of Micro Inertial Measurement Unit (IMU) in Sensing Basketball Dynamics. *Sens. Actuators A Phys.* **2019**, *293*, 7–13. [CrossRef]
18. Manchado, C.; Martínez, J.T.; Pueo, B.; Tormo, J.M.C.; Vila, H.; Ferragut, C.; Sánchez, F.S.; Busquier, S.; Amat, S.; Ríos, L.J.C. High-performance handball player's time-motion analysis by playing positions. *Int. J. Environ. Res. Public Health* **2020**, *17*, 6768. [CrossRef]
19. Hicks, D.; Drummond, C.; Williams, K.; Van den Tillaar, R. Investigating Vertical and Horizontal Force-Velocity Profiles in Club-Level Field Hockey Athletes: Do Mechanical Characteristics Transfer Between Orientation of Movement? *Int. J. Strength Cond.* **2023**, *3*, 1–14. [CrossRef]
20. Ishac, K.; Bourahmoune, K.; Carmichael, M. An IoT Sensing Platform and Serious Game for Remote Martial Arts Training. *Sensors* **2023**, *23*, 7565. [CrossRef] [PubMed]
21. Rusdiana, A. Analysis of Wearable Smartphone-Based Technologies for the Measuresment of Barbell Velocity in Different Resistance Training, Exercises. *Songklanakarin J. Sci. Technol.* **2021**, *43*, 448–452. [CrossRef]
22. Balsalobre-Fernández, C.; Marchante, D.; Baz-Valle, E.; Alonso-Molero, I.; Jiménez, S.L.; Muñóz-López, M. Development of agility, coordination, and reaction time training device with infrared sensor and WiFi module Arduino in badminton. *Front. Physiol.* **2017**, *8*, 649. [CrossRef] [PubMed]
23. Balsalobre-Fernández, C.; Marchante, D.; Baz-Valle, E.; Alonso-Molero, I.; Jiménez, S.L.; Muñóz-López, M. Effects of Plyometric Training Program on Speed and Explosive Strength of Lower Limbs in Young Athletes. *J. Phys. Educ. Sport* **2018**, *18*, 2476–2482. [CrossRef]

24. bin Abdullah, M.N.; Yeong, C.F.; Ming, E.S.L.; Ramli, M.S.B.; Khor, K.X.; Yang, C. Development of ReactRun: Advanced Universal IoT System for Agility Training in Sports and Healthcare. In Proceedings of the 2020 IEEE-EMBS Conference on Biomedical Engineering and Sciences (IECBES), Langkawi Island, Malaysia, 1–3 March 2021; pp. 286–291. [CrossRef]

25. Turna, B. The Effect of Agility Training on Reaction Time in Fencers. *J. Educ. Learn.* **2020**, *9*, 127–135. [CrossRef]

26. Witkowski, M.; Karpowicz, K.; Tomczak, M.; Bronikowski, M. A loss of precision of movements in fencing due to increasing fatigue during physical exercise. *J. Fisiopatol. Dello Sport* **2019**, *72*, 331–343. [CrossRef]

27. Balkó, S.; Rous, M.; Balkó, I.; Hnízdil, J.; Borysiuk, Z. Influence of a 9-week training intervention on the reaction time of fencers aged 15 to 18 years. *Phys. Act. Rev.* **2017**, *5*, 146–154. [CrossRef]

28. Nita, V.A.; Magyar, P. Smart IoT device for measuring body angular velocity and centralized assesing of balance and control in fencing. In Proceedings of the 2023 International Symposium on Signals, Circuits and Systems (ISSCS), Iasi, Romania, 13–14 July 2023; pp. 1–4. [CrossRef]

29. Wojciechowski, Z. *This Is Fencing! Advanced Training and Performance Principles for Foil*; The Crowood Press: Marlborough, MA, USA, 2019.

30. Niță, V.A.; Popa, V. Bringing Technology into Fencing Training. The art of Counterattacking. In Proceedings of the 2019 11th International Symposium on Advanced Topics in Electrical Engineering (ATEE), Bucharest, Romania, 28–30 March 2019; pp. 1–4. [CrossRef]

31. Available online: https://www.thingiverse.com/thing:3999211 (accessed on 10 April 2023).

32. Inuiguchi, M.; Sakawa, M.; Ushiro, S. Mean-absolute-deviation-based fuzzy linear regression analysis by level sets automatic deduction from data. In Proceedings of the 6th International Fuzzy Systems Conference, Barcelona, Spain, 5 July 1997; Volume 2, pp. 829–834. [CrossRef]

33. Selim, S.Z.; Ismail, M.A. K-Means-Type Algorithms: A Generalized Convergence Theorem and Characterization of Local Optimality. *IEEE Trans. Pattern Anal. Mach. Intell.* **1984**, *PAMI-6*, 81–87. [CrossRef] [PubMed]

34. Wayne, D.W. Kruskal–Wallis One-Way Analysis of Variance by Ranks. In *Applied Nonparametric Statistics*, 2nd ed.; PWS-Kent: Boston, MA, USA, 1990; pp. 226–234, ISBN 0-534-91976-6.

35. Hettmansperger, T.P.; McKean, J.W. Robust nonparametric statistical methods. In *Kendall's Library of Statistics*; Edward Arnold: London, UK; John Wiley and Sons, Inc.: New York, NY, USA, 1998; Volume 5, p. 467, ISBN 978-0-340-54937-7.

36. Dunn, O.J. Multiple Comparisons Among Means. *J. Am. Stat. Assoc.* **1961**, *56*, 52–64. [CrossRef]

37. Noether, G.E. Introduction to Wilcoxon (1945) Individual Comparisons by Ranking Methods. In *Breakthroughs in Statistics: Methodology and Distribution*; Kotz, S., Johnson, N.L., Eds.; Springer: Berlin/Heidelberg, Germany, 1992; pp. 191–195. [CrossRef]

38. Gonzalez, R.; Dabove, P. Performance Assessment of an Ultra Low-Cost Inertial Measurement Unit for Ground Vehicle Navigation. *Sensors* **2019**, *19*, 3865. [CrossRef]

sensors

Article

Myotendinous Thermoregulation in National Level Sprinters after a Unilateral Fatigue Acute Bout—A Descriptive Study

Alessio Cabizosu [1,2], Cristian Marín-Pagán [2,*], Antonio Martínez-Serrano [2], Pedro E. Alcaraz [2] and Francisco Javier Martínez-Noguera [2]

1   THERMHESC Group, Chair of Molina Ribera Hospital, C. Asociación, S/N, 30500 Molina de Segura, Spain; acabizosu@ucam.edu

2   Research Center for High Performance Sport, Catholic University of Murcia (UCAM), Campus de los Jerónimos, Nº 135, 30107 Murcia, Spain; amartinez30@ucam.edu (A.M.-S.); palcaraz@ucam.edu (P.E.A.); fjmartinez3@ucam.edu (F.J.M.-N.)

*   Correspondence: cmarin@ucam.edu; Tel.: +34-968-278-566

Abstract: In the last decade there has been a growing interest in infrared thermography in the field of sports medicine in order to elucidate the mechanisms of thermoregulation. The aim of this study was to describe bilateral variations in skin temperature of the anterior thigh and patellar tendon in healthy athletes and to provide a model of baseline tendon and muscle thermoregulation in healthy sprinters following a unilateral isokinetic fatigue protocol. Fifteen healthy national-level sprinters (eleven men and four women), with at least 3 years of athletic training experience of 10–12 h/week and competing in national-level competitions, underwent unilateral isokinetic force testing and electrostimulation in which their body temperature was measured before, during, and after the protocol using an infrared thermographic camera. ANOVA detected a significant difference in the time $\times$ side interaction for patellar temperature changes ($p \leq 0.001$) and a significant difference in the time/side interaction for quadriceps temperature changes ($p \leq 0.001$). The thermal challenge produces homogeneous changes evident in quadriceps areas, but not homogeneous in tendon areas. These data show that metabolic and blood flow changes may depend on the physical and mechanical properties of each tissue. Future research could be conducted to evaluate the predictive value of neuromuscular fatigue in the patellar tendon and quadriceps after exercise in order to optimize post-exercise recovery strategies.

Keywords: thermography; tendons; quadriceps muscle; body temperature regulation

Citation: Cabizosu, A.; Marín-Pagán, C.; Martínez-Serrano, A.; Alcaraz, P.E.; Martínez-Noguera, F.J. Myotendinous Thermoregulation in National Level Sprinters after a Unilateral Fatigue Acute Bout—A Descriptive Study. *Sensors* **2023**, *23*, 9330. https://doi.org/10.3390/s23239330

Academic Editor: Stefano Sfarra

Received: 21 September 2023
Revised: 17 November 2023
Accepted: 20 November 2023
Published: 22 November 2023

1. Introduction

Since it has been observed that cutaneous infrared radiation changes may be evidence of acute physiological responses, thermography has been the subject of numerous studies in sports medicine, due to the possibility of obtaining immediate data about the functional state of the studied structures is a great advantage over other techniques [1,2].

Because it is inexpensive, non-invasive, and free of contraindication tools [3,4], in order to clarify the thermoregulatory processes and the physiological and metabolic responses to exercise, different studies have been carried out in many disciplines using infrared thermography [5–7].

According to the microscopic concept of temperature, it is known that this measure corresponds to the average value of the kinetic energy of all the particles that constitute matter [8]. It is known that substances are composed of numerous particles that have a disordered movement, and the temperature indicates the degree of molecular agitation. So, the greater the molecular agitation, the greater the kinetic energy, and this is reflected in a higher temperature [9]. If we consider that, in humans, there are several conditions that can generate an increase or decrease in cellular activity, it is easy to understand that thermography can be a very useful tool for the assessment of the physiological state of the human being [10].

However, it should be noted that there are many other biomedical fields, in addition to sports medicine, in which this technique is being implemented exponentially. In health sciences for example the use of this technique has been extended to different medical areas such as neurology [11], cardiology [12], endocrinology [13], and dermatology [14].

Although it is currently considered mainly as a complementary diagnostic tool, there are already numerous studies that attempt to classify certain diseases on the basis of the differences in thermographic patterns obtained in an attempt to validate the thermographic tool as a reliable diagnostic or monitoring technique [15,16].

In fact, associating, for example, the use of this technique to the Doppler signal, recent studies have demonstrated its great usefulness in the analysis of patellar tendinitis in young athletes [17], or studies comparing CT images with thermographic images have allowed important advances in the field of rare neurodegenerative diseases [3], however, it should be noted that more and more research is demonstrating good inter- and intra-examiner reliability [18] and the diagnostic validity of this tool [19].

The fact that this diagnostic technique is easily transportable, due to its weight and size of approximately 1 kg (2.2 lbs) and that it is also a quick and easy-to-use technique, has encouraged its use progressively in sports medicine, being currently a very reliable and objective technique of metabolic response, before, during, and immediately after competitions.

Initially, sports such as soccer and basketball have been the subject of numerous studies [20–23]. However, there is currently a growing increase in the literature on running sports due to the interest generated by the different thermoregulatory responses under prolonged muscular stress [24].

It is known that high-performance training generates important physiological changes at a systemic level, so some studies carried out on runners analyze the role of thermography in relation to sports performance and muscular response [25–27] so that assessing the thermoregulatory response before, during, and after exercise is essential to more accurately establish workloads in individual athletes. To this end, the analysis of differences or associations between skin temperature and musculoskeletal response is currently a priority in the scientific field related to competitive running sports, with bilateral metabolic responses being analyzed mostly in the days following physical exercise, but not exhaustively in the short term [28,29].

In this regard, it should be noted that despite the effort being made to generate new knowledge in the field of thermography, either in mathematical analysis methods of thermal patterns [30] or in the automation and standardization of the diagnostic process with infrared biomedical images [31], there are still many aspects that need to be investigated and clarified.

Due to the difficulty of reproducing measurement protocols and environmental and metabolic conditions [32], studies of analysis and interpretations of normal thermographic data are quantitatively scarce, so research is currently focused on analyzing thermographic values after immediate exertion or 24–48 h after exertion, in the unilaterality/bilaterality relationship, especially in healthy runners [28,29].

In this sense, previous studies have shown that depending on the exercise performed, the exercise intensity, and the relationship with certain biochemical markers related to muscle damage, the skin responds by altering its basal thermal state [33–35]. Therefore, the technique of skin surface temperature assessment can be considered as a thermoregulatory indicator of acute metabolic stress and fatigue at the muscle level.

Therefore, the aim of this study was to describe bilateral variations in skin temperature of the anterior thigh and patellar tendon in healthy athletes and to provide a model of baseline tendon and muscle thermoregulation in healthy sprinters following, in situ, a unilateral isokinetic fatigue protocol.

2. Material and Method

2.1. Study Design

This descriptive study consisted of 2 laboratory visits. Visit 1 consisted of a medical examination to determine health status and an isokinetic familiarization session. In visit 2, athletes performed the protocol described in Section 2.3.2, where patellar skin surface temperature was measured at baseline (B), after warm-up (W), after 1st electrostimulation (1st ELEC), after isokinetic (ISO), and after 2nd electrostimulation (2nd ELEC). In the quadriceps, skin surface temperature was measured at BA after 1st ELEC, after ISO, and after 2nd ELEC. In addition, before each testing session (visit 2), the nutritionist prescribed a standardized breakfast 2 h before the rectangular test, which consisted of 1.30 g/BW carbohydrate, 0.43 g/BW protein, and 0.57 g/BW fat.

2.2. Participants

Fifteen healthy national-level sprinters were recruited (eleven men and four women). All subjects had to meet the following inclusion criteria: (a) age 18–30 years; (b) BMI 19.0–25.5 kg·m^2; and (c) at least 3 years of athletic training experience of 10–12 h/week and compete in national level competitions. Subjects were excluded if they: (a) had any metabolic or cardiovascular pathology or abnormality; (b) smoked or drank regularly; (c) took any supplements or medication in the previous 2 weeks; (d) had suffered any injury in the last 6 months; (e) had received physiotherapy 24 h before the thermographic measurements; (f) had a pathological or metabolic disease that could alter the thermographic results; and (g) were taking any supplement or medication that could affect thermoregulation. The study was conducted in accordance with the Declaration of Helsinki for Human Research [36] and was approved by the Ethics Committee of the Catholic University of Murcia (CE102201). All participants were informed of the study procedures and signed informed consent forms; in the case of underage athletes, their relatives were informed and signed the informed consent forms.

2.3. Assessments

2.3.1. Medical Exam and Familiarization Session

A medical examination included a health history, a resting electrocardiogram, and a cardiorespiratory examination (auscultation, blood pressure, etc.), and confirmed that the volunteer was healthy to participate in the study.

In the familiarization session, participants were placed in the assessment position (isokinetic chair device) and electrostimulations were applied to familiarize them with the stimulus.

2.3.2. Thermography Protocol and Muscular Stress Test

We used a Flir E75 camera (Wilsonville, OR, EE.UU) with an infrared resolution of 320 × 240 pixels and thermal sensitivity of <0.04 °C. This infrared camera offers wide measuring ranges (from −20 to 120 °C, from 0 to 650 °C, and from 300 °C to 1000 °C), with the range from −20 to 120 °C having been used in this study. The emissivity was set to 0.98 according to the bibliographic indications regarding the skin study [37,38]. The spectral range was 7.4–14.0 micrometer (μm), and the detector type was a non-refrigerated microbolometer of 17 μm. Measurement accuracy and frame rate were ±2 °C or ±2% of the reading, the ambient temperature was from 15 to 35 °C, and the object temperature was higher than 0 °C and 30 Hz. The minimum focal length was 0.5 m.

The athletes were positioned with shorts on a 1.5 m thick cotton pad to avoid direct contact with the floor and, thus, generate temperature changes in the lower limbs. To allow the correct configuration of the machine in the room, the principal investigator turned on the machine one hour before the first recording was made.

The thermographic protocol was carried out in different phases following the TISSEM protocol [38]. Data were collected by the sex, height, weight, and BMI of the participants. In addition, information was collected on the last physiotherapy and training sessions performed. High-intensity sessions could not have been performed 48 h before the study to avoid alterations in the metabolic responses of the muscular system. Participants were informed beforehand that they could not sunbathe, use chemicals on the skin, use drugs, or drink coffee in the 5 h prior to the study. The diet prior to the thermographic test was organized by the nutritionist in relation to the sports requirements of the athletes and the thermographic recommendations.

Image processing was performed by 2 blinded investigators using Flir IR research software following the previous analysis methods described by Cabizosu et al. [37]. For all measurements, the regions of interest (ROI) in the body were defined using conventional anatomical and scientific bibliography [39]. The thermographic measurements were carried out both in bipedal and seated positions, analyzing the quadricipital regions in the bipedal position and the patellar regions in the seated position. In order to be precise and reliable in the analysis of the anatomical regions, the anatomical limits of the patellar tendon were marked with a reflective sticker, placing a reflective sticker on the lower edge of the patella and another on the anterior tibial tuberosity. In this way, it was ensured that in the thermographic images, the anatomical limits of this structure had a different infrared emission gradient with respect to the skin temperature.

Phase 1: The athletes were acclimatized for a period of 20 min at 21–23 °C, a humidity of 40 ($\pm$0.8)%, and an atmospheric pressure of 1 ATM.

Phase 2 (baseline (B)): After the acclimatization period, the first thermographic acquisition was performed with the patients in a standing position for the anterior thigh region and in a seated position for the patellar region. In both cases, the athletes were located 1 m away from the thermograph, which was positioned on a fixed tripod.

Phase 3 (warm-up (W)): After the thermographic acquisition, the athletes completed a lower body ballistic warm-up exercise, recording again the temperatures of the regions studied.

Phase 4 (1st ELEC): Muscle peripheral properties of the participants' right leg were assessed using a high-voltage (400 V) constant current research stimulator (DS7R, Digitimer, Hertfordshire, United Kingdom). Square-wave stimuli of 2 ms were applied to the quadriceps muscle using bipolar electrodes (9 $\times$ 5 cm). Electrodes were positioned to ensure that rectus femoris, vastus medialis, and vastus lateralis were stimulated. The position of the electrodes was marked on the skin with a permanent marker. Firstly, single twitches were applied with progressive increases of the current (10 mA) until the evoked twitch peak torque reached a plateau. The current was then increased by 20% to ensure that a supramaximal stimulation was applied. Consequently, this current intensity was used to apply two doublets (10 and 100 Hz).

Phase 5 (isokinetic): Participants performed two knee extension maximum voluntary isometric contractions (MVIC) at 90° knee flexion. They were asked to apply and maintain the maximal force for 5 s. Resting time between repetitions was 1 min. They had visual feedback of the torque-time series, and the signal was reviewed by the researcher to determine if there was a countermovement at the beginning of the contraction. After determining MVIC, participants performed a fatiguing test that consisted of maximal isokinetic repetitions of the knee extensors at 60°/s in concentric mode with a 90° range of motion (ROM). They were asked to apply maximum force in all repetitions and throughout the full ROM and to cease force production in knee flexion so that the hamstring muscles were not involved. The test ended when the maximal force achieved during the test decreased by 50% compared to the best repetition.

Phase 6 (2nd ELEC): After the fatiguing task, two more doublets with the same intensity at 10 and 100 Hz were applied, as well as two knee extensors MVIC to assess the presence of neuromuscular fatigue (Figure 1).

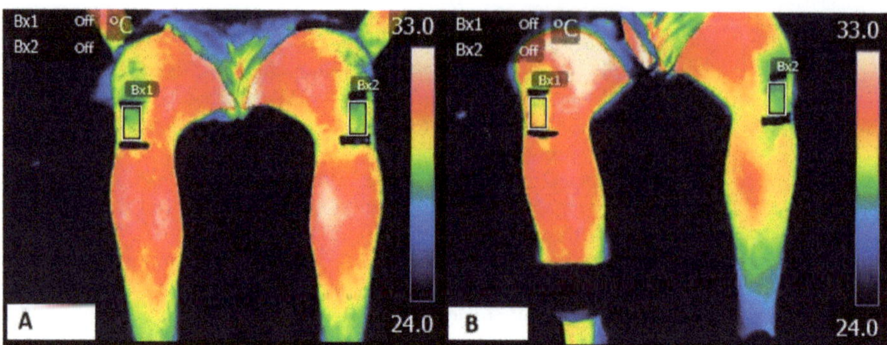

Figure 1. Patellar thermography during protocol. (**A**) Basal thermography and (**B**) final thermography.

2.3.3. Anthropometry

A certified ISAK Level-1 researcher (FJMN) performed the anthropometric measurements. Height and body weight were measured using a digital scale with a stadiometer for clinical use (SECA 780; Vogel & Halke GmbH & Co., Hamburg, Germany). Skinfold thickness was measured using Holtain Skinfold Calipers (Holtain, Ltd., Crymych Pembrokeshire, UK) in accordance with the International Society for the Advancement of Kinanthropometry guidelines [40]. The percentage of body fat was determined using the Faulkner equation [41], while the percentage of muscle mass was calculated using the modified Matiegka equation [42]. The sum of the eight skinfolds (triceps, subscapular, bicep, iliac crest, supraspinal, abdominal, thigh, and calf) was also calculated.

2.3.4. Statistical Analysis

IBM Social Sciences software (SPSS, v.21.0, Chicago, IL, USA) was used for statistical analysis. Data are presented as mean $\pm$ SD. Homogeneity and normality of the data were checked with the Levene and Shapiro–Wilk tests, respectively. For each ROI variable, a three-way repeated-measures ANOVA with time factor (BA vs. CAL vs. 1st ELEC vs. ISO vs. 2nd ELEC), sex factor (men vs. female), and side factor (right (R) and left (L)) were performed. Tukey's post hoc analysis was carried out if significance was found in the ANOVA models. Partial eta squared (ηp2) was also calculated as the effect size for the interaction of all variables in the ANOVA analysis. Partial eta square thresholds were used as follows: <0.01, irrelevant; $\geq$0.01, small; $\geq$0.059, moderate; $\geq$0.138, large [43,44]. The different correlations between the parameters were evaluated using Pearson's or Spearman's correlation (r). Significance level was set at $p \leq 0.05$.

3. Results

Assessing skin surface temperature changes by thermography can help to understand certain physiological processes that occur during exercise. Therefore, it is important to be able to quantify these changes in order to establish control criteria.

In relation to temperature changes in the patella (Figure 2), we found a significant difference in the time $\times$ side interaction ($p \leq 0.001$; η^2p = 0.366). In addition, during the protocol (intra-group analysis), post hoc Tukey's found no significant changes from B to CAL (L: -1.1 °C; $p = 0.408$ and R: -0.9 °C; $p = 0.536$), from CAL to 1st ELEC (L: 0.04 °C; $p = 1.000$ and R: 0.46 °C; $p = 0.966$), and from 1st ELEC to ISO (L: -0.33 °C; $p = 1.000$ and R: -0.27 °C; $p = 1.000$), but, we did find a significant increase from ISO to 2nd ELEC in the left patellar (L: 1.7 °C; $p = 0.007$ and R: 2.6 °C; $p = 0.256$). On the other hand, post hoc Tukey's showed significant differences in 1st ELEC ($p = 0.016$) and ISO ($p = 0.003$) between both sides. In addition, a significant positive correlation was found in the 2nd ELEC (r = 0.975; $p \leq 0.001$) between the right and left sides. Moreover, a significant positive correlation was found in 1st ELEC (r = 0.824; $p \leq 0.001$), ISO (r = 0.786; $p \leq 0.001$), and 2nd ELEC (r = 1.000; $p \leq 0.001$) between the right and left side.

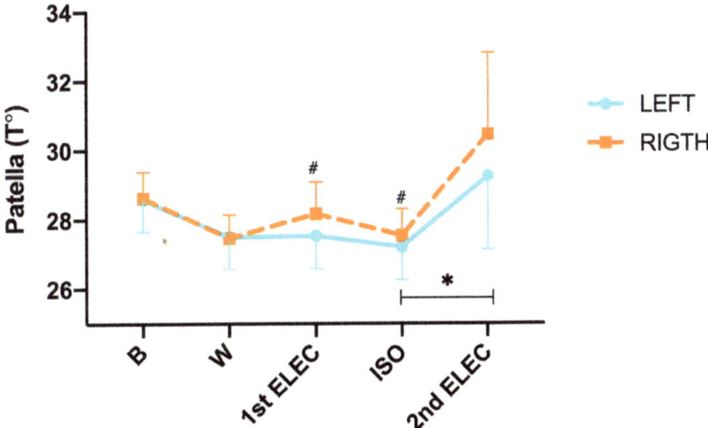

Figure 2. Patellar skin surface temperature changes were measured by infrared thermography during the exercise protocol. B = rest; W = warm-up; 1st ELEC = first electrostimulation; ISO = isokinetic and 2nd ELEC = second electrostimulation. * = $p \leq 0.05$ between ISO and 2nd ELEC on the left leg. # = $p \leq 0.05$ between both sides.

Regarding quadriceps temperature changes (Figure 3), we found a significant difference in the time × side interaction ($p \leq 0.001$; $\eta^2 p = 0.786$). In addition, during the protocol (intra-group analysis) post hoc Tukey´s found no significant changes from B to CAL (L: −1.6 °C; $p = 0.422$ and R: −1.6 °C; $p = 0.347$), but we did find a significant decrease in the T° from CAL to 2nd ELEC in QUA left (L: −2.2 °C; $p = 0.005$ and R: −0.8 °C; $p = 0.684$) and from B to 2nd ELEC (L: −3.2 °C; $p \leq 0.001$ and R: −1.7 °C; $p = 0.018$). On the other hand, post hoc Tukey's showed significant differences in 2nd ELEC ($p \leq 0.001$) between both sides. In addition, a significant positive correlation was found in the 2nd ELEC (r = 0.975; $p \leq 0.001$) between the right and left sides.

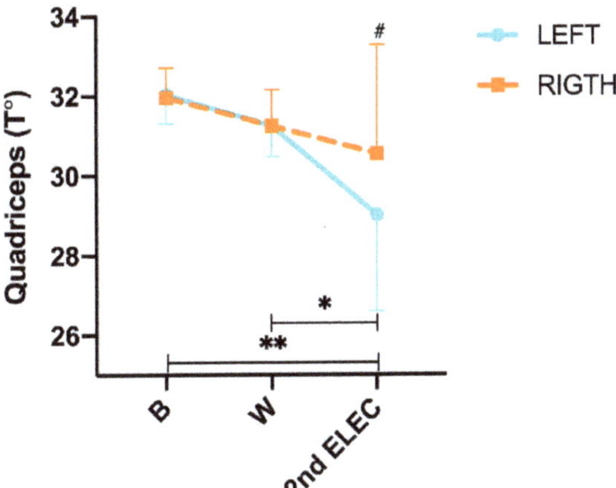

Figure 3. Quadriceps skin surface temperature changes were measured by infrared thermography during the exercise protocol. B = rest; W = warm-up; 1st ELEC = first electrostimulation; ISO = isokinetic and 2nd ELEC = second electrostimulation. * = $p \leq 0.05$ between W and 2nd ELEC in the left leg. ** = $p \leq 0.05$ between B and 2nd ELEC in both legs. # = $p \leq 0.05$ between both sides.

4. Discussion

The aim of this study was to observe and describe the bilateral skin temperature variation in national-level sprinters, in the anterior thigh region and patellar tendon, after the application of a unilateral muscle fatigue protocol carried out with electrostimulation and isokinetic knee extension exercises. The main findings found in this study generally show a decrease in temperature in muscular and tendon regions, although it should be noted that the tendon portion does not follow a homogeneous pattern of thermal regulation.

In the anterior thigh region, a progressive decrease in temperature was observed bilaterally as muscle fatigue increased, becoming more evident and statistically significant only in the left thigh at the end of the protocol, $-3.2\,^{\circ}\text{C}$; $p < 0.001$. These data confirm the results previously described by [24,45,46], which observed a reduction in temperature from baseline to the end of the muscle fatigue protocols performed. However, these findings contrast with those obtained by other authors, since, in these cases, an increase in temperature after fatigue in the most stressed region was observed [33,47,48]. Exercise physiology foresees that, during muscle contraction, due to the metabolic response of the structures involved, thermoregulatory and vascular changes are generated from the least stressed regions to the most stressed regions [49,50]. This process of blood redistribution is reflected in a variation in skin temperature, due to superficial vasoconstriction, as opposed to deep vasodilatation, regulated by cardiac output and arterial pressure according to oxygen demand at a deeper level [51,52].

According to some authors [53,54], the result of a reduction in circulation at the superficial level to favor an increase at the deep level, from the thermographic view, translates into a decrease in skin temperature, while for others [55–57], the result of deep overheating produced by muscle contraction produces an increase in skin temperature, affirming that there is a transmission effect of this heat at the superficial level, from the depth, which generates a release of heat by dissipation at the end of the exercise. In line with these authors, the results obtained in this study would be evidence that, on the one hand, thermoregulation is a global and not a local process. So, the decrease in temperature is observed bilaterally, and on the other hand, that the lower thermal decrease of the stressed limb at the end of the left protocol $-3.2\,^{\circ}\text{C}$ and right, $-1.7\,^{\circ}\text{C}$, could be the result of unilateral overheating produced by muscle contraction.

The variation of the contralateral temperature could also be justified by an aspect that has been widely studied from the physiological point of view, but not thermographically, which is the crossover effect [58,59]. It is known that, due to the activation of commissural interneurons at the spinal level, it is possible to observe an activation of certain physiological and functional parameters in the contralateral limb, so that the temperature variation could represent a deep activation of the unsolicited system, not only from the point of view of blood flow redirection, but also of nervous activation [60]. Since the physiological processes that generate this response are not conclusively known, different authors have approached this phenomenon advancing several hypotheses. A possible explanation could be represented by the overflow of the neuronal electrical impulse that would generate contralateral activations or possible neuromuscular adaptations that would influence the contralateral side since improvements in the strength of the untrained side have been observed in unilateral training [61]. However, it should be noted that, from a thermographic point of view, a consensus has not yet been reached regarding the cross-over effect, since, on the one hand, as Dindorf et al. [49] observed, there is a statistical decrease in the temperature of the non-exercised limb at the end of the unilateral effort in different anatomical regions area: 1 $p < 0.001$; area 3 $p < 0.001$. On the other hand, Escamilla et al. [49] described an increase in temperature of up to $+1\,^{\circ}\text{C}$ after the end of the muscle fatigue protocol in poorly trained and highly trained individuals.

In this regard, it should be noted that the results obtained in this work were measured immediately after exercise, as were the results obtained by Dindorf et al. [62], while Escamilla et al. [58] performed the measurements 30 and 60 min after the end of the protocol. Based on previous studies, we then speculate on a model that predicts a progressive bilateral cooling process of skin temperature which will occur as deep energy expenditure increases. This decrease will be significant and progressive until it reaches a peak, at which point it will tend to stabilize until the end of the test. At the end of the test, we can observe in the following 10–15 min a process of return to the superficial thermal normality with progressive warming of the temperature with respect to the final phase of the effort [47,63,64] until an increase in basal temperature in the following 24–48 h due to systemic inflation related to the muscle repair process [65–67]. In order to confirm this model, it would be interesting to continuously record the variation of skin temperature for 48 h after a muscular fatigue protocol to confirm if and how long the thermal reorganization process lasts after exercise.

In relation to the tendon our, results show a first phase of thermoregulatory behavior similar to the muscular system with a decrease from B to W in the left tendon of $-1.1\,^{\circ}$C and on the right of $-0.9\,^{\circ}$C. However, this decrease is followed by an increase in temperature after the first electrotherapy, followed by a second decrease in temperature after the isokinetic, and finally followed by a final overheating left ($+1.7\,^{\circ}$C) and right ($2.6\,^{\circ}$C). Since the patellar tendon is a much more superficial structure with respect to the muscular bellies of the thigh, and since fat at this level is much more present than at the tendon level, our results could be evidence of the thermoregulatory model proposed above since, in a first phase of active heating, the blood would be directed towards the depth to meet the physiological demands of the tissue, generating a thermal decrease at the cutaneous level. In the second phase, after the 1st electro, we would be observing this superficial overheating since, in this phase, the protocol foresees the absence of motor activity in relation to the tendon, which becomes again a thermal decrease after the motor effort performed in the isokinetic. Immediately after the isokinetic, there is a cessation of motor activity at the tendon level, so the overheating observed in this last phase is the result of the heat generated and transmitted with the isokinetic. In this sense, our results represent an exclusivity since, to our knowledge, there are no known previous studies or reference values in professional runners that evaluate the thermoregulatory response in tendons after protocols that alternate isokinetic quadriceps exercises to electrostimulation under normal environmental conditions.

Although no similar previous studies are known that could justify the thermal challenge obtained in this work, the temperature increase observed in the second phase could be justified with the results proposed by other authors, since it is known that the responses to load, as in the muscular system, include mechanical and metabolic-circulatory variations at the tendon level [68]. According to some authors, even with short-duration exercises, changes in tendon microcirculation are generated, increasing the total hemoglobin values and intra-structural oxygen saturation as physiological stress increases [69]. From a thermographic point of view, this could translate into a thermal increase due to the increase in vascular demand generated after the initial heating process. However, it should be noted that this theory could contrast with the results obtained by other authors who correlated a thermal increase by thermographic imaging in relation to a process of tendon vascular restriction by echo-Doppler in patients with patellar tendinitis [17]. An increased thermal state in tendon pathology could represent the first state of autonomous tissue repair, as proposed in muscle structures [65]. In fact, as in the muscular system, this could be due to a previous thermal decrease as a result of overexertion. If we consider that the increase in vascularization, in non-pathological processes, coincides with a decrease in tendon stiffness and a greater elongation of the tendons, this could be due to a previous thermal decrease as a result of overexertion [70], and that in pathological processes an intra-tendinous vascular resistance is generated which results in an increase in stiffness and a decrease in its extensibility, it would be of extreme interest for future research, to observe by means of clinical thermography, the influence of different loads at the tendon level to relate them successively with recovery processes of the pathological state.

According to these results, thermography could be a useful tool to determine the metabolic and functional response obtained after a short-duration, high-intensity physiological muscle and tendon stress protocol in national-level runners. However, it should be considered that the small sample used in this study could represent a limiting factor. Future studies are needed in which the importance of the crossover effect on thermoregulatory processes is also analyzed, as it could influence the response to short-duration exercise. This would help to clarify differences in functional activation levels in relation to skin surface temperature that are not yet entirely clear. The observation of the metabolic response in all phases of exercise could be interesting in predicting the athlete's response to high-intensity, short-duration work.

5. Conclusions

Our results confirmed that muscle exposure to high-intensity stress can generate significant changes in temperature patterns in the lower limbs, both in the muscle and tendon portions. In addition, we can observe that muscle activation during exercise affects both the homolateral and contralateral sides and is reflected in a change in skin temperature response. In addition, according to our results, the thermoregulatory effects at the tendon level appear much faster than at the muscle level, which could be of extreme interest not only to assess workloads in athletes but also for future research focused on the recovery and rehabilitation of the tendon system. Future research could be conducted to evaluate the predictive value of neuromuscular fatigue in the patellar tendon and quadriceps after exercise in order to optimize post-exercise recovery strategies. This study has also shown that, due to its easy handling and transport, thermography can be a good tool to follow and monitor instantaneously and, in situ, the muscle and tendon metabolic response in professional runners.

6. Limitations of the Study

One of the main limitations of this study was the small number of participants and the lack of similar research with which to compare results. However, although there is currently great controversy regarding thermoregulatory responses and the consequent cutaneous response in relation to muscle and tendon stress, our study may be of great use for future research in this field. Another limitation of the study could be represented by the type of thermograph used. In this regard, it should be noted that this tool is an industrial device not specifically manufactured for research with human beings. It is pertinent to point out that there is no updated registration by the Food and Drug Administration (FDA) or CE mark that classifies it as a medical device for research, although it has been validated in different studies as a diagnostic tool in humans.

Author Contributions: Conceptualization, A.C., C.M.-P., P.E.A. and F.J.M.-N.; Methodology, A.C., C.M.-P., A.M.-S. and F.J.M.-N.; Software, A.C.; Formal analysis, C.M.-P., A.M.-S. and F.J.M.-N.; Investigation, A.C., C.M.-P., A.M.-S. and F.J.M.-N.; Writing – original draft, A.C.; Writing – review & editing, C.M.-P., P.E.A. and F.J.M.-N.; Funding acquisition, P.E.A. All authors have read and agreed to the published version of the manuscript.

Funding: This research received no external funding.

Institutional Review Board Statement: We confirm that we have read the Journal's position on issues involved in ethical publication and affirm that this report is consistent with those guidelines. The study was conducted in accordance with the Declaration of Helsinki, and approved by the "Institutional Ethics Committee of Catholic University of Murcia (CE102201 28th of October 2022)" for studies involving humans.

Informed Consent Statement: Informed consent was obtained from all subjects involved in the study.

Data Availability Statement: The data is available contacting with the corresponding author by email.

Conflicts of Interest: The authors declare no conflict of interest.

References

1. Menezes, P.; Rhea, M.R.; Herdy, C.; Simão, R. Effects of Strength Training Program and Infrared Thermography in Soccer Athletes Injuries. *Sports* **2018**, *6*, 148. [CrossRef]
2. Sánchez-Jiménez, J.L.; Tejero-Pastor, R.; Calzadillas-Valles, M.D.C.; Jimenez-Perez, I.; Cibrián Ortiz de Anda, R.M.; Salvador-Palmer, R.; Priego-Quesada, J.I. Chronic and Acute Effects on Skin Temperature from a Sport Consisting of Repetitive Impacts from Hitting a Ball with the Hands. *Sensors* **2022**, *22*, 8572. [CrossRef]
3. Cabizosu, A.; Carboni, N.; Figus, A.; Vegara-Meseguer, J.M.; Casu, G.; Hernández Jiménez, P.; Martinez-Almagro Andreo, A. Is Infrared Thermography (IRT) a Possible Tool for the Evaluation and Follow up of Emery-Dreifuss Muscular Dystrophy? A Preliminary Study. *Med. Hypotheses* **2019**, *127*, 91–96. [CrossRef]
4. Ring, E.F.J.; Ammer, K. Infrared Thermal Imaging in Medicine. *Physiol. Meas.* **2012**, *33*, R33–R46. [CrossRef]
5. Amaro, A.M.; Paulino, M.F.; Neto, M.A.; Roseiro, L. Hand-Arm Vibration Assessment and Changes in the Thermal Map of the Skin in Tennis Athletes during the Service. *Int. J. Environ. Res. Public Health* **2019**, *16*, 5117. [CrossRef]
6. Lino-Samaniego, Á.; de la Rubia, A.; Sillero-Quintana, M. Acute Effect of Auxotonic and Isometric Contraction Evaluated by Infrared Thermography in Handball Players. *J. Therm. Biol.* **2022**, *109*, 103318. [CrossRef]
7. Pérez-Guarner, A.; Priego-Quesada, J.I.; Oficial-Casado, F.; Cibrián Ortiz de Anda, R.M.; Carpes, F.P.; Palmer, R.S. Association between Physiological Stress and Skin Temperature Response after a Half Marathon. *Physiol. Meas.* **2019**, *40*, 034009. [CrossRef]
8. Hu, H.; Yang, X.; Zhai, F.; Hu, D.; Liu, R.; Liu, K.; Sun, Z.; Dai, Q. Far-Field Nanoscale Infrared Spectroscopy of Vibrational Fingerprints of Molecules with Graphene Plasmons. *Nat. Commun.* **2016**, *7*, 12334. [CrossRef]
9. Wu, T.; Luo, Y.; Wei, L. Mid-Infrared Sensing of Molecular Vibrational Modes with Tunable Graphene Plasmons. *Opt. Lett.* **2017**, *42*, 2066–2069. [CrossRef]
10. Naviaux, R.K. Metabolic Features of the Cell Danger Response. *Mitochondrion* **2014**, *16*, 7–17. [CrossRef]
11. Pérez-Buitrago, S.; Tobón-Pareja, S.; Gómez-Gaviria, Y.; Guerrero-Peña, A.; Díaz-Londoño, G. Methodology to Evaluate Temperature Changes in Multiple Sclerosis Patients by Calculating Texture Features from Infrared Thermography Images. *Quant. InfraRed Thermogr. J.* **2022**, *19*, 1–11. [CrossRef]
12. Maki, K.A.; Griza, D.S.; Phillips, S.A.; Wolska, B.M.; Vidovich, M.I. Altered Hand Temperatures Following Transradial Cardiac Catheterization: A Thermography Study. *Cardiovasc. Revascularization Med. Mol. Interv.* **2019**, *20*, 496–502. [CrossRef] [PubMed]
13. da Rosa, S.E.; Neves, E.B.; Martinez, E.C.; Marson, R.A.; Machado de Ribeiro dos Reis, V.M. Association of Metabolic Syndrome Risk Factors with Activation of Brown Adipose Tissue Evaluated by Infrared Thermography. *Quant. InfraRed Thermogr. J.* **2023**, *20*, 1–17. [CrossRef]
14. Verstockt, J.; Verspeek, S.; Thiessen, F.; Tjalma, W.A.; Brochez, L.; Steenackers, G. Skin Cancer Detection Using Infrared Thermography: Measurement Setup, Procedure and Equipment. *Sensors* **2022**, *22*, 3327. [CrossRef]
15. Bardhan, S.; Nath, S.; Debnath, T.; Bhattacharjee, D.; Bhowmik, M.K. Designing of an Inflammatory Knee Joint Thermogram Dataset for Arthritis Classification Using Deep Convolution Neural Network. *Quant. InfraRed Thermogr. J.* **2022**, *19*, 145–171. [CrossRef]
16. Özdil, A.; Yilmaz, B. Medical Infrared Thermal Image Based Fatty Liver Classification Using Machine and Deep Learning. *Quant. InfraRed Thermogr. J.* **2023**, *20*, 1–18. [CrossRef]
17. Molina-Payá, F.J.; Ríos-Díaz, J.; Carrasco-Martínez, F.; Martínez-Payá, J.J. Infrared Thermography, Intratendon Vascular Resistance, and Echotexture in Athletes with Patellar Tendinopathy: A Cross-Sectional Study. *Ultrason. Imaging* **2023**, *45*, 47–61. [CrossRef]
18. Molina-Payá, J.; Ríos-Díaz, J.; Martínez-Payá, J. Inter and Intraexaminer Reliability of a New Method of Infrared Thermography Analysis of Patellar Tendon. *Quant. InfraRed Thermogr. J.* **2021**, *18*, 127–139. [CrossRef]
19. Viana, J.R.; Campos, D.; Ulbricht, L.; Sato, G.Y.; Ripka, W.L. Thermography for the Detection of Secondary Raynaud's Phenomenon by Means of the Distal-Dorsal Distance. In Proceedings of the 42nd Annual International Conference of the IEEE Engineering in Medicine & Biology Society (EMBC), Montreal, QC, Canada, 20–24 July 2020; pp. 1528–1531. [CrossRef]
20. Gómez-Carmona, P.; Fernández-Cuevas, I.; Sillero-Quintana, M.; Arnaiz-Lastras, J.; Navandar, A. Infrared Thermography Protocol on Reducing the Incidence of Soccer Injuries. *J. Sport Rehabil.* **2020**, *29*, 1222–1227. [CrossRef]
21. Majano, C.; García-Unanue, J.; Hernandez-Martin, A.; Sánchez-Sánchez, J.; Gallardo, L.; Felipe, J.L. Relationship between Repeated Sprint Ability, Countermovement Jump and Thermography in Elite Football Players. *Sensors* **2023**, *23*, 631. [CrossRef] [PubMed]
22. Rodrigues Júnior, J.L.; Duarte, W.; Falqueto, H.; Andrade, A.G.P.; Morandi, R.F.; Albuquerque, M.R.; de Assis, M.G.; Serpa, T.K.F.; Pimenta, E.M. Correlation between Strength and Skin Temperature Asymmetries in the Lower Limbs of Brazilian Elite Soccer Players before and after a Competitive Season. *J. Therm. Biol.* **2021**, *99*, 102919. [CrossRef]
23. Yeste-Fabregat, M.; Baraja-Vegas, L.; Vicente-Mampel, J.; Pérez-Bermejo, M.; Bautista González, I.J.; Barrios, C. Acute Effects of Tecar Therapy on Skin Temperature, Ankle Mobility and Hyperalgesia in Myofascial Pain Syndrome in Professional Basketball Players: A Pilot Study. *Int. J. Environ. Res. Public Health* **2021**, *18*, 8756. [CrossRef]
24. Martínez-Noguera, F.J.; Cabizosu, A.; Marín-Pagan, C.; Alcaraz, P.E. Body Surface Profile in Ambient and Hot Temperatures during a Rectangular Test in Race Walker Champions of the World Cup in Oman 2022. *J. Therm. Biol.* **2023**, *114*, 103548. [CrossRef]
25. Racinais, S.; Ihsan, M.; Taylor, L.; Cardinale, M.; Adami, P.E.; Alonso, J.M.; Bouscaren, N.; Buitrago, S.; Esh, C.J.; Gomez-Ezeiza, J.; et al. Hydration and Cooling in Elite Athletes: Relationship with Performance, Body Mass Loss and Body Temperatures during the Doha 2019 IAAF World Athletics Championships. *Br. J. Sports Med.* **2021**, *55*, 1335–1341. [CrossRef]

26. Racinais, S.; Havenith, G.; Aylwin, P.; Ihsan, M.; Taylor, L.; Adami, P.E.; Adamuz, M.-C.; Alhammoud, M.; Alonso, J.M.; Bouscaren, N.; et al. Association between Thermal Responses, Medical Events, Performance, Heat Acclimation and Health Status in Male and Female Elite Athletes during the 2019 Doha World Athletics Championships. *Br. J. Sports Med.* **2022**, *56*, 439–445. [CrossRef]

27. Rodriguez-Sanz, D.; Losa-Iglesias, M.E.; Becerro-de-Bengoa-Vallejo, R.; Dorgham, H.A.A.; Benito-de-Pedro, M.; San-Antolín, M.; Mazoteras-Pardo, V.; Calvo-Lobo, C. Thermography Related to Electromyography in Runners with Functional Equinus Condition after Running. *Phys. Ther. Sport Off. J. Assoc. Chart. Physiother. Sports Med.* **2019**, *40*, 193–196. [CrossRef]

28. Priego-Quesada, J.I.; Pérez-Guarner, A.; Gandia-Soriano, A.; Oficial-Casado, F.; Galindo, C.; Cibrián Ortiz de Anda, R.M.; Piñeiro-Ramos, J.D.; Sánchez-Illana, Á.; Kuligowski, J.; Gomes Barbosa, M.A.; et al. Effect of a Marathon on Skin Temperature Response After a Cold-Stress Test and Its Relationship with Perceptive, Performance, and Oxidative-Stress Biomarkers. *Int. J. Sports Physiol. Perform.* **2020**, *15*, 1467–1475. [CrossRef] [PubMed]

29. Priego-Quesada, J.I.; Catalá-Vilaplana, I.; Bermejo-Ruiz, J.L.; Gandia-Soriano, A.; Pellicer-Chenoll, M.T.; Encarnación-Martínez, A.; Cibrián Ortiz de Anda, R.; Salvador-Palmer, R. Effect of 10 Km Run on Lower Limb Skin Temperature and Thermal Response after a Cold-Stress Test over the Following 24 h. *J. Therm. Biol.* **2022**, *105*, 103225. [CrossRef]

30. Serantoni, V.; Jourdan, F.; Louche, H.; Sultan, A. Proposal for a Protocol Using an Infrared Microbolometer Camera and Wavelet Analysis to Study Foot Thermoregulation. *Quant. InfraRed Thermogr. J.* **2021**, *18*, 73–91. [CrossRef]

31. Özdil, A.; Yılmaz, B. Automatic Body Part and Pose Detection in Medical Infrared Thermal Images. *Quant. InfraRed Thermogr. J.* **2022**, *19*, 223–238. [CrossRef]

32. Machado, Á.S.; Priego-Quesada, J.I.; Jimenez-Perez, I.; Gil-Calvo, M.; Carpes, F.P.; Perez-Soriano, P. Influence of Infrared Camera Model and Evaluator Reproducibility in the Assessment of Skin Temperature Responses to Physical Exercise. *J. Therm. Biol.* **2021**, *98*, 102913. [CrossRef]

33. Formenti, D.; Ludwig, N.; Gargano, M.; Gondola, M.; Dellerma, N.; Caumo, A.; Alberti, G. Thermal Imaging of Exercise-Associated Skin Temperature Changes in Trained and Untrained Female Subjects. *Ann. Biomed. Eng.* **2013**, *41*, 863–871. [CrossRef]

34. Rojas-Valverde, D.; Tomás-Carús, P.; Timón, R.; Batalha, N.; Sánchez-Ureña, B.; Gutiérrez-Vargas, R.; Olcina, G. Short-Term Skin Temperature Responses to Endurance Exercise: A Systematic Review of Methods and Future Challenges in the Use of Infrared Thermography. *Life* **2021**, *11*, 1286. [CrossRef]

35. Čoh, M.; Širok, B. Use of the Thermovision Method in Sport Training. *Phys. Educ. Sport* **2007**, *5*, 85–94.

36. World Medical Association. World Medical Association Declaration of Helsinki: Ethical Principles for Medical Research Involving Human Subjects. *JAMA* **2013**, *310*, 2191–2194. [CrossRef]

37. Cabizosu, A.; Carboni, N.; Martínez-Almagro Andreo, A.; Casu, G.; Ramón Sánchez, C.; Vegara-Meseguer, J.M. Relationship between Infrared Skin Radiation and Muscular Strength Tests in Patients Affected by Emery-Dreifuss Muscular Dystrophy. *Med. Hypotheses* **2020**, *138*, 109592. [CrossRef]

38. Moreira, D.G.; Costello, J.T.; Brito, C.J.; Adamczyk, J.G.; Ammer, K.; Bach, A.J.E.; Costa, C.M.A.; Eglin, C.; Fernandes, A.A.; Fernández-Cuevas, I.; et al. Thermographic Imaging in Sports and Exercise Medicine: A Delphi Study and Consensus Statement on the Measurement of Human Skin Temperature. *J. Therm. Biol.* **2017**, *69*, 155–162. [CrossRef]

39. Bouzas Marins, J.C.; de Andrade Fernandes, A.; Gomes Moreira, D.; Souza Silva, F.; Magno, A.; Costa, C.; Pimenta, E.M.; Sillero-Quintana, M. Thermographic profile of soccer players' lower limbs. *Rev. Andal. Med. Deporte* **2014**, *7*, 1–6. [CrossRef]

40. Stewart, A.; Marfell-Jones, M.; Olds, T.; De Ridder, J. *International Standards for Anthropometric Assessment*; ISAK: Brasilia, Brazil, 2011; Volume 137, ISBN 978-0-620-36207-8.

41. Faulkner, J.A. Physiology of Swimming. *Res. Q.* **1966**, *37*, 41–54. [CrossRef]

42. Norton, K. *Antropometrica*; Spanish Version of Anthropometrica; Norton, K., Olds, T., Eds.; BIOSYSTEM Servicio Educativo: Rosario, Argentina, 1995.

43. Cohen, J. *Statistical Power Analysis for the Behavioral Sciences*, 2nd ed.; Reprint; Psychology Press: New York, NY, USA, 2009; ISBN 978-0-8058-0283-2.

44. Cuddy, J.S.; Hailes, W.S.; Ruby, B.C. A Reduced Core to Skin Temperature Gradient, Not a Critical Core Temperature, Affects Aerobic Capacity in the Heat. *J. Therm. Biol.* **2014**, *43*, 7–12. [CrossRef]

45. Adamczyk, J.G.; Boguszewski, D.; Siewierski, M. Thermographic Evaluation of Lactate Level in Capillary Blood During Post-Exercise Recovery. *Kinesiology* **2014**, *46*, 186–193.

46. Rynkiewicz, M.; Korman, P.; Zurek, P.; Rynkiewicz, T. Application of Thermovisual Body Image Analysis in the Evaluation of Paddling Effects on a Kayak Ergometer. *Med. Dello Sport* **2015**, *68*, 31–42.

47. Priego Quesada, J.I.; Martinez, N.; Salvador-Palmer, R.; Psikuta, A.; Annaheim, S.; Rossi, R.; Corberan, J.; Cibrian, R.; Perez-Soriano, P. Effects of the Cycling Workload on Core and Local Skin Temperatures. *Exp. Therm. Fluid Sci.* **2016**, *77*, 91–99. [CrossRef]

48. Weigert, M.; Nitzsche, N.; Kunert, F.; Lösch, C.; Baumgärtel, L.; Schulz, H. Acute Exercise-Associated Skin Surface Temperature Changes after Resistance Training with Different Exercise Intensities. *Int. J. Kinesiol. Sports Sci.* **2018**, *6*, 12–18. [CrossRef]

49. Fernández-Cuevas, I.; Torres, G.; Sillero-Quintana, M.; Navandar, A. Thermographic Assessment of Skin Response to Strength Training in Young Participants. *J. Therm. Anal. Calorim.* **2023**, *148*, 3407–3415. [CrossRef]

50. Robles Dorado, V. Variaciones termométricas en la planta del pie y piernas valorada en corredores antes y después de correr 30 km. *Rev. Int. Cienc. Podol.* **2016**, *10*, 31–40. [CrossRef]

51. Joyner, M.J.; Casey, D.P. Regulation of Increased Blood Flow (Hyperemia) to Muscles during Exercise: A Hierarchy of Competing Physiological Needs. *Physiol. Rev.* **2015**, *95*, 549–601. [CrossRef]
52. Rowell, L.B. Blood Pressure Regulation during Exercise. *Ann. Med.* **1991**, *23*, 329–333. [CrossRef]
53. Fritzsche, R.G.; Coyle, E.F. Cutaneous Blood Flow during Exercise Is Higher in Endurance-Trained Humans. *J. Appl. Physiol.* **2000**, *88*, 738–744. [CrossRef]
54. Neves, E.; Cunha, R.; Rosa, C.; Antunes, N.; Felisberto, I.; Alves, J.; Reis, V. Correlation between Skin Temperature and Heart Rate during Exercise and Recovery, and the Influence of Body Position in These Variables in Untrained Women. *Infrared Phys. Technol.* **2016**, *75*, 70–76. [CrossRef]
55. Al-Nakhli, H.H.; Petrofsky, J.S.; Laymon, M.S.; Berk, L.S. The Use of Thermal Infra-Red Imaging to Detect Delayed Onset Muscle Soreness. *J. Vis. Exp.* **2012**, *59*, 3551. [CrossRef]
56. Priego Quesada, J.I.; Carpes, F.P.; Bini, R.R.; Salvador Palmer, R.; Pérez-Soriano, P.; Cibrián Ortiz de Anda, R.M. Relationship between Skin Temperature and Muscle Activation during Incremental Cycle Exercise. *J. Therm. Biol.* **2015**, *48*, 28–35. [CrossRef]
57. Silva, Y.A.; Santos, B.H.; Andrade, P.R.; Santos, H.H.; Moreira, D.G.; Sillero-Quintana, M.; Ferreira, J.J.A. Skin Temperature Changes after Exercise and Cold Water Immersion. *Sport Sci. Health* **2017**, *13*, 195–202. [CrossRef]
58. Escamilla-Galindo, V.L.; Estal-Martínez, A.; Adamczyk, J.G.; Brito, C.J.; Arnaiz-Lastras, J.; Sillero-Quintana, M. Skin Temperature Response to Unilateral Training Measured with Infrared Thermography. *J. Exerc. Rehabil.* **2017**, *13*, 526–534. [CrossRef]
59. Valdes, O.; Ramirez, C.; Perez, F.; Garcia-Vicencio, S.; Nosaka, K.; Penailillo, L. Contralateral Effects of Eccentric Resistance Training on Immobilized Arm. *Scand. J. Med. Sci. Sports* **2021**, *31*, 76–90. [CrossRef]
60. Benito-Martínez, E.; Senovilla-Herguedas, D.; de la Torre-Montero, J.C.; Martínez-Beltrán, M.J.; Reguera-García, M.M.; Alonso-Cortés, B. Local and Contralateral Effects after the Application of Neuromuscular Electrostimulation in Lower Limbs. *Int. J. Environ. Res. Public Health* **2020**, *17*, 9028. [CrossRef]
61. Carroll, T.J.; Herbert, R.D.; Munn, J.; Lee, M.; Gandevia, S.C. Contralateral Effects of Unilateral Strength Training: Evidence and Possible Mechanisms. *J. Appl. Physiol.* **2006**, *101*, 1514–1522. [CrossRef]
62. Dindorf, C.; Bartaguiz, E.; Janowicz, E.; Fröhlich, M.; Ludwig, O. Effects of Unilateral Muscle Fatigue on Thermographic Skin Surface Temperature of Back and Abdominal Muscles-A Pilot Study. *Sports* **2022**, *10*, 41. [CrossRef]
63. Binek, M.; Drzazga, Z.; Teresa, S.; Pokora, I. Do Exist Gender Differences in Skin Temperature of Lower Limbs Following Exercise Test in Male and Female Cross-Country Skiers? *J. Therm. Anal. Calorim.* **2022**, *147*, 7373–7383. [CrossRef]
64. Oliveira, S.; Oliveira, F.; Marins, J.; Gomes, A.; Silva, A.; Brito, C.; Gomes Moreira, D.; Quintana, M. Original Article Measuring of Skin Temperature via Infrared Thermography after an Upper Body Progressive Aerobic Exercise. *J. Phys. Educ. Sport* **2018**, *18*, 184–192. [CrossRef]
65. Alburquerque Santana, P.V.; Alvarez, P.D.; Felipe da Costa Sena, A.; Serpa, T.K.; de Assis, M.G.; Pimenta, E.M.; Costa, H.A.; Sevilio de Oliveira Junior, M.N.; Torres Cabido, C.E.; Veneroso, C.E. Relationship between Infrared Thermography and Muscle Damage Markers in Physically Active Men after Plyometric Exercise. *J. Therm. Biol.* **2022**, *104*, 103187. [CrossRef]
66. da Silva, W.; Machado, Á.S.; Souza, M.A.; Kunzler, M.R.; Priego-Quesada, J.I.; Carpes, F.P. Can Exercise-Induced Muscle Damage Be Related to Changes in Skin Temperature? *Physiol. Meas.* **2018**, *39*, 104007. [CrossRef]
67. Rojas-Valverde, D.; Gutiérrez-Vargas, R.; Sánchez-Ureña, B.; Gutiérrez-Vargas, J.C.; Priego-Quesada, J.I. Relationship between Skin Temperature Variation and Muscle Damage Markers after a Marathon Performed in a Hot Environmental Condition. *Life* **2021**, *11*, 725. [CrossRef]
68. Kjaer, M.; Langberg, H.; Miller, B.F.; Boushel, R.; Crameri, R.; Koskinen, S.; Heinemeier, K.; Olesen, J.L.; Døssing, S.; Hansen, M.; et al. Metabolic Activity and Collagen Turnover in Human Tendon in Response to Physical Activity. *J. Musculoskelet. Neuronal Interact.* **2005**, *5*, 41–52.
69. Kubo, K.; Ikebukuro, T. Blood Circulation of Patellar and Achilles Tendons during Contractions and Heating. *Med. Sci. Sports Exerc.* **2012**, *44*, 2111. [CrossRef]
70. Kubo, K.; Ikebukuro, T.; Yaeshima, K.; Kanehisa, H. Effects of Different Duration Contractions on Elasticity, Blood Volume, and Oxygen Saturation of Human Tendon in Vivo. *Eur. J. Appl. Physiol.* **2009**, *106*, 445–455. [CrossRef]

sensors

Article

The Role of the Specific Strength Test in Handball Performance: Exploring Differences across Competitive Levels and Age Groups

Luis J. Chirosa-Ríos [1], Ignacio J. Chirosa-Ríos [1,*], Isidoro Martínez-Marín [2], Yolanda Román-Montoya [3] and José Fernando Vera-Vera [3]

[1] Physical Education and Sports Department, Faculty of Sport Sciences, University of Granada, 18011 Granada, Spain; lchirosa@ugr.es

[2] Sports Science Department, Faculty of Sport Sciences, University of Leon, 24004 Leon, Spain; imarm@unileon.es

[3] Statistics and Operational Research Department, Faculty of Sciences, University of Granada, 18071 Granada, Spain; yroman@ugr.es (Y.R.-M.); jfvera@ugr.es (J.F.V.-V.)

* Correspondence: ichirosa@ugr.es

Abstract: The aim of this study was to determine if specific physical tests are sufficiently discriminant to differentiate players of similar anthropometric characteristics, but of different playing levels. Physical tests were conducted analyzing specific strength, throwing velocity, and running speed tests. Thirty-six male junior handball players ($n = 36$; age 19.7 ± 1.8 years; 185.6 ± 6.9 cm; 83.1 ± 10.3 kg; 10.6 ± 3.2 years of experience) from two different levels of competition participated in the study: NT = 18 were world top-level elite players, belonging to the Spanish junior men's national team (National Team = NT) and A = 18 players of the same age and anthropometric conditions, who were selected from Spanish third league men's teams (Amateur = A). The results showed significant differences ($p < 0.05$) between the two groups in all physical tests, except for two-step-test velocity and shoulder internal rotation. We conclude that a battery combining the Specific Performance Test and the Force Development Standing Test is useful in identifying talent and differentiating between elite and sub-elite players. The current findings suggest that running speed tests and throwing tests are essential in selecting players, regardless of age, sex, or type of competition. The results shed light on the factors that differentiate players of different levels and can help coaches in selecting players.

Keywords: team handball; specific physical tests; players selection; field tests

Citation: Chirosa-Ríos, L.J.; Chirosa-Ríos, I.J.; Martínez-Marín, I.; Román-Montoya, Y.; Vera-Vera, J.F. The Role of the Specific Strength Test in Handball Performance: Exploring Differences across Competitive Levels and Age Groups. *Sensors* **2023**, *23*, 5178. https://doi.org/10.3390/s23115178

Academic Editor: Javad Foroughi

Received: 20 April 2023
Revised: 23 May 2023
Accepted: 26 May 2023
Published: 29 May 2023

1. Introduction

Handball is a high-intensity team sport, where explosive actions with high strength involvement, such as jumping, sprinting, throwing, and changing direction, are fundamental for sport performance [1–6]. Players must perform frequent periods of high-intensity activity with highly variable recoveries over time, combined with moments of low intensity [4,7,8]. Further, for top-level players, decision making, as well as other conditional and biometric factors will determine the level of the player (reference). The first ones could be defined as internal factors (cognitive) and the second ones as external factors (conditional and biometric).

Although handball is a team sport, as any other sport, where the causes of performance are multifactorial, strength is one of the most determining external factors due to the high involvement it has in fundamental gestures such as throwing, jumping, changes in direction or impacts-contacts [9–12]. When assessing strength, it is necessary to apply tests capable of both discriminating analytical improvements in athlete strength production, as well as its transfer and application to the game. The way to be able to evaluate it is a matter of concern for researchers of Sport Sciences, especially when they want to use it to classify and discriminate athletes of different levels, categories, sex, etc. A good evaluation protocol should differentiate a good player from a not so good player, in addition to identifying the rates of improvement during a season.

It can be complicated to choose the type of test intended to define performance and classify players for selection in a team or national team [13,14]. Among the most commonly used factors in handball evaluation is anthropometry [15] along with strength assessment tests [16] and field application tests [17]. In the literature of the last decade, several evaluation proposals have appeared, more or less specific, combining assessment of anthropometric characteristics together with their capacity for force production and application in handball players of different levels, age and sex [18–22]. Previous work has shown that there are differences between elite and amateur handball players in terms of strength manifestations, such as maximal dynamic strength and muscle power development, as well as anthropometric characteristics [2,23]. Several studies have observed how anthropometric differences are determinants in the performance of handball players [15,21,24–26], favoring the application of explosive strength to determinant gestures in the game such as throwing [27]; in fact, it seems clear that greater anthropometry is clearly related to an increase in game performance and that it is a differential factor in any category [18,25,28].

Although the evidence for performance between anthropometry (external factor) and level of play is clear, there is an underlying question little studied in our sport that is related exclusively to the level of play and the production of force, what happens when players of different levels, but of the same age have the same anthropometric characteristics? Are there still differences in their physical ability to manifest and apply force? Recent studies in other sports, such as soccer, have partially answered this question [29] and these works have shown that players of different competitive levels with similar anthropometric conditions have different physical performances, concluding that elite soccer players show better performance indicators in the strength variables studied [29].

Studies on male handball players have shown that both physiological and physical characteristics differentiate players according to category, sex and level [26,28,30–33]. The problem is that in all these investigations, the factors of age, sex and anthropometry are not kept fixed in order to analyze exclusively the physical factors. Therefore, it would be of great interest to conduct a comparison between players with different training experiences, keeping age and anthropometry fixed so that they cannot act as contaminating variables. It would also be important to know whether a given battery of tests, in which the application of strength is an essential component, is able to discriminate the level of the players by itself without other factors being present and thus avoid test redundancies.

Consequently, the present study aims to examine the differences in physical performance between non-elite and elite junior handball players, keeping anthropometry and age stable. The aim is to determine whether the force production capacity and its application in field tests are sufficient to differentiate or classify selected junior players at the highest level from their non-elite peers, regardless of their anthropometry. At the same time, the aim is to check which test can be more determinant to differentiate players who should be eligible for selection from those who are not.

2. Materials and Methods

Thirty-six male junior handball players ($n = 36$; age 19.7 ± 1.8 years; 185.6 ± 6.9 cm; 83.1 ± 10.3 kg; 10.6 ± 3.2 years of experience) from two different levels of competition participated in this study: National Team = 18 were world top-level elite players ($n = 18$; age 20.2 ± 0.8 years; 185.6 ± 7.5 cm; 83.1 ± 10.5 kg), belonging to the Spanish junior men's national team (National Team = NT) and A = 18 players of the same age and anthropometric conditions ($n = 18$; age 19.7 ± 1.0 years; 185.6 ± 4.5 cm; 83.1 ± 10.2 kg), who were selected from Spanish third league men's teams that had not been selected as amateur players (Amateur = A). This study adopts a Quasi-Experimental Design to compare and assess differences between the groups under investigation.

The measurements were taken before the start of the competition season; all the tests were performed on the handball court. This study was approved by the Ethics Committee of the University of Granada, Spain, and was conducted in accordance with the requirements established in the Declaration of Helsinki (2/2020). The participants were informed about the procedures to be performed and written informed consent was obtained from all of them. Information on age, training experience, supplementation and presence of injuries was collected by means of a personalized interview. All the players had undergone previous medical tests; only those athletes who were not injured at the time of the test participated in this study.

To reduce the influence of uncontrolled variables, all participants were instructed to maintain their typical lifestyle and dietary habits before and during the study. Subjects were told not to exercise the day before the test and to consume their last meal (without caffeine) at least 3 h before the scheduled test. In addition, they drank at least 0.5 L of pure water during the last hour before the test. They were also asked to sleep regularly before the protocol. During all performance-based tests, athletes were instructed to perform at their maximum capacity.

2.1. Procedures

The subjects were familiar with the testing protocol, for which all tests had passed a reliability test (Hopkins, 2000) prior to the familiarization period, with the ICC being above 0.9 and CV less than 10%, thus it was considered that the tests were learned by the players.

The physical test battery consists of two different blocks:

The Specific Performance Test: Two tests of strength application related to the game—for the upper body, the throwing velocity that deals with throws at maximum speed with three steps (Chirosa-Ríos et al., 2020); and for the lower body, the 30 m sprint test (30 mST) was used, which is a maximum speed run of 30 m taking the time in 10–20–30 m.

The Functional Dynamometric Strength Test: Two tests that measure the manifestation of strength in a more specific way through the use of a functional electromechanical dynamometer (DEMF) [34]. An isometric test for the lower body is the shoulder rotation test [35] and the two-step test (TST) [36], which is a dynamic test that evaluates the maximal manifestation of strength in the lower body. In our case, we have chosen, from the multitude of data, the speed of displacement (two-step test—velocity, TSTv) and the power of the performance of the gesture (two-step test—force, TSTf).

2.1.1. Anthropometric Parameters

Prior to testing, participants were assessed for height and body mass. Height was assessed to the nearest 0.001 m, using a stadiometer (Holtain Ltd., Crymych, UK). Body mass and body fat percentage were measured to the nearest 0.1 kg, using an electronic scale (Seca Instruments Ltd., Hamburg, Germany). To determine the participants' body mass index (BMI), the measured values were used in a standard calculation.

2.1.2. Handball Test Protocol

Prior to the performance of the test battery, the players were given verbal instructions and shown videos about the different tests in which they were to participate. Each participant was carefully instructed and verbally encouraged to give their maximum performance in the tests. Each test was supervised by a handball specialist, and only those that complied were recorded.

The warm-up was standardized, consisting of soft running, static and dynamic range of motion, 5 progressive speed runs of 30 m, 2 series of 10 push-ups and 2 series of 6 throws progressively looking for maximum speed. The test battery was divided into two test blocks: the Specific Performance Test and the Functional Dynamometric Strength Test. Each of these parts had two tests in order not to fatigue the players and not to intervene too much in the preparation of the teams, the order of application was randomized.

2.1.3. Throwing Velocity

The participants performed the running throw as executed in the game, behind the throwing line at a distance of 9 m.

Throws were performed with an official ball according to IHF/EHF regulations (ball weight 425 g, −475 g, ball radius = 58–60 cm), and subjects were allowed to use resin according to convenience.

Five throws were measured—of the five throws, the last four were scored for this study. Each throw was measured in kilometers per hour (km/h) using a radar (Stalker ATR, Professional radar, Applied Concepts Inc., Plano, TX, USA), with an accuracy of ±0.1 km/h and a sampling frequency of 100 HZ within a field of action of 10° where the pistol was placed behind the goal on the thrower's axis at the height of his arm. The participants were placed in the corresponding position and performed the throw, with a rest between throws of 1 min. High test–retest reliability was found (ICC = 0.97, CV = 4%).

2.1.4. The 30 m Sprint Test

Each player was asked to perform a sprint as fast as possible standing with both feet shoulder-width apart, 5 cm behind the first timing gate. To measure sprint time, 3 light beams (Brower Timing System CM L5, Brower, UT, USA) placed at 10, 20 and 30 m from the test distance were used. Each subject had to repeat the sprint test twice, with 2 min recovery between tests. The fastest sprint time of 30 m was used for the calculation. High test–retest reliability was found (ICC = 0.93, CV = 2%).

The Functional Dynamometric Strength Test:

2.1.5. Standing Shoulder Internal Rotation

Subjects were positioned standing and supporting the dominant upper limb on a subjection system of own manufacture. The subjection system was regulated, taking into account the subject's height with a variation of ±1 cm. The humerus was fixed with a cinch at 2/3 of the distance between the lateral epicondyle and the acromion. Position was determined with a baseline goniometer (Gymna hoofdzetel, Bilzen, Belgium). The position consisted of a 90° adduction of the glenohumeral joint and a 90° flexion of the humeroulnar joint. For the glenohumeral joint, the fulcrum was positioned in the acromion with the vertical arm stable and the arm movable along the humerus with the lateral epicondyle as a point of reference. For the humeroulnar articulation, the fulcrum was positioned in the lateral epicondyle with the arm stable in horizontal and the arm movable along the forearm with the processus styloideus ulnae as a reference point.

Participants first attend (four subjects each time) in a well-rested condition at the start of each testing session of 45 min with the FEMD. The protocol consisted of a general warm-up for both test session consisted on 5 min of jogging (beats per minute < 130; measurement with a Polar M400), 5 min of joint mobility and 2 sets of 6 s of internal rotation and external rotation in the previous stablished position. After the warm-up of familiarization protocol, participants rested for 5 min before the initiation. The test consisted of two series of 6 s of shoulder internal and external rotators. The rest between sets was a three-minute. The mean and peak force were taken to calculate the mean dynamic force for each participant, (ICC = 0.97, CV = 3%).

2.1.6. The Two-Step Test

Participants stood with their feet shoulder-width apart. They had to perform a two-step forward movement, touching at the end a person simulating an opponent. The final position was standing with the second leg they moved forward and holding an "opponent". Any initial or final position was previously established based on one's own freedom of movement. An appropriate belt was used to avoid injury when performing the explosive forward steps. A free range of motion was established without taking any measurements.

The test consisted of two sets of six consecutive maximum repetitions, with 15% body weight overload with free range of motion. Participants were required to perform a gesture similar to the forward movement of holding an opponent in two steps in handball at the maximum possible speed. There was a three-minute pause between sets. Only the maximum velocity, TSTv, and the power of each repetition, TSTf, were taken as variables.

Body displacement velocity was evaluated with a FEMD (Dynasystem Health, Symotech, Granada, Spain) with a precision of 3 mm for displacement, 100 g for a sensed load, and a range of velocities between 0.05 m·s^{-1} and 2.80 m·s^{-1}, coupled with a standard bench, an appropriate hip belt, a pulley system and a subjection system (ICC = 0.96, CV = 4%).

3. Statistical Analysis

All data were analyzed considering the group to which each player belongs. Descriptive measures (mean, standard deviation and 95% confidence intervals) were calculated for the anthropometric variables, testing the difference between the mean values of each group using a general linear model. Differences between means were considered statistically significant if the *p*-values are less than 0.05 and the effect size (eta) takes values greater than 0.10.

For the performance variables, the Strength Transfer Field Test and the Specific Strength Test, a general linear model was used to analyze the differences between the two teams by looking at the results of both tests. The mean values obtained by each player after the series of repetitions of the different tests are considered. The normality of the observations was previously verified. The differences between the mean values of the results obtained were considered statistically significant if the *p*-values are less than 0.05 and the effect size (eta) takes values greater than 0.10. The confidence intervals (at 95% confidence) of the mean values were also included.

Linear discriminant analysis was used to determine the most influential variables in classifying a player as belonging or not to the national team, as well as to determine a linear discriminant rule to determine the classification of any new player to one of the two groups. Taking into account the sample size, the hypotheses of the model were checked for each variable by means of Shapiro–Wilks and Levené contrasts. The hypotheses of equality of means and multivariate homoscedasticity were tested with Wilks' Lambda test and Box's test. The most determinant variables were selected using Wilks' method and Mahalanobis' distance, and the ranking results for the players in the sample were also analyzed with a cross-validation procedure to determine the ranking results. All analyses were performed using SPSS Statistics 27 (IBM Corp., Armonk, NY, USA).

3.1. Comparative Analysis

Analyzing the anthropometric characteristics of the players according to the team to which they belong, it is observed that there are no significant differences between the main variables studied, considering the confidence intervals and *p*-value associated with the contrast of equality of mean values between groups (Table 1).

Table 1. Anthropometric characteristics of the National Team (*n* = 18) and amateur players (*n* = 18).

Anthropometric Characteristics	National Team			Amateur			P	ES
	Means	SD	SE	Means	SD	SE		
Age (yrs)	20.22	0.80	0.19	19.72	2.51	0.59	0.428	−0.268
Body Height (cm)	185.61	7.51	1.71	185.44	6.74	1.59	0.945	−0.023
Body Weigh (kg)	86.78	10.52	2.48	85.56	10.22	2.48	0.288	−0.360
Body Fat (%)	14.55	4.16	0.98	13.70	3.58	0.84	0.233	−0.324
Body Mass (Kg)	74.43	17.22	4.05	73.08	9.35	2.73	0.141	0.502

ES = Coen´s d

In view of the results, regarding the physical tests performed, there were significant differences in the two groups of tests of the battery used, the specific performance test (SPT) and the Functional Dynamometric Strength Test (FDST) (Figure 1).

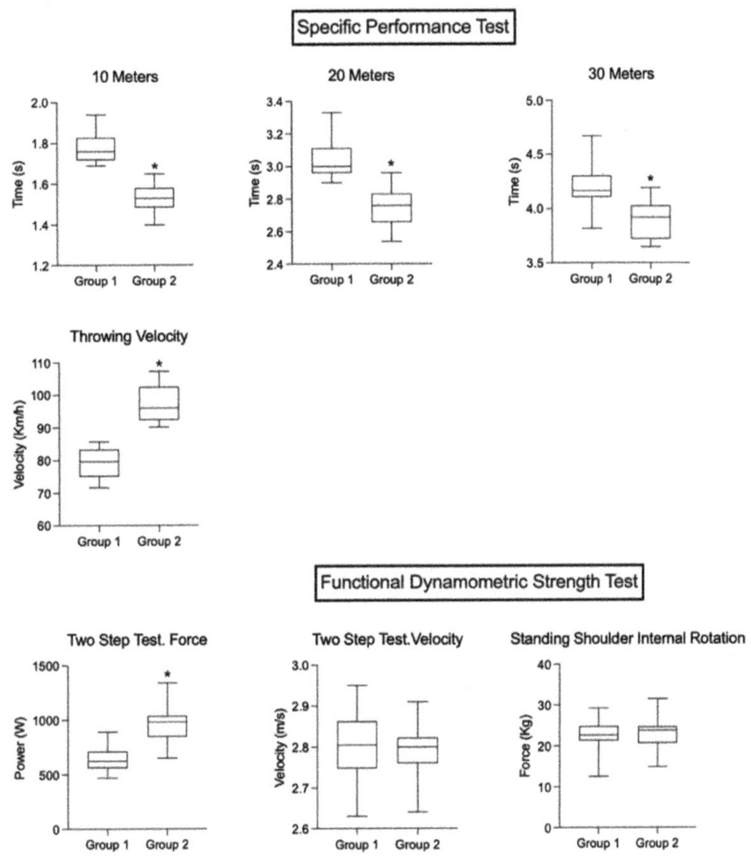

Figure 1. Physical fitness relationship of the National Team ($n = 18$) vs. the amateur player ($n = 18$). * An asterisk means statistically significant differences between groups ($p < 0.001$).

In the SPT, the displacement velocity (m/s) in the first section 10 m speed was 1.49 ± 0.06 NT, versus 1.74 ± 0.07 A ($p < 0.001$; ES = 3.43); in the second section of 20 m speed it was 2. 71 ± 0.10 NT, versus 3.01 ± 0.10 A ($p < 0.001$; ES = 2.56); in the third stretch of 30 m speed it was 3.86 ± 0.15 NT, versus 4.11 ± 0.34 A ($p < 0.008$; ES = 0.93). In ball displacement speed (KM/h) the results were 102.44 ± 5.49 NT, versus 81.00 ± 4.95 A ($p < 0.001$; ES= -4.09).

In the FDST, the two-step test revealed a significant difference in power (W) between the force results: 1128.72 ± 172.7 NT versus 817.00 ± 141.87 A ($p < 0.001$; ES = -1.97). The two-step test velocity results were 2.65 ± 0.06 NT, versus 2.65 ± 0.01 A ($p < 0.001$; ES = -1.01 A), 2.65 ± 0.01 A ($p = 0.984$; ES = 0.01) and in the shoulder internal rotation the results were 22.98 ± 4.9 NT, versus 22.26 ± 4.47 A ($p = 0.0621$; ES = -0.17).

3.2. Discriminant Analysis

The influence of the variables, BMI, body height, body weight, body fat, body mass, 10 m speed, 20 m speed, 30 m speed, throwing velocity, two-step test—force, two-step test—velocity, shoulder internal rotation to determine the membership of a player to the selection has been analyzed using linear discriminant analysis with SPSS. The variables were first standardized by their range to avoid measurement differences.

The usual assumption for the linear discriminant analysis is performed. First, the normality and homoscedasticity of each variable among the two groups of players (1 = selection, 2 = national club) were analyzed. The Shapiro–Wilks normality test was significant at the level 0.05 for the variables BMI ($p = 0.027$), body fat ($p = 0.005$) and 30 m speed ($p = 0.004$), both for the group of the national selection players, while the Levené's test indicates lack of homoscedasticity for total water ($p = 0.002$) and shoulder internal rotation med ($p = 0.026$). Hence, the influence of variables body mass, 10 m speed, 20 m speed, throwing velocity, two-step test—force and shoulder internal rotation max was analyzed to determine a discriminant rule for the membership of a player to the national selection. Significant differences are identified for each variable in terms of the Wilks' Lambda test of equality of groups means, except for body mass ($p = 0.257$), two-step test—force ($p = 1$) and shoulder internal rotation max ($p = 0.621$). Second, the Box's test of equality of covariance matrices showed a p-value of 0.915 which does not contradict the multivariate homoscedasticity hypothesis, and a Wilks' Lambda value of 0.088 indicates that almost all the variance is explained by group differences, which related to a p-value of 0, indicates highly significant differences between the two group centroids (Chi-square value of 74.047).

A stepwise procedure was used to determine the influent variables in the linear discriminant procedure. The variables related to the small value of the Wilks' Lambda, which in this data sets also coincides which that maximizing the Mahalanobis distance between the two closets groups, were throwing velocity, 10 m speed and 20 m speed, for Wilks' Lambda values of 0.202, 0.146 and 0.149, respectively, while the remaining variables were not included in the analysis. The standardized canonical discriminant function coefficient were -1.622 for 10 m speed, -1.119 for 10 m 20 m speed and 0.750 for throwing velocity, which shows the SPRINT10 and SPRINT20 variables appear to have the greatest impacts.

Considering the above results, linear discriminant analysis is performed for the resulting variables. For each group, the discrimination functions to classifying a player in the group related to the highest score (NT = National Team or A = club amateur) are:

$$NT = -32.770 \times 10 \text{ m speed} + 208.50 \times 20 \text{ m speed} + 127.212 \times \text{throwing velocity} - 490.280$$

$$A = -102.380 \times 10 \text{ m speed} + 251.599 \times 20 \text{ m speed} + 158.714 \times \text{throwing velocity} - 513.625$$

For the analyzed sample, the overall success of the three variables in the model for classifying cases into one of the two groups is 100%. Since the sample is small, cross-validated classification results (for one leave-out player) also showed a 100% of correct results.

4. Discussion

The main objective of this study was to determine whether the force production and application capacity of junior elite handball players were different from that of their non-selected peers, independently of their anthropometry, one of the most studied factors together. At the same time, we wanted to know if the physical tests are sufficiently discriminant to differentiate players of similar anthropometric characteristics, but of different playing levels. The results have clearly shown that there are significant differences $p < 0.05$ between the two groups in all the physical tests analyzed, both for the lower and upper body, except in the two-step test velocity and the shoulder internal rotation.

To the best of our knowledge, this study represents the first analysis of strength capabilities among handball players of varying levels, while controlling for age and anthropometric factors. We have sought to know the differences in two types of tests: on the one hand the SPT, close to the specific skills related to performance and the player's ability to apply force, for this purpose we chose the throwing velocity and the 30 m speed which are the most used by coaches and researchers and on the other hand we applied the FDST, with the standing shoulder internal rotation and the two-step test for the upper and lower body, respectively, these tests were chosen because they are the most related to the gestures of competition.

After the analysis of the results it is clear that a battery combining the SPT and the FDST, is useful to facilitate the discrimination of elite players from sub-elite players, which gives a basis for identifying talent [18,19,28,37]. It was known that anthropometry is a factor that plays in favor of performance in handball players in important gestures such as throwing and displacement and that it is also a differentiating factor between levels of play [19,26,27]. However, to date, it was unknown what happened in force production capacity and its application to gestures close to those of competition in players of similar anthropometric characteristics and age. The selected variables enable discrimination between an elite group and another. Based on these variables, it is more likely to predict that an individual is part of the elite group when they have high values in stronger, faster, throw, and speed. Furthermore, these differences could be attributed to the more qualitative and targeted training undertaken by high-level athletes, although the specific quantification of these values was not conducted in this study, which would be an area of significant interest for future research. The discriminant analysis confirms these findings as well. Our research is in line with recent work carried out in soccer, where it has been shown that players of different competitive levels (elite and non-elite), with similar anthropometric conditions had very different physical performances [29].

Partially analyzing the results, as the most outstanding data, indicates that in the SPT, the two tests applied to the selected players were significantly superior to the amateur group ($p < 0.001$), over 20% faster in the speed tests and 21% in the throwing tests. These results are in line with other investigations with similar objectives to ours, [28] for example, found large differences between categories of play in the German handball league (professional and amateur level) in the assessment of throwing speed, concluding that strength, power and throwing speed are important and discriminating factors in professional handball. Similar results have been given in other studies related either to throwing ability or to speed, but differences in the type of sample or in the test battery applied make direct comparison difficult [11,15,19,26].

It can be suggested after the analysis of the related literature and with the results of our study that in the selection of players, regardless of age, sex and type of competition travel speed tests and throwing tests, such as the ones performed here, are tests that allow differentiating levels of play and therefore should be used by coaches to know their players [11,15,19,28]. The reasons for these differences may be diverse, as they may be related to intra- and intermuscular synchronization, to greater motor coordination on the part of the players, to maturity itself, etc., which clearly should be a reason to study for new research to shed light in this field.

On the other hand, analyzing the effects of the tests related to the FDST, it should be noted that, in this case, only the power in the two-step test shows significant differences between groups. This is a test clearly related to the measurement of force production capacity in a basic handball gesture that is used both in attack and defense and that allows direct measurement of specific displacement power thanks to the use of the DEMF. What is relevant in our results is that the elite players almost doubled their amateur peers (68% difference), maintaining practically the same execution velocities (two-step test—velocity = 2.65 m/s). Since it is a free gesture similar to the one used in the game in an acceleration of unmarking or in a defensive action, if more force is produced at the same velocity, it is clearly indicating that the player can apply more power. To our knowledge, this is the first study where the DEMF has been applied to FDSTs, which makes it very difficult to compare with other works performed in handball, since these usually use generic tests, such as bench press or squats and other technological measurement devices [38,39]. Regardless of the type of test, if we only take into consideration force production, the results coincide with other investigations that see large differences between 20 and 40% between elite and amateurs in the ability to manifest force [19,31,38,40].

The second objective of this research was to test which tests could be more determinant to differentiate selectable players from non-selectable players. This analysis, in addition to being novel for junior players, is important because it allows coaches, if necessary, to reduce the number of tests to be applied or to prioritize the result of some tests over others when establishing a test battery. Especially in junior players, as in this case, that although they are still in a training process, their level of biological maturation is high and therefore anthropometric factors will not undergo major changes or be affected by age and maturity as happens in other lower categories [41,42].

After the analysis, being very cautious due to the size of the sample, we can say that from the battery of tests applied, the tests that clearly discriminate the players are the 3PT, being the 10 m speed the most discriminating, followed by the 20 m speed and the throwing velocity (the coefficient of the standardized canonical discriminant function was −1.622 for 10 m speed, −1.119 for 20 m speed and 0.750 for throwing velocity). It should be noted that the overall accuracy of the three variables of the two classification models used was 100%, having a small sample it was decided to refute the case model with the cross-validation model, as can be seen the results are convincing when it comes to discriminating the players. It is likely that the large differences in the level of the selected sample are what facilitate the discrimination.

Data on discriminant variables in handball players are scarce, so it is relatively difficult to compare the results—there is little research and the variables studied for discriminant analysis and the type of samples differ from our study [26,32,43,44]. In a similar study in which they included part of the variables studied here, but in younger players and therefore exposed to clearer maturational processes, they also concluded that throwing velocity had the strongest relationship with the discriminant function (Wilk's λ = 0.502, χ^2 = 57.21, p < 0.001); however, displacement did not come out as a discriminant factor in this function [22]—the difference with our work may be in the type of sample analyzed, higher biological maturation and the influence of sex.

Although as we have commented that it is not possible to make direct comparisons when discriminating performance factors that can influence the selection of players, due to the disparity of sample and procedures, in light of the literature reviewed, it is evident that throwing and/or pitching are discriminating factors between players of different levels or sex and can be used as evidence for this purpose and this is undoubtedly an important requirement to be among the elite [21,38,39,41,45,46]. It also appears that conditioning variables that have to do with the application of force, such as throwing and running, are more discriminant for boys than for girls [14,46].

To conclude, we agree with [18], who evidenced that most of the batteries and tests that are applied to collective sports players are conducted outside of real game situations, thus eliminating possible sources of very valid information to really achieve the objective of facilitating the discrimination of players. It would be of interest to further expand this line of research by incorporating control variables within the actual competition setting. The utilization of micro-technology sensors [4–6] already enables the possibility of employing discriminant analysis to select players based on factors such as displacement speed, accelerations, and throwing speed, which are crucial indicators of performance. These variables could effectively discriminate between different levels of play, even when keeping anthropometric values constant, as demonstrated in this particular case.. More studies along these lines are needed to facilitate and clarify these talent selection processes, using the fewest number of variables.

5. Conclusions

With this study, it is demonstrated, with a small sample, but of great sporting level, that not only is anthropometry the key to reaching elite levels, but other factors such as the ability to manifest strength in specific skills will play a factor, concluding that:

- There are clear significant differences in the ability to produce and apply force in the throwing velocity, the 30 m speed and the two-step test between elite and amateur players of the same age and anthropometry.
- The tests that discriminate junior elite handball players are the SPT, with the 10 m speed the most discriminating, followed by the 20 m speed and the throwing velocity.

6. Practical Applications

Our research has shown that the use of specific strength assessment tests, the FDST, together with field tests, the SPT, is useful for the trainer or researcher to know the operational status of the athlete at a given time without deviating too much from the specific preparation objective, since these tests are suitable to evaluate and improve performance at the same time. In this research, FDSTs could be applied thanks to the use of the DEMF, a new device that allows evaluating and training strength in natural conditions [34,47–50].

Based on these data and the studies analyzed, we can indicate that players with high speeds over short distances, no more than 20 m, and with power in the throw will have an advantage over the rest. Undoubtedly, these are going to be key factors to reach higher levels of performance in handball and therefore should be considered when choosing a player and designing training sessions. Our findings provide information for the assessment and evaluation of talents, allow differentiating players of the same anthropometry and age, and indicate that the batteries that combine strength production, especially with gestures similar to those of competition, and field tests are useful to facilitate the decision to choose players.

A more concrete contribution is that it is possible to discriminate the level of the elite handball player according to his maximum speed in the first 10 or 20 m as well as his throwing ability. These data will allow those in charge of selecting players in the final stages of maturation, such as juniors, to focus their attention on explosive tests as close as possible to the competitive game and to reduce, if necessary, the number of tests to be applied.

7. Limitations

The main limitation of this study is the size of the sample, and this limitation has been caused by the type of sample we have sought—elite players of the highest level compared with their not so successful peers. Therefore, these data should be interpreted with caution and in comparison with similar research.

Despite the inherent methodological limitations, the use of a cross-sectional comparative study, when using a top-level sample, can have important implications in the sporting arena, guiding coaches in their daily work and providing scientists with meaningful information to develop future research.

Author Contributions: Conceptualization, L.J.C.-R. and I.J.C.-R.; Methodology; Formal analysis, J.F.V.-V.; Data curation, Y.R.-M. and J.F.V.-V.; Writing—original draft, L.J.C.-R.; Writing—review & editing, I.J.C.-R.; Supervision, I.M.-M. and J.F.V.-V. All authors have read and agreed to the published version of the manuscript.

Funding: This study was supported by FEDER/ Ministry of Science, Innovation and Universities—State Research Agency (Dossier number: RTI2018-099723-B-I00).

Institutional Review Board Statement: This study was approved by the Ethics Committee of the University of Granada, Spain, and was conducted in accordance with the requirements established in the Declaration of Helsinki (2/2020).

Informed Consent Statement: Written informed consent has been obtained from the participant(s) to publish this paper.

Data Availability Statement: Data is unavailable due to privacy or ethical restrictions of the Spanish Handball Federation.

Conflicts of Interest: The authors declare no conflict of interest.

References

1. Chelly, M.S.; Ghenem, M.A.; Abid, K.; Hermassi, S.; Tabka, Z.; Shephard, R.J. Effects of in-season short-term plyometric training program on leg power, jump-and sprint performance of soccer players. *J. Strength Cond. Res.* **2010**, *24*, 2670–2676. [CrossRef]
2. Granados, C.; Izquierdo, M.; Ibáñez, J.; Ruesta, M.; Gorostiaga, E.M. Are there any differences in physical fitness and throwing velocity between national and international elite female handball players? *J. Strength Cond. Res.* **2013**, *27*, 723–732. [CrossRef] [PubMed]
3. Hermassi, S.; Van Den Tillaar, R.; Khlifa, R.; Chelly, M.S.; Chamari, K. Comparison of In-Season-Specific Resistance vs. A Regular Throwing Training Program on Throwing Velocity, Anthropometry, and Power Performance in Elite Handball Players. *J. Strength Cond. Res.* **2015**, *29*, 2105–2114. [CrossRef] [PubMed]
4. Manchado, C.; Martínez, J.T.; Pueo, B.; Tormo, J.M.C.; Vila, H.; Ferragut, C.; Sánchez, F.S.; Busquier, S.; Amat, S.; Ríos, L.J.C. High-performance handball player's time-motion analysis by playing positions. *Int. J. Environ. Res. Public Health* **2020**, *17*, 6768. [CrossRef] [PubMed]
5. Manchado, C.; Pueo, B.; Chirosa-Rios, L.J.; Tortosa-Martínez, J. Time-motion analysis by playing positions of male handball players during the european championship 2020. *Int. J. Environ. Res. Public Health* **2021**, *18*, 2787. [CrossRef]
6. Pueo, B.; Tortosa-Martínez, J.; Chirosa-Rios, L.J.; Manchado, C. Throwing performance by playing positions of male handball players during the European Championship 2020. *Scand. J. Med. Sci. Sport.* **2022**, *32*, 588–597. [CrossRef]
7. Michalsik, L.B.; Aagaard, P.; Madsen, K. Locomotion characteristics and match-induced impairments in physical performance in male elite team handball players. *Int. J. Sport. Med.* **2013**, *34*, 590–599. [CrossRef]
8. Wagner, H.; Gierlinger, M.; Adzamija, N.; Ajayi, S.; Bacharach, D.W.; Von Duvillard, S.P. Specific physical training in elite male team handball. *J. Strength Cond. Res.* **2017**, *31*, 3083–3093. [CrossRef]
9. Aguilar-martínez, D.; Chirosa-Ríos, L.J.; Martín-Tamayo, I.; Chirosa-Rios, I.J.; Cuadrado-Reyes, J. Efecto del entrenamiento de la potencia sobre la velocidad de lanzamiento en balonmano/Effect of power training in throwing velocity in team handball. *Rev. Int. Med. En Cienc. La Act. Física Y Del Deport.* **2012**, *12*, 729–744.
10. Chelly, M.S.; Hermassi, S.; Shephard, R.J. Relationships between power and strength of the upper and lower limb muscles and throwing velocity in male handball players. *J. Strength Cond. Res.* **2010**, *24*, 1480–1487. [CrossRef]
11. Hammami, M.; Gaamouri, N.; Ramirez-Campillo, R.; Shephard, R.J.; Bragazzi, N.L.; Chelly, M.S.; Knechtle, B.; Gaied, S. Effects of high-intensity interval training and plyometric exercise on the physical fitness of junior male handball players. *Eur. Rev. Med. Pharmacol. Sci.* **2021**, *25*, 7380–7389. [CrossRef] [PubMed]
12. Hermassi, S.; van den Tillaar, R.; Bragazzi, N.L.; Schwesig, R. The Associations Between Physical Performance and Anthropometric Characteristics in Obese and Non-obese Schoolchild Handball Players. *Front. Physiol.* **2021**, *11*, 580991. [CrossRef]
13. Bautista, I.J.; Chirosa, I.J.; Robinson, J.E.; Van Der Tillaar, R.; Chirosa, L.J.; Martín, I.M. A new physical performance classification system for elite handball players: Cluster analysis. *J. Hum. Kinet.* **2016**, *50*, 131–142. [CrossRef]
14. Saavedra, J.M.; Halldórsson, K.; Porgeirsson, S.; Einarsson, I.; Guðmundsdóttir, M.L. Prediction of Handball Players' Performance on the Basis of Kinanthropometric Variables, Conditioning Abilities, and Handball Skills. *J. Hum. Kinet.* **2020**, *73*, 229–239. [CrossRef] [PubMed]
15. Hammami, M.; Hermassi, S.; Gaamouri, N.; Aloui, G.; Comfort, P.; Shephard, R.J.; Chelly, M.S. Field Tests of Performance and Their Relationship to Age and Anthropometric Parameters in Adolescent Handball Players. *Front. Physiol.* **2019**, *10*, 01124. [CrossRef] [PubMed]
16. Ortega-Becerra, M.; Pareja-Blanco, F.; Jiménez-Reyes, P.; Cuadrado-Peñafiel, V.; González-Badillo, J.J. Determinant factors of physical performance and specific throwing in handball players of different ages. *J. Strength Cond. Res.* **2018**, *32*, 1778–1786. [CrossRef]
17. Nuño, A.; Chirosa, I.J.; Van Den Tillaar, R.; Guisado, R.; Martín, I.; Martinez, I.; Chirosa, L.J. Effects of fatigue on throwing performance in experienced team handball players. *J. Hum. Kinet.* **2016**, *54*, 103–113. [CrossRef] [PubMed]
18. Farley, J.B.; Stein, J.; Keogh, J.W.L.; Woods, C.T.; Milne, N. The Relationship Between Physical Fitness Qualities and Sport-Specific Technical Skills in Female, Team-Based Ball Players: A Systematic Review. *Sport. Med.—Open* **2020**, *6*, 18. [CrossRef] [PubMed]
19. Haugen, T.A.; Tønnessen, E.; Seiler, S. Physical and physiological characteristics of male handball players: Influence of playing position and competitive level. *J. Sport. Med. Phys. Fitness* **2016**, *56*, 19–26.
20. Massuça, L.; Fragoso, I. Morphological characteristics of adult male handball players considering five levels of performance and playing position. *Coll. Antropol.* **2015**, *39*, 109–118.
21. Moss, S.L.; McWhannell, N.; Michalsik, L.B.; Twist, C. Anthropometric and physical performance characteristics of top-elite, elite and non-elite youth female team handball players. *J. Sport. Sci.* **2015**, *33*, 1780–1789. [CrossRef] [PubMed]
22. Naisidou, S.; Kepesidou, M.; Kontostergiou, M.; Zapartidis, I. Differences of physical abilities between successful and less successful young female athletes. *J. Phys. Educ. Sport* **2017**, *17*, 294–299. [CrossRef]
23. Hermassi, S.; Laudner, K.; Schwesig, R. Playing level and position differences in body characteristics and physical fitness performance among male team handball players. *Front. Bioeng. Biotechnol.* **2019**, *7*, 149. [CrossRef] [PubMed]
24. Camacho-Cardenosa, A.; Camacho-Cardenosa, M.; González-Custodio, A.; Martínez-Guardado, I.; Timón, R.; Olcina, G.; Brazo-Sayavera, J. Anthropometric and Physical Performance of Youth Handball Players: The Role of the Relative Age. *Sports* **2018**, *6*, 47. [CrossRef] [PubMed]

25. Karcher, C.; Buchheit, M. On-court demands of elite handball, with special reference to playing positions. *Sport. Med.* **2014**, *44*, 797–814. [CrossRef] [PubMed]

26. Lijewski, M.; Burdukiewicz, A.; Stachoń, A.; Pietraszewska, J. Differences in anthropometric variables and muscle strength in relation to competitive level in male handball players. *PLoS ONE* **2021**, *16*, e0261141. [CrossRef]

27. Ferragut, C.; Vila, H.; Abraldes, J.A.; Manchado, C. Influence of Physical Aspects and Throwing Velocity in Opposition Situations in Top-Elite and Elite Female Handball Players. *J. Hum. Kinet.* **2018**, *63*, 23–32. [CrossRef]

28. Fieseler, G.; Hermassi, S.; Hoffmeyer, B.; Schulze, S.; Irlenbusch, L.; Bartels, T.; Delank, K.S.; Laudner, K.G.; Schwesig, R. Differences in anthropometric characteristics in relation to throwing velocity and competitive level in professional male team handball: A tool for talent profiling. *J. Sport. Med. Phys. Fit.* **2017**, *57*, 985–992. [CrossRef]

29. Tereso, D.; Paulo, R.; Petrica, J.; Duarte-Mendes, P.; Gamonales, J.M.; Ibáñez, S.J. Assessment of Body Composition, Lower Limbs Power, and Anaerobic Power of Senior Soccer Players in Portugal: Differences According to the Competitive Level. *Int. J. Environ. Res. Public Health* **2021**, *18*, 8069. [CrossRef]

30. Fernandez-Fernandez, J.; Granacher, U.; Martinez-Martin, I.; Garcia-Tormo, V.; Herrero-Molleda, A.; Barbado, D.; Garcia-Lopez, J. Physical fitness and throwing speed in U13 versus U15 male handball players. *BMC Sport. Sci. Med. Rehabil.* **2022**, *14*, 113. [CrossRef]

31. Granados, C.; Izquierdo, M.; Ibañez, J.; Bonnabau, H.; Gorostiaga, E.M. Differences in physical fitness and throwing velocity among elite and amateur female handball players. *Int. J. Sport. Med.* **2007**, *28*, 860–867. [CrossRef] [PubMed]

32. Lido, R.; Falk, B.; Arnol, M.; Cohen, M.; Segal, G.; Lander, Y. Measuremet of talent in team handball. *J. Strength Cond. Res.* **2005**, *19*, 318–325. [CrossRef]

33. Wagner, H.; Fuchs, P.; Fusco, A.; Fuchs, P.; Bell, J.W.; von Duvillard, S.P. Physical Performance in Elite Male and Female Team Handball Players. *Int. J. Sport. Physiol. Perform.* **2018**, *14*, 60–67. [CrossRef] [PubMed]

34. Rodriguez-Perea, Á.; Jerez-Mayorga, D.; García-Ramos, A.; Martínez-García, D.; Chirosa-Ríos, L.J. Reliability and concurrent validity of a functional electromechanical dynamometer device for the assessment of movement velocity. *J. Sport. Eng. Technol.* **2021**, *235*, 1754337120984883. [CrossRef]

35. Martínez-García, D.; Rodríguez-Perea, Á.; Huerta-Ojeda, Á.; Jerez-Mayorga, D.; Aguilar-Martínez, D.; Chirosa-Rios, I.; Ruiz-Fuentes, P.; Chirosa-Rios, L.J. Effects of Pre-Activation with Variable Intra-Repetition Resistance on Throwing Velocity in Female Handball Players: A Methodological Proposal. *J. Hum. Kinet.* **2021**, *77*, 235–244. [CrossRef]

36. Chirosa-Ríos, I.; Ruiz-Orellana, L.; Jerez-Mayorga, D.; Chirosa-Ríos, L.; del-Cuerpo, I.; Martínez-Martín, I.; Rodríguez-Perea, Á.; Pelayo-Tejo, I.; Martinez-Garcia, D. Defensive Two-Step Test in Handball Players: Reliability of a New Test for Assessing Displacement Velocity [Prueba defensiva de dos pasos en jugadores de balonmano: Fiabilidad de una nueva prueba para evaluar la velocidad de desplazamiento]. *E-Balonmano.com Rev. Cienc. Deporte* **2022**, *18*, 233–244.

37. Renshaw, I.; Chow, J.Y. A constraint-led approach to sport and physical education pedagogy. *Phys. Educ. Sport Pedagog.* **2018**, *24*, 103–116. [CrossRef]

38. Gorostiaga, E.M.; Granados, C.; Ibáñez, J.; Izquierdo, M. Differences in physical fitness and throwing velocity among elite and amateur male handball players. *Int. J. Sport. Med.* **2005**, *26*, 225–232. [CrossRef]

39. Pereira, L.A.; Nimphius, S.; Kobal, R.; Kitamura, K.; Turisco, L.A.L.; Orsi, R.C.; Abad, C.C.C.; Loturco, I. Relationship Between Change of Direction, Speed, and Power in Male and Female National Olympic Team Handball Athletes. *J. Strength Cond. Res.* **2018**, *32*, 2987–2994. [CrossRef]

40. Pereira, M.I.R.; Gomes, P.S.C. Effects of isotonic resistance training at two movement velocities on strength gains. *Rev. Bras. Med. Esporte* **2007**, *13*, 91–96. [CrossRef]

41. Koopmann, T.; Lath, F.; Büsch, D.; Schorer, J. Predictive Value of Technical Throwing Skills on Nomination Status in Youth and Long-Term Career Attainment in Handball. *Sport. Med.—Open* **2022**, *8*, 6. [CrossRef] [PubMed]

42. Tuquet, J.; Zapardiel, J.C.; Saavedra, J.M.; Jaén-Carrillo, D.; Lozano, D. Relationship between Anthropometric Parameters and Throwing Speed in Amateur Male Handball Players at Different Ages. *Int. J. Environ. Res. Public Health* **2020**, *17*, 7022. [CrossRef] [PubMed]

43. Mohamed, H.; Vaeyens, R.; Matthys, S.; Multael, M.; Lefevre, J.; Lenoir, M.; Philppaerts, R. Anthropometric and performance measures for the development of a talent detection and identification model in youth handball. *J. Sport. Sci.* **2009**, *27*, 257–266. [CrossRef] [PubMed]

44. Nikolaidis, P.T.; Ingebrigtsen, J. Physical and Physiological Characteristics of Elite Male Handball Players from Teams with a Different Ranking. *J. Hum. Kinet.* **2013**, *38*, 115. [CrossRef]

45. Palamas, A.; Zapartidis, I.; Kidou, Z.K.; Tsakalou, L.; Natsis, P.; Kokaridas, D. The Use of Anthropometric and Skill Data to Identify Talented Adolescent Team Handball Athletes. *J. Phys. Educ. Sport. Manag.* **2015**, *2*, 174–183. [CrossRef]

46. Wagner, H.; Sperl, B.; Bell, J.W.; Von Duvillard, S.P. Testing Specific Physical Performance in Male Team Handball Players and the Relationship to General Tests in Team Sports. *J. Strength Cond. Res.* **2019**, *33*, 1056–1064. [CrossRef]

47. Jerez-Mayorga, D.; Huerta-Ojeda, Á.; Chirosa-Ríos, L.J.; Guede-Rojas, F.; Guzmán-Guzmán, I.P.; Intelangelo, L.; Miranda-Fuentes, C.; Delgado-Floody, P. Test–Retest Reliability of Functional Electromechanical Dynamometer on Five Sit-to-Stand Measures in Healthy Young Adults. *Int. J. Environ. Res. Public Health* **2021**, *18*, 6829. [CrossRef]

48. Reyes-Ferrada, W.; Chirosa-Rios, L.; Martinez-Garcia, D.; Rodríguez-Perea, Á.; Jerez-Mayorga, D. Reliability of trunk strength measurements with an isokinetic dynamometer in non-specific low back pain patients: A systematic review. *J. Back Musculoskelet. Rehabil.* **2022**, *35*, 937–948. [CrossRef]

49. Reyes-Ferrada, W.; Rodríguez-Perea, Á.; Chirosa-Ríos, L.; Martínez-García, D.; Jerez-Mayorga, D. Muscle Quality and Functional and Conventional Ratios of Trunk Strength in Young Healthy Subjects: A Pilot Study. *Int. J. Environ. Res. Public Health* **2022**, *19*, 2673. [CrossRef]

50. Sánchez-Sánchez, A.J.; Chirosa-Ríos, L.J.; Chirosa-Ríos, I.J.; García-Vega, A.J.; Jerez-Mayorga, D. Test-retest reliability of a functional electromechanical dynamometer on swing eccentric hamstring exercise measures in soccer players. *PeerJ* **2021**, *9*, e11743. [CrossRef]

Article

Level of Agreement between the MotionMetrix System and an Optoelectronic Motion Capture System for Walking and Running Gait Measurements

Diego Jaén-Carrillo [1], Felipe García-Pinillos [2,3,4,*], José M. Chicano-Gutiérrez [3], Alejandro Pérez-Castilla [5,6], Víctor Soto-Hermoso [2,3], Alejandro Molina-Molina [7] and Santiago A. Ruiz-Alias [2,3]

[1] Department of Sport Science, Universität Innsbruck, 6020 Innsbruck, Austria
[2] Department of Physical Education and Sports, Faculty of Sport Sciences, University of Granada, 18016 Granada, Spain
[3] Sport and Health University Research Institute (iMUDS), University of Granada, 18007 Granada, Spain
[4] Department of Physical Education, Sports and Recreation, Universidad de La Frontera, Temuco 1145, Chile
[5] Department of Education, Faculty of Education Sciences, University of Almería, 04120 Almería, Spain
[6] SPORT Research Group (CTS-1024), CERNEP Research Center, University of Almería, 04120 Almería, Spain
[7] Department of Physiotherapy, Universidad San Jorge, 50830 Zaragoza, Spain
* Correspondence: fgpinillos@ugr.es; Tel.: +34-660062066

Citation: Jaén-Carrillo, D.; García-Pinillos, F.; Chicano-Gutiérrez, J.M.; Pérez-Castilla, A.; Soto-Hermoso, V.; Molina-Molina, A.; Ruiz-Alias, S.A. Level of Agreement between the MotionMetrix System and an Optoelectronic Motion Capture System for Walking and Running Gait Measurements. *Sensors* 2023, 23, 4576. https://doi.org/10.3390/s23104576

Academic Editors: Edward Sazonov, Michael Beigl and Roozbeh Ghaffari

Received: 23 February 2023
Revised: 31 March 2023
Accepted: 5 May 2023
Published: 9 May 2023

Abstract: Markerless motion capture systems (MCS) have been developed as an alternative solution to overcome the limitations of 3D MCS as they provide a more practical and efficient setup process given, among other factors, the lack of sensors attached to the body. However, this might affect the accuracy of the measures recorded. Thus, this study is aimed at evaluating the level of agreement between a markerless MSC (i.e., MotionMetrix) and an optoelectronic MCS (i.e., Qualisys). For such purpose, 24 healthy young adults were assessed for walking (at 5 km/h) and running (at 10 and 15 km/h) in a single session. The parameters obtained from MotionMetrix and Qualisys were tested in terms of level of agreement. When walking at 5 km/h, the MotionMetrix system significantly underestimated the stance and swing phases, as well as the load and pre-swing phases ($p < 0.05$) reporting also relatively low systematic bias (i.e., ≤ -0.03 s) and standard error of the estimate (SEE) (i.e., ≤ 0.02 s). The level of agreement between measurements was perfect (r > 0.9) for step length left and cadence and very large (r > 0.7) for step time left, gait cycle, and stride length. Regarding running at 10 km/h, bias and SEE analysis revealed significant differences for most of the variables except for stride time, rate and length, swing knee flexion for both legs, and thigh flexion left. The level of agreement between measurements was very large (r > 0.7) for stride time and rate, stride length, and vertical displacement. At 15 km/h, bias and SEE revealed significant differences for vertical displacement, landing knee flexion for both legs, stance knee flexion left, thigh flexion, and extension for both legs. The level of agreement between measurements in running at 15 km/h was almost perfect (r > 0.9) when comparing Qualisys and MotionMetrix parameters for stride time and rate, and stride length. The agreement between the two motion capture systems varied for different variables and speeds of locomotion, with some variables demonstrating high agreement while others showed poor agreement. Nonetheless, the findings presented here suggest that the MotionMetrix system is a promising option for sports practitioners and clinicians interested in measuring gait variables, particularly in the contexts examined in the study.

Keywords: analysis; biomechanics; gait; markerless; motion; reliability

1. Introduction

There are various technologies for analyzing gait and related parameters, including 3D motion capture systems (MCS), image processing, and wearable sensors [1–3]. MCS use various methods such as optical, magnetic, or inertial sensors to measure the position and orientation of markers placed on the objects or individuals. In the field of human biomechanics, the tracking of three-dimensional (3D) motion is usually accomplished through the attachment of retro-reflective markers to participants, which are then monitored using infrared cameras. This technique can detect marker locations with high precision, down to sub-millimeter levels within a specific capture volume. It is widely regarded as the closest alternative to the highly precise fluoroscopy method, which is considered the gold standard in this field [1]. The resulting data is used to create a digital representation of movement and can be used in various fields such as biomechanics, sports science, animation, or in clinical settings [?]. However, commercially available systems for this analysis often have limitations that make their practicality challenging.

The use of marker-based MCS has become widespread in both the research and diagnosis fields, but its application in places like patient homes, sports fields, and public spaces is limited by the need for very specific recording settings (i.e., number, placement, and features of cameras, calibration process, reflected markers attached to the individual's body, and data analysis). Motion capture aims to measure the movement of the skeletal system. However, due to factors such as clothing, skin, and soft tissue movement, the retro-reflective markers placed on key bony landmarks like the lateral malleolus may move in relation to the underlying bones [3]. Various approaches have been employed to reduce this error, but they cannot completely eliminate it [4]. Therefore, to minimize this error source, participants are usually instructed to wear tight or minimal clothing. Additionally, during long data collection periods or when participants sweat excessively (e.g., during a maximal aerobic capacity running test), the likelihood of markers moving or falling off increases, which can compromise data accuracy.

The practicality and clinical adoption of marker-based motion capture is hindered by the considerable amount of time it takes to securely fix markers to participants. Even with a skilled researcher, this process can take between 20 to 30 min, not including the time needed to prepare the markers before data collection. Additionally, the presence of the markers on the participant's body or clothing may make them aware of their presence, possibly affecting their natural movement.

Markerless MCS share similarities with marker-based systems as they aim to accurately and reliably measure the 3D motion of human segments. Both systems aim for the representation of the skeletal system with a simplified biomechanical model. Nonetheless, Markerless systems do not require markers to be placed on the participant, instead using synchronized 2D video cameras to achieve a 3D reconstruction. Markerless MCS have been developed as an alternative solution to overcome the limitations of marker-based MCS [5–9]. These systems eliminate the need for markers, providing a time-efficient and user-friendly option for capturing human motion data. Unlike marker-based MCS, markerless systems rely on image-based tracking and machine-learning algorithms to estimate the movement of body joints and segments [5–9]. This provides a more natural experience for the subject being recorded, as well as a more practical and less time-consuming setting-up process for the practitioner. Recent technological advancements in computational speed and a growing commercial demand for markerless MCS have led to the development of several software packages that are available for purchase, such as Theia3D [10], DariMotion, and MotionMetrix. Additionally, some markerless software packages, such as OpenPose [9] and OpenCap, are available as open-source software [11].

One commercially available markerless system is the MotionMetrix software (MotionMetrix AB, Lidingö, Sweden). The reliability of the system has been tested by evaluating the test-retest results of all variables during walking (at 5 km/h) and running (at 10 and 15 km/h), and the results suggest that the system may be useful in settings where more sophisticated systems (such as 3D MCS) are not accessible, taking into account the limitations of the software itself (i.e., the reliability of some parameters may vary depending on walking and running velocity) [12]. Additionally, the agreement between spatio-temporal parameters measured by MotionMetrix to high-speed videoanalysis and OptoGait measurements has been previously assessed [13]. Valid measures for step frequency and step length were reported, but the MotionMetrix system tended to overestimate contact time (CT) and underestimate flight time (FT) [13].

Despite its popularity among researchers and widespread use by clinicians and sports practitioners, there is still doubt about the accuracy and consistency of the MotionMetrix system compared to a recognized standard 3D MCS. Therefore, this study aims to evaluate the level of agreement between the MotionMetrix system and a 3D MCS (Qualisys AB, Göteborg, Sweden) for walking (at 5 km/h) and running (at 10 and 15 km/h) in healthy adults.

2. Materials and Methods

2.1. Participants

Twenty-four recreationally active young adults (16 men, 8 women; age = 22.7 $\pm$ 2.6 years; body height = 1.72 $\pm$ 0.10 m; body mass = 69.1 $\pm$ 11.7 kg; weekly training = 6.9 $\pm$ 2.4 h/week) [14] were included in the study. They declared to participate voluntarily in the study and being familiar with running on a treadmill and free from injuries and health problems. Each participant signed an informed consent form after being informed of the study's objectives and procedures, and it was made clear that they could leave at any time. The study was conducted in accordance with the Declaration of Helsinki (2013) and was approved by the local university's Ethics Board (No. 2546/CEIH/2022).

2.2. Procedures

Subjects were asked to avoid any strenuous physical activity for at least 48 h before the data collection and came to the laboratory only once. During the test, they wore their typical running clothes and shoes and went through a walking and running protocol on a treadmill (WOODWAY Pro XL, Woodway, Inc., Waukesha, WI, USA) [12]. Before data collection, subjects had a minimum 8 min accommodation period on the treadmill at a self-selected pace [15]. Immediately after, the subjects went through a protocol where they walked and ran for one minute at speeds of 5, 10, and 15 km/h [12]. Data were only collected during the last 30 s of each bout to ensure the subjects had adjusted to the speed.

2.3. Materials and Testing

The body height (m) and body mass (kg) of each subject were measured employing a stadiometer (SECA 222, SECA, Corp., Hamburg, Germany) and a bioimpedance scale (Inbody 230, Inbody, Seoul, Republic of Korea), respectively.

The MotionMetrix system was used in combination with two Kinect sensors (version 1.0, Microsoft, Washington, DC, USA) positioned on either side of the treadmill following a specific configuration (see Figure 1) as suggested by the manufacturer. The MotionMetrix system calculates various kinetic parameters based on the movement being analyzed (i.e., walking or running gait) [12]. The Kinect sensors, which have a depth sensor, allow for monitoring of 3D movements by recognizing 20 body joints in 3D space at a rate of 30 Hz. When both sensors track the same point simultaneously, this rate increases to 60 Hz. To ensure accurate data collection, the manufacturer's guidelines were followed, including software calibration, wearing fitted clothing without shiny black fabric or reflective surfaces, securing shoelaces, tucking away hair, avoiding direct sunlight, and ensuring that treadmill parts do not obstruct the subject's entire view. For a full description of how the MotionMetrix software provides measurements of the variables, the reader is pointed toward a previous study [12].

Eight Qualisys Oqus cameras (Qualisys AB, Gothenburg, Sweden) operating at 250 Hz and meticulously positioned to allow a full view of the treadmill (Figure 1) were employed for 3D motion capture analysis as the measure of reference. Before collecting data, the testing area was properly calibrated using a dynamic wand. After, 40 retro-reflective markers were attached to the subjects' bodies to track 3D movement (Figure 2). Markers were placed on specific anatomical locations such as the right/left ilium crest tubercle, right/left posterior superior iliac spine, right/left femur greater trochanter, right/left anterior superior iliac spine, right/left femur lateral epicondyle, right/left femur medial epicondyle, right/left fibula apex of the lateral malleolus, right/left tibia apex of the medial malleolus, right/left head of the fifth metatarsals, right/left head of the first metatarsus, and right/left posterior surface of the calcaneus. Additionally, two marker sets were attached to the thigh and shank. After static calibration, the subjects performed the accommodation period to the treadmill and the entire protocol described above.

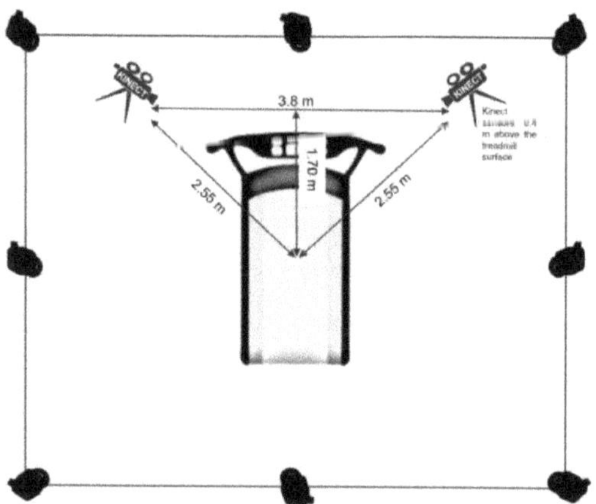

Figure 1. Diagram of the protocol setting for data acquisition displaying Kinect sensors and Qualisys cameras placement.

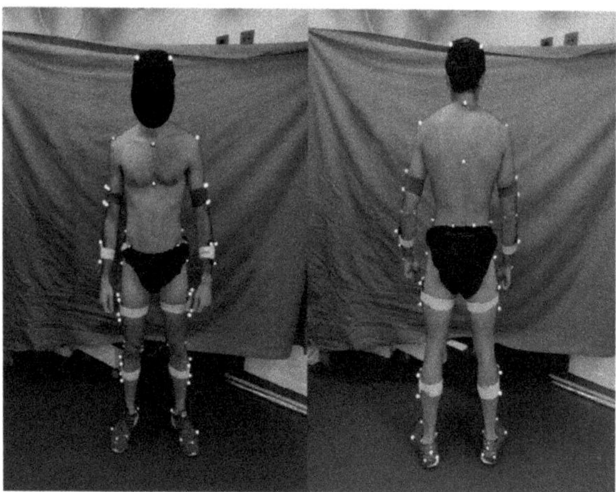

Figure 2. Marker set for data collection.

Visual 3D software (C-Motion Inc., Germantown, MD, USA) was used to process static and kinetic data. The motion data was processed using a low-pass filter with a cut-off frequency of 8 Hz to eliminate high-frequency noise. Joint angles were calculated using the x-y-z Cardan sequence, which represents flexion/extension, abduction/adduction, and axial rotation, without normalizing them to a static standing position. The laboratory frame was set up using the right-hand rule, with the positive y-direction facing forward, positive x-direction to the left, and positive z-direction upward. Variables of interest included the different phases of the gait cycle (walking at 5 km/h), stride rate and length, vertical displacement, spine angle, and landing, stance, and swing knee flexion (at 10 and 15 km/h).

2.4. Statistical Analysis

Descriptive data are presented as mean (±SD) and 95% confidence intervals (CI) The normal distribution of the data and equal distribution of variance were confirmed through the Shapiro–Wilk test and Levene's test, respectively. A *t*-test pairwise mean comparison was performed to compare data obtained from MotionMetrix and Qualisys MCS. To assess the level of agreement between the two systems, Pearson's product-moment correlation coefficient (r) was calculated between each variable. The correlation between measurements was interpreted using established criteria: <0.1 (trivial), 0.1–0.3 (small), 0.3–0.5 (moderate), 0.5–0.7 (large), 0.7–0.9 (very large), 0.9–1.0 (almost perfect) [16]. Additionally, intraclass correlation coefficients (ICC) were calculated. The authors followed the guidelines outlined by Koo and Li [17] and applied a "two-way random-effects" model (ICC [2,1]), using a "single measurement" type and "absolute agreement" definition for the ICC calculation. The benchmarks from Koo and Li [17] were used to interpret the ICC results: ICC values below 0.5 indicate 'poor' reliability, values between 0.5–0.75 represent 'moderate' reliability, 0.75–0.90 indicate 'good' reliability, and values above 0.90 signify 'excellent' reliability. The level of agreement between the MotionMetrix and Qualisys MCS was also evaluated through the systematic bias and the standard error of the estimate (SEE) from linear regression analysis. Statistical analysis was performed using the software package SPSS (IBM SPSS version 25.0, Chicago, IL, USA), and a significance level of $p < 0.05$ was established.

3. Results

3.1. Walking at 5 km/h

The pairwise comparison between the data obtained from the MotionMetrix and 3D MCS revealed significant differences for most of the variables, although the systematic bias and SEE were low (Table 1).

Table 1. Descriptive data (mean ± SD), bias, Pearson's product-moment correlation coefficients (r), standard error of the estimate (SEE), and intraclass correlation coefficients (ICC [2,1]) for comparison between values obtained from MotionMetrix and Qualisys motion capture systems walking at 5 km/h.

Variable	Qualisys	MotionMetrix	Bias (95% CI) ^	r	SEE	ICC (95% CI)
Stance phase left (s)	0.68 (0.04)	0.65 (0.01)	−0.03 (−0.05 to −0.01) *	0.205	0.01	0.101 (−0.479 to 0.542)
Stance phase right (s)	0.68 (0.03)	0.65 (0.01)	−0.03 (−0.05 to −0.02) *	−0.029	0.04	−0.013 (−0.524 to 0.439)
Swing phase left (s)	0.36 (0.02)	0.35 (0.01)	−0.01 (−0.02 to 0.00) *	0.372	0.01	0.334 (−0.271 to 0.690)
Swing phase right (s)	0.36 (0.02)	0.35 (0.01)	−0.01 (−0.02 to −0.01) *	0.337	0.02	0.257 (−0.298 to 0.634)
Load response left (s)	0.16 (0.02)	0.15 (0.01)	−0.00 (−0.01 to 0.00)	0.365	0.02	0.446 (−0.290 to 0.767)
Load response right (s)	0.16 (0.01)	0.15 (0.02)	−0.01 (−0.02 to −0.00) *	0.180	0.01	0.173 (−0.347 to 0.576)
Pre-swing left (s)	0.16 (0.02)	0.15 (0.01)	−0.01 (−0.02 to 0.00) *	−0.005	0.02	−0.04 (−1.09 to 0.527)
Pre-swing right (s)	0.16 (0.01)	0.15 (0.01)	−0.01 (−0.02 to −0.00) *	0.149	0.02	0.186 (−0.581 to 0.629)
Total double support (s)	0.32 (0.03)	0.30 (0.01)	−0.02 (−0.03 to −0.00) *	0.236	0.03	0.247 (−0.440 to 0.654)
Step time left (s)	0.52 (0.02)	0.52 (0.02)	0.00 (−0.00 to 0.00)	0.920 **	0.01	0.960 (0.901 to 0.984)
Step time right (s)	0.52 (0.03)	0.52 (0.02)	0.00 (−0.01 to 0.01)	0.697 **	0.02	0.808 (0.531 to 0.922)
Gait cycle (s)	1.05 (0.04)	1.07 (0.12)	0.02 (−0.02 to 0.06)	0.785 **	0.03	0.686 (0.236 to 0.872)
Step length left (m)	0.87 (0.04)	0.73 (0.03)	−0.15 (−0.16 to −0.13) *	0.746 **	0.03	0.176 (−0.04 to 0.548)
Step length right (m)	0.57 (0.04)	0.72 (0.03)	0.15 (0.13 to 0.17) *	0.510 *	0.04	0.101 (−0.053 to 0.390)
Stride length (m)	1.44 (0.08)	1.46 (0.06)	0.01 (−0.01 to 0.03)	0.713 **	0.06	0.817 (0.563 to 0.924)
Cadence (spm)	114.5 (4.84)	114.4 (4.85)	−0.06 (−0.30 to 0.17)	0.994 **	0.56	0.997 (0.993 to 0.999)
Step width (m)	0.01 (0.00)	0.16 (0.04)	0.15 (0.13 to 0.16) *	−0.006	0.00	0.00 (−0.04 to 0.09)

%GT: percentage of gait cycle; spm: number steps per minute. ^ calculated by pairwise mean comparison (*t*-test). * $p < 0.05$. ** $p < 0.001$.

Specifically, the MotionMetrix system significantly underestimated the stance and swing phases, as well as the load and pre-swing phases ($p < 0.05$). The systematic bias was relatively low (Bias ≤ -0.03 s) as well as the SEE (≤ 0.02 s).

However, the level of agreement between measurements when walking at 5 km/h (Table 1) was perfect (r > 0.9) for step length left and cadence. Very large (r > 0.7) when comparing Qualisys and MotionMetrix parameters for step time left, gait cycle, and stride length. Large agreements (r > 0.5) were revealed for step time right and step length right.

Moreover, a moderate correlation (r > 0.4) was found for landing knee flexion for both legs. A small correlation (r > 0.1) for CT left, thigh extension left and thigh flexion right, and spine angle. Finally, a trivial correlation (r < 0.1) was found for CT right, step width, thigh flexion left, and thigh extension right. The ICCs also found a "good to excellent" association between MotionMetrix and Qualisys MCS measurements (ICCs > 0.75) for stride time, stride rate, stride length, and stance knee flexion left. Moderate agreements (ICCs > 0.642) were found for vertical displacement, stance knee flexion right, and swing knee flexion right. For the other variables, poor agreement (ICC < 0.5) was exhibited between both MCS.

3.2. Running at 10 km/h

Bias, obtained by pairwise comparison between data, and SEE for differences among MotionMetrix and Qualisys MCS revealed significant differences for most of the variables except for stride time, rate and length, swing knee flexion for both legs, and thigh flexion left (Table 2).

Table 2. Descriptive data (mean ± SD), bias, Pearson's product-moment correlation coefficients (r), standard error of the estimate (SEE), and intraclass correlation coefficients (ICC [2,1]) for comparison between values obtained from MotionMetrix and Qualisys running at 10 km/h.

Variable	Qualisys	MotionMetrix	Bias (95% CI) ^	r	SEE	ICC (95% CI)
Stride time (s)	0.73 (0.04)	0.73 (0.05)	0.00 (−0.01 to 0.02)	0.770 **	0.03	0.873 (0.694 to 0.948)
Stride rate (spm)	82.22 (4.88)	82.13 (5.28)	−0.09 (−1.72 to 1.54)	0.741 **	3.64	0.855 (0.648 to 0.940)
Stride length (m)	2.03 (0.12)	2.04 (0.13)	0.00 (−0.04 to 0.04)	0.766 **	0.08	0.872 (0.688 to 0.947)
Contact time left (s)	0.25 (0.03)	0.30(0.03)	0.05 (0.03 to 0.07) *	0.107	0.03	0.096 (−0.285 to 0.473)
Contact time right (s)	0.25 (0.04)	0.28 (0.02)	0.03 (0.01 to 0.04) *	0.010	0.02	0.013 (−0.286 to 0.359)
Step width (m)	0.03 (0.01)	0.05 (0.02)	0.03 (0.01 to 0.04) *	−0.384	0.02	−0.419 (−1.169 to 0.335)
Vertical displacement (m)	0.10 (0.01)	0.08 (0.02)	−0.02 (−0.02 to −0.01) *	0.741 **	0.01	0.642 (−0.214 to 0.882)
Landing knee flexion left (°)	18.22 (6.84)	13.39 (2.94)	−4.83 (−7.53 to −2.12) *	0.452 *	2.69	0.382 (−0.225 to 0.719)
Landing knee flexion right (°)	15.24 (6.66)	18.42 (3.34)	3.17 (0.53 to 5.82) *	0.448 *	3.06	0.475 (−0.130 to 0.771)
Stance knee flexion left (°)	41.06 (5.60)	39.09 (4.11)	−1.97 (−3.8 to −0.13) *	0.673 **	3.11	0.752 (0.401 to 0.897)
Stance knee flexion right (°)	40.23 (4.87)	44.43 (4.71)	4.20 (2.49 to 5.91) *	0.676 **	3.55	0.661 (−0.132 to 0.883)
Swing knee flexion left (°)	90.46 (10.72)	89.87 (13.50)	−0.59 (−5.16 to 3.98)	0.622 *	10.18	0.803 (0.508 to 0.920)
Swing knee flexion right (°)	90.74 (9.80)	91.83 (15.90)	1.09 (−4.97 to 7.15)	0.519 *	13.93	0.643 (0.124 to 0.853)
Thigh flexion left (°)	18.94 (3.44)	23.98 (7.99)	5.04 (1.21 to 8.86)	0.023	8.18	0.026 (−0.782 to 0.534)
Thigh extension left (°)	−12.19 (14.00)	−26.48 (3.68)	−14.29 (−20.26 to −8.31) *	0.268	3.63	0.128 (−0.297 to 0.512)
Thigh flexion right (°)	17.93 (3.90)	24.31 (7.70)	6.38 (2.73 to 10.03) *	0.112	7.84	0.113 (−0.468 to 0.546)
Thigh extension right (°)	−17.28 (3.55)	−25.97 (4.20)	−8.68 (−11.04 to −6.33) *	0.070	4.30	0.039 (−0.169 to 0.328)
Spine angle (°)	4.11 (2.11)	6.99 (2.12)	2.88 (1.66 to 4.09) *	0.165	2.15	0.161 (−0.286 to 0.544)

Stpm: number of strides per minute; ^ calculated by pairwise mean comparison (*t*-test). * $p < 0.05$. ** $p < 0.001$.

The level of agreement between measurements when running at 10 km/h (Table 2) was very large (r > 0.7) when comparing Qualisys and MotionMetrix parameters for stride time and rate, stride length, and vertical displacement. Large agreements (r > 0.5) were revealed for stance knee flexion and swing knee flexion for both legs. Moreover, a moderate correlation (r > 0.4) was found for landing knee flexion for both legs. A small correlation (r > 0.1) for CT left, thigh extension left and thigh flexion right, and spine angle. Finally, a trivial correlation (r < 0.1) was found for CT right, step width, thigh flexion left, and thigh extension right. The ICCs found a good association between MotionMetrix and Qualisys MCS measurements (ICCs > 0.75) for stride time, stride rate, stride length, and stance knee flexion left. Moderate agreements (ICCs > 0.642) were found for vertical displacement, stance knee flexion right, and swing knee flexion right. For the other variables, poor agreement (ICC < 0.5) was exhibited between both MCS.

3.3. Running at 15 km/h

Bias and SEE obtained by pairwise comparison between MotionMetrix and Qualisys MCS revealed significant differences for vertical displacement, landing knee flexion for both legs, stance knee flexion left, thigh flexion, and extension for both legs (Table 3).

Table 3. Descriptive data (mean ± SD), bias, Pearson's product-moment correlation coefficients (r), standard error of the estimate (SEE), and intraclass correlation coefficients (ICC [2,1]) for comparison between values obtained from MotionMetrix and Qualisys running at 15 km/h.

Variable	Qualisys	MotionMetrix	Bias (95% CI) ^	r	SEE	ICC (95% CI)
Stride time (s)	0.67 (0.05)	0.67 (0.05)	0.00 (−0.00 to 0.00)	0.994 **	0.00	0.997 (0.993 to 0.999)
Stride rate (spm)	90.05 (6.50)	89.93 (6.50)	−0.11 (−0.40 to 0.17)	0.995 **	0.65	0.998 (0.994 to 0.999)
Stride length (m)	2.79 (0.21)	2.79 (0.20)	0.00 (−0.01 to 0.01)	0.994 **	0.02	0.997 (0.993 to 0.999)
Contact time left (s)	0.21 (0.02)	0.23 (0.01)	0.02 (0.01 to 0.03) *	0.387	0.01	0.313 (−0.251 to 0.673)
Contact time right (s)	0.21 (0.02)	0.22 (0.01)	0.02 (0.01 to 0.02) *	0.373	0.01	0.337 (−0.264 to 0.691)
Step width (m)	0.03 (0.02)	0.05 (0.03)	0.01 (0.00 to 0.03)	−0.533 *	0.02	−1.580 (−5.508 to −0.031)
Vertical displacement (m)	0.09 (0.01)	0.07 (0.02)	−0.02 (−0.01 to 0.03) *	0.747 **	0.01	0.617 (−0.238 to 0.877)
Landing knee flexion left (°)	18.41 (8.40)	14.31 (3.59)	−4.10 (−7.75 to −0.44) *	0.256	3.56	0.274 (−0.478 to 0.674)
Landing knee flexion right (°)	17.17 (8.40)	18.05 (2.91)	0.87 (−2.83 to 4.57) *	0.192	2.93	0.218 (−0.957 to 0.680)
Stance knee flexion left (°)	42.40 (5.37)	37.97 (4.73)	−4.42 (−6.34 to −2.50) *	0.639 **	3.73	0.634 (−0.116 to 0.868)
Stance knee flexion right (°)	41.51 (4.49)	42.93 (4.66)	1.42 (−0.43 to 3.27)	0.586 *	3.87	0.725 (0.359 to 0.884)
Swing knee flexion left (°)	112.93 (13.0)	112.42 (12.81)	−0.51 (−3.72 to 2.71)	0.842 **	3.87	0.918 (0.801 to 0.966)
Swing knee flexion right (°)	112.91 (13.08)	111.67 (10.59)	−1.24 (−5.18 to 2.70)	0.737 **	7.33	0.842 (0.620 to 0.934)
Thigh flexion left (°)	21.25 (10.92)	32.38 (5.88)	11.03 (6.41 to 15.64) *	0.354	5.63	0.288 (−0.250 to 0.653)
Thigh extension left (°)	−21.62 (10.35)	−34.57 (3.53)	−12.95 (−17.64 to −8.25) *	0.102	3.60	0.052 (−0.272 to 0.412)
Thigh flexion right (°)	22.66 (3.85)	34.03 (6.00)	11.37 (8.56 to 14.19) *	0.229	5.99	0.112 (−0.149 to 0.430)
Thigh extension right (°)	−20.75 (10.72)	−35.45 (4.49)	−14.69 (−20.09 to −9.29) *	−0.134	4.56	−0.078 (−0.424 to 0.322)
Spine angle (°)	4.69 (2.56)	6.72 (3.45)	2.03 (−0.11 to 4.17)	−0.277	3.40	−0.588 (−2.364 to 0.303)

Stpm: number of strides per minute; ^ calculated by pairwise mean comparison (*t*-test). * $p < 0.05$. ** $p < 0.001$.

The level of agreement between measurements in running at 15 km/h (Table 3) was almost perfect (r > 0.9) when comparing Qualisys and MotionMetrix parameters for stride time and rate and stride length. Very large agreements (r > 0.7) were revealed for vertical displacement and swing knee flexion for both legs. In addition, large correlations (r > 0.5) were found in the consistency between measurements for stance knee flexion in both legs. Moreover, a moderate correlation (r > 0.35) was found for CT for both legs and thigh flexion left. A small correlation (r > 0.1) for landing knee flexion for both legs, thigh extension left and thigh flexion right was revealed. Finally, a trivial correlation (r < 0.1) was found for step width, thigh extension right, and spine angle. The ICCs also found a "good to excellent" association between MotionMetrix and Qualisys MCS measurements (ICCs > 0.84) for Swing knee flexion for both legs, stride time, stride rate, and stride length. Moderate agreements were found for stance, knee flexion for both legs, and vertical displacement (ICCs > 0.617). For the other variables, poor agreement (ICCs < 0.5) was reflected between both MCS.

4. Discussion

This study is the first, to the best of the authors' knowledge, to assess the level of agreement between the MotionMetrix system and a 3D MCS for walking and running kinetic parameters. The study involved 24 healthy young adults to achieve this aim. The MotionMetrix system demonstrated the highest level of agreement, indicating excellent agreement (ICC > 0.9), for the left step time and cadence, compared to Qualisys MCS, while walking at a speed of 5 km/h. Regarding the agreement of other variables, the text highlights that stride rate and length showed good agreement, with ICCs above 0.75, while gait cycle exhibited a moderate level of agreement, with ICCs exceeding 0.5. However, the other variables demonstrated poor agreement, with ICCs below 0.5. Concerning running at a speed of 10 km/h, both MCSs demonstrated good levels of agreement for stride time, stride rate, and knee flexion angles (stance and swing) on the left side. The agreement was also good for stride length. Conversely, the other variables had moderate to poor levels of agreement. At a faster running speed of 15 km/h, both MCSs demonstrated excellent levels of agreement for stride time, stride rate, stride length, and swing knee flexion on the left side. A good level of agreement was also observed for swing knee flexion on the right side, while moderate agreement was found for vertical displacement and knee flexion angles on the stance side. However, the ICCs for the other variables indicated poor levels of agreement.

Assessing walking and running gait can be important for a variety of reasons, including diagnosing and monitoring certain conditions and injuries, evaluating the effectiveness of interventions, and tracking changes over time. However, the accuracy and reliability of gait measurements depend on the selection of appropriate variables and the quality of the measurement instruments used.

Previous research has shown that the MotionMetrix system provides reliable measures for walking at 5 km/h of all parameters except for step width [12]. Our study supports these findings, as we found that step width measurements obtained with the MotionMetrix system showed poor reliability (ICC = 0.000) and significant differences ($p < 0.05$) when compared to the gold standard system (i.e., Qualisys MCS) during walking at 5 km/h. It is important to note that the MotionMetrix system utilized two Kinect cameras and software to evaluate walking and running gait in this study. Prior research has demonstrated that the Kinect cameras have underestimated step time and step length by 16% and 1.7%, respectively, during walking [18]. Our findings partially refute this statement since no significant differences were observed for step time ($p > 0.05$, Bias = 0.00 s). Although significant differences were detected for step length when measured unilaterally, no significant differences were observed when measured as stride length ($p > 0.05$, Bias = 0.00 m), that is, one step after the other.

Additionally, the aforementioned study investigated the test-retest reliability of MotionMetrix for analyzing running at 10 and 15 km/h [12]. These authors reported reliable measures for all its parameters at 10 km/h, except for thigh flexion, landing knee flexion, and step width, which exhibited coefficients of variation (CV) of 16.26%, 10.14%, and 10.72%, respectively [12]. Consistent with the earlier investigation, the current findings showed significant differences ($p < 0.05$) between MotionMetrix and Qualisys for thigh flexion in both legs (Bias = 5.04° and 6.38°, left and right legs, respectively), landing knee flexion (Bias = −4.83° and 3.17°, left and right legs, respectively), and step width (Bias = 0.03 m) at 10 km/h. At 15 km/h, the previous study found that MotionMetrix provided reliable values for all parameters except spine angle and step width, which exhibited CVs of 23.27% and 22.22%, respectively [12]. In the current investigation, large SEE values were found between MotionMetrix and 3D MCS for spine angle (SEE = 3.40) and step width (SEE = 0.02).

When interpreting the findings here reported, readers must be aware of certain limitations. It is important to note that the treadmill protocol employed in this study aimed to minimize gait and running variability due to inexperience or fatigue [19,20]. Previous studies have shown that novice treadmill runners and healthy young adults require a minimum of 6 to 8 min to adjust to the treadmill's locomotion. Running on a treadmill may not be the same as running on an overground outdoor surface, but they are largely comparable [15]. The present study was limited by its sample of healthy, active young adults, and as such, caution should be taken when generalizing the findings to other populations.

Despite the limitations exposed above, the study examined several variables that are commonly used to analyze gait patterns in walking and running, including step time, cadence, stride rate, stride length, and knee flexion angles. The study's findings indicate that the MotionMetrix system is a reliable system for measuring these variables in both walking and running gait. These variables offer valuable information about the temporal and spatial aspects of gait, such as the duration of different phases, the frequency and distance of steps, and the angles of the joints involved in the movement. Accurate and reliable measurement of these variables can provide insights into the mechanics and efficiency of gait and help identify abnormalities or deviations from normal patterns. This information can guide the development of interventions, such as exercise programs or orthotics, to address gait impairments and improve functional outcomes. Therefore, considering these variables when assessing walking gait and running can provide valuable information for both research and clinical purposes, leading to a better understanding, diagnosis, and treatment of gait-related conditions and injuries.

5. Conclusions

After comparing the measurements of gait and running variables provided by the MotionMetrix markerless software against a gold standard system (i.e., Qualisys 3D MCS), it is concluded that the agreement between the two motion capture systems varied for different variables and speeds of locomotion, with some variables demonstrating high agreement while others showed poor agreement. Nonetheless, the findings presented here suggest that the MotionMetrix system is a promising option for sports practitioners and clinicians interested in measuring gait variables, particularly in the contexts examined in the study. However, further research is needed to examine the agreement between different systems for other variables and in different contexts.

Author Contributions: Conceptualization, D.J.-C. and F.G.-P.; methodology, D.J.-C. and S.A.R.-A.; software, A.P.-C. and A.M.-M.; validation, S.A.R.-A. and F.G.-P.; formal analysis, D.J.-C. and A.M.-M.; investigation, J.M.C.-G.; resources, V.S.-H. and J.M.C.-G.; data curation, V.S.-H.; writing—original draft preparation, D.J.-C.; writing—review and editing, F.G.-P.; visualization, S.A.R.-A.; supervision, V.S.-H. and A.P.-C.; project administration, F.G.-P.; funding acquisition, V.S.-H. and S.A.R.-A. All authors have read and agreed to the published version of the manuscript.

Funding: This study has been partially supported by the State Research Agency (SRA) and European Regional Development Fund (ERDF) with the project EDUSPORT (REF: PID2020-115600RB-C21).

Institutional Review Board Statement: The study was conducted in accordance with the Declaration of Helsinki, and approved by the Institutional Review Board (or Ethics Committee) of the University of Granada (protocol code 2546/CEIH/2022, 18 January 2022).

Informed Consent Statement: Informed consent was obtained from all subjects involved in the study.

Data Availability Statement: The data presented in this study are available upon request from the corresponding author. The data are not publicly available due to participant privacy.

Conflicts of Interest: The authors declare no conflict of interest.

References

1. Colyer, S.L.; Evans, M.; Cosker, D.P.; Salo, A.I.T. A Review of the Evolution of Vision-Based Motion Analysis and the Integration of Advanced Computer Vision Methods Towards Developing a Markerless System. *Sports Med.-Open* **2018**, *4*, 24. [CrossRef] [PubMed]
2. Lam, W.W.T.; Fong, K.N.K. The application of markerless motion capture (MMC) technology in rehabilitation programs: A systematic review and meta-analysis. *Virtual Real.* **2022**, 1–16. [CrossRef]
3. Lucchetti, L.; Cappozzo, A.; Cappello, A.; Della Croce, U. Skin movement artefact assessment and compensation in the estimation of knee-joint kinematics. *J. Biomech.* **1998**, *31*, 977–984. [CrossRef] [PubMed]
4. Leardini, A.; Chiari, L.; Della Croce, U.; Cappozzo, A. Human movement analysis using stereophotogrammetry: Part 3. Soft tissue artifact assessment and compensation. *Gait Posture* **2005**, *21*, 212–225. [CrossRef] [PubMed]
5. Knippenberg, E.; Verbrugghe, J.; Lamers, I.; Palmaers, S.; Timmermans, A.; Spooren, A. Markerless motion capture systems as training device in neurological rehabilitation: A systematic review of their use, application, target population and efficacy. *J. Neuroeng. Rehabil.* **2017**, *14*, 61. [CrossRef] [PubMed]
6. Armitano-Lago, C.; Willoughby, D.; Kiefer, A.W. A SWOT Analysis of Portable and Low-Cost Markerless Motion Capture Systems to Assess Lower-Limb Musculoskeletal Kinematics in Sport. *Front. Sports Act. Living* **2022**, *3*, 809898. [CrossRef] [PubMed]
7. Schmitz, A.; Ye, M.; Shapiro, R.; Yang, R.; Noehren, B. Accuracy and repeatability of joint angles measured using a single camera markerless motion capture system. *J. Biomech.* **2014**, *47*, 587–591. [CrossRef] [PubMed]
8. van der Kruk, E.; Reijne, M.M. Accuracy of human motion capture systems for sport applications; state-of-the-art review. *Eur. J. Sport Sci.* **2018**, *18*, 806–819. [CrossRef] [PubMed]
9. Nakano, N.; Sakura, T.; Ueda, K.; Omura, L.; Kimura, A.; Iino, Y.; Fukashiro, S.; Yoshioka, S. Evaluation of 3D Markerless Motion Capture Accuracy Using OpenPose With Multiple Video Cameras. *Front. Sports Act. Living* **2020**, *2*, 50. [CrossRef] [PubMed]
10. Kanko, R.M.; Laende, E.; Selbie, W.S.; Deluzio, K.J. Inter-session repeatability of markerless motion capture gait kinematics. *J. Biomech.* **2021**, *121*, 110422. [CrossRef] [PubMed]
11. Wade, L.; Needham, L.; McGuigan, P.; Bilzon, J. Applications and limitations of current markerless motion capture methods for clinical gait biomechanics. *PeerJ* **2022**, *10*, e12995. [CrossRef] [PubMed]
12. Jaén-Carrillo, D.; Ruiz-Alias, S.A.; Chicano-Gutiérrez, J.M.; Ruiz-Malagón, E.J.; Roche-Seruendo, L.E.; García-Pinillos, F. Test-Retest Reliability of the MotionMetrix Software for the Analysis of Walking and Running Gait Parameters. *Sensors* **2022**, *22*, 3201. [CrossRef] [PubMed]
13. García-Pinillos, F.; Jaén-Carrillo, D.; Hermoso, V.S.; Román, P.L.; Delgado, P.; Martinez, C.; Carton, A.; Seruendo, L.R. Agreement Between Spatiotemporal Gait Parameters Measured by a Markerless Motion Capture System and Two Reference Systems—A Treadmill-Based Photoelectric Cell and High-Speed Video Analyses: Comparative Study. *JMIR mHealth uHealth* **2020**, *8*, e19498. [CrossRef] [PubMed]

14. McKay, A.K.; Stellingwerff, T.; Smith, E.S.; Martin, D.T.; Mujika, I.; Goosey-Tolfrey, V.L.; Sheppard, J.; Burke, L.M. Defining Training and Performance Caliber: A Participant Classification Framework. *Int. J. Sports Physiol. Perform.* **2022**, *17*, 317–331. [CrossRef] [PubMed]

15. Van Hooren, B.; Fuller, J.T.; Buckley, J.D.; Miller, J.R.; Sewell, K.; Rao, G.; Barton, C.; Bishop, C.; Willy, R.W. Is Motorized Treadmill Running Biomechanically Comparable to Overground Running? A Systematic Review and Meta-Analysis of Cross-Over Studies. *Sports Med.* **2020**, *50*, 785–813. [CrossRef] [PubMed]

16. Hopkins, W.; Marshall, S.; Batterham, A.; Hanin, J. Progressive statistics for studies in sports medicine and exercise science. *Med. Sci. Sports Exerc.* **2009**, *41*, 3. [CrossRef] [PubMed]

17. Koo, T.K.; Li, M.Y. A Guideline of Selecting and Reporting Intraclass Correlation Coefficients for Reliability Research. *J. Chiropr. Med.* **2016**, *15*, 155–163. [CrossRef] [PubMed]

18. Clark, R.A.; Bower, K.J.; Mentiplay, B.F.; Paterson, K.; Pua, Y.-H. Concurrent validity of the Microsoft Kinect for assessment of spatiotemporal gait variables. *J. Biomech.* **2013**, *46*, 2722–2725. [CrossRef] [PubMed]

19. Lavcanska, V.; Taylor, N.F.; Schache, A.G. Familiarization to treadmill running in young unimpaired adults. *Hum. Mov. Sci.* **2005**, *24*, 544–557. [CrossRef] [PubMed]

20. Schieb, D.A. Kinematic accommodation of novice treadmill runners. *Res. Q. Exerc. Sport* **1986**, *57*, 1–7. [CrossRef]

Article

RunScribe Sacral Gait Lab™ Validation for Measuring Pelvic Kinematics during Human Locomotion at Different Speeds

Emilio J. Ruiz-Malagón [1,2], Felipe García-Pinillos [1,2,3,*], Alejandro Molina-Molina [4], Víctor M. Soto-Hermoso [1,2] and Santiago A. Ruiz-Alias [1,2]

1   Department of Physical Education and Sports, Faculty of Sport Sciences, University of Granada, 18071 Granada, Spain
2   Sport and Health University Research Institute (iMUDS), University of Granada, 18071 Granada, Spain
3   Department of Physical Education, Sports and Recreation, Universidad de La Frontera, Temuco 4811230, Chile
4   Campus Universitario, Universidad San Jorge, Villanueva de Gállego, 50830 Zaragoza, Spain
*   Correspondence: fgpinillos@ugr.es

Abstract: Optoelectronic motion capture systems are considered the gold standard for measuring walking and running kinematics parameters. However, these systems prerequisites are not feasible for practitioners as they entail a laboratory environment and time to process and calculate the data. Therefore, this study aims to evaluate the validity of the three-sensor RunScribe Sacral Gait Lab™ inertial measurement unit (IMU) in measuring pelvic kinematics in terms of vertical oscillation, tilt, obliquity, rotational range of motion, and the maximum angular rates during walking and running on a treadmill. Pelvic kinematic parameters were measured simultaneously using an eight-camera motion analysis system (Qualisys Medical AB, GÖTEBORG, Sweden) and the three-sensor RunScribe Sacral Gait Lab™ (Scribe Lab. Inc. San Francisco, CA, USA) in a sample of 16 healthy young adults. An acceptable level of agreement was considered if the following criteria were met: low bias and SEE (<0.2 times the between-subject differences SD), almost perfect ($r > 0.90$), and good reliability (ICC > 0.81). The results obtained reveal that the three-sensor RunScribe Sacral Gait Lab™ IMU did not reach the validity criteria established for any of the variables and velocities tested. The results obtained therefore show significant differences between the systems for the pelvic kinematic parameters measured during both walking and running.

Keywords: sacrum; vertical oscillation; tilt; obliquity; rotation

Citation: Ruiz-Malagón, E.J.; García-Pinillos, F.; Molina-Molina, A.; Soto-Hermoso, V.M.; Ruiz-Alias, S.A. RunScribe Sacral Gait Lab™ Validation for Measuring Pelvic Kinematics during Human Locomotion at Different Speeds. *Sensors* **2023**, *23*, 2604. https://doi.org/10.3390/s23052604

Academic Editor: Carlo Ricciardi

Received: 28 January 2023
Revised: 21 February 2023
Accepted: 23 February 2023
Published: 27 February 2023

1. Introduction

Human movement is a complex task that requires correct intersegmental coordination and human locomotion is concerned in particular with the forward propulsion of the body [1]. In this sense, pelvic kinematics play an important role in both maximizing athletic performance and in minimizing the risk of injury in runners during human locomotion [1–3]. Hence, it was suggested that their analysis in all the planes (sagittal, frontal, and transverse) could reveal important information for practitioners [4–7]. However, the accurate measurement of pelvic kinematics constitutes an essential element for clinicians or trainers working in human locomotion, in order to avoid misinterpretations [8].

Optoelectronic motion capture systems are the gold standard for measuring the kinematics parameters of walking and running. In fact, these systems have improved in recent years and the current measurement error can be less than <0.5 mm [9]. However, optoelectronic motion capture systems require laboratory settings in addition to the time and knowledge for data processing, making them unfeasible systems for some practitioners [10,11]. In response to these limitations, Inertial Measurement Units (IMUs) are a low-cost alternative that allows data to be collected in the field [12]; reports are provided immediately [13]. Nowadays, there is a wide variety of IMUs that allow kinematic monitoring of various joints in everyday tasks. Different levels of agreement have been reported according to the task performed and the joint or axis measured [14–17]. Therefore, a more detailed analysis of the new IMUs entering the market is required.

The RunScribe™ IMU system (Scribe Labs Inc., San Francisco, CA, USA) is an example of such a device widely used in the field of human locomotion analysis, consisting of two IMU devices (one on each foot). Each IMU is based on a nine-axis (three-axis gyroscope, three-axis accelerometer, three-axis magnetometer) with a sampling rate of 500 Hz. Both allow the measurement of several spatiotemporal parameters of running [18] and foot-strike patterns [19] for each foot which have previously been validated. The hardware device weighs 15 g and measures 35 × 25 × 7.5 mm. Recently, the same manufacturer launched a three-sensor system (RunScribe Sacral Gait Lab™) that includes an additional device to be placed on the runner's sacrum that allows the measurement of the pelvic kinematics in the frontal, sagittal, and transverse planes synchronously with the two previous sessions that were placed on both feet. This would allow the already validated spatiotemporal parameters of the feet [18] to be combined with the pelvic motion, providing health and sports professionals with a complete analysis of their patients and athletes. Therefore, the aim of this study is to examine the validity of the RunScribe Sacral Gait Lab™ to measure pelvic kinematics in terms of vertical oscillation, tilt, obliquity, rotational range of motion, and maximal angular rates during walking and running on a treadmill.

2. Materials and Methods

2.1. Experimental Design

Pelvic kinematic parameters were simultaneously recorded using an eight-camera motion analysis system (Qualisys Medical AB, Sweden) and the three-sensor RunScribe Sacral Gait Lab™ IMU (Scribe Lab. Inc. San Francisco, CA, USA) during a walking and running treadmill protocol (WOODWAY Pro XL, Woodway, Inc., Waukesha, WI, USA). To acclimate to the treadmill, participants began with a 10-min warm-up at a self-selected comfortable speed [20,21]. After familiarization, participants completed 3 sets of 1 min at 5, 10, and 15 km·h^{-1}. The last 30 s were used for recording with both systems. Participants were instructed to refrain from strenuous activity for at least 48 h prior to data collection.

2.2. Participants

A group of 16 healthy young male adults (age = 22.7 ± 2.6 years; body mass = 69.1 ± 11.7 kg; height = 1.72 ± 0.10 m; weekly training = 6.9 ± 2.4 h/week) participated in the study. All subjects were required to meet the following inclusion criteria: (i) be between 18 and 30 years of age, (ii) not have suffered any injury within six months prior to data collection, and (iii) be physically active according to the guidelines of the American College of Sports Medicine (ACSM) [22]. All the subjects were informed of the purpose and procedures of the study before signing a written consent form. The study protocol adhered to the tenets of the Declaration of Helsinki and was approved by the Institutional Review Board (No. 2546/CEIH/2022).

2.3. Procedures

The height (m) and body mass (kg) of the participants were obtained using the stadiometer SECA 222 (SECA, Corp., Hamburg, Germany) and the bioimpedance meter Inbody 230 (Inbody Seúl, Corea), respectively.

Pelvic kinematics were assessed using two different systems (i.e., optoelectronic motion capture system versus IMU). Three-dimensional kinematics were acquired using an eight-camera motion analysis system (Qualisys Medical AB, Sweden) with a sampling rate of 250 Hz. The cameras were positioned to provide a complete view of the treadmill location. Safety bars were removed to avoid any potential masking of the markers by the structures. Prior to the data collection, the volume of the test space was calibrated using a dynamic T-wand, and the origin and axes of the coordinate system were established by placing an L-frame on the treadmill. Subjects were then fitted with a lower body marker model. A total of 40 markers were placed by two experienced researchers based on the palpation of appropriate anatomical landmarks (Figure 1). One of the researchers was responsible for placing the reflective markers and the other researcher was responsible for checking their placement, thus providing a double check. The anatomical locations of the markers were the right and left iliac crest tubercle, the right and left posterior superior iliac spine, the right and left femur greater trochanter, the right and left anterior superior iliac spine, the right and left femur lateral epicondyle, the right and left femur medial epicondyle, the right and left fibula apex of the lateral malleolus, the right and left tibia apex of the medial malleolus, the right and left fifth metatarsal head, the right and left first metatarsal head, and the right and left posterior surface of the calcaneus. In addition, two cluster marker sets (a group of four retro-reflective markers attached to a lightweight rigid plastic shell) were also placed on the thigh and on the shank. Once in position, a static test was performed with the participants positioned in an anatomical position prior to the start of the treadmill running protocol.

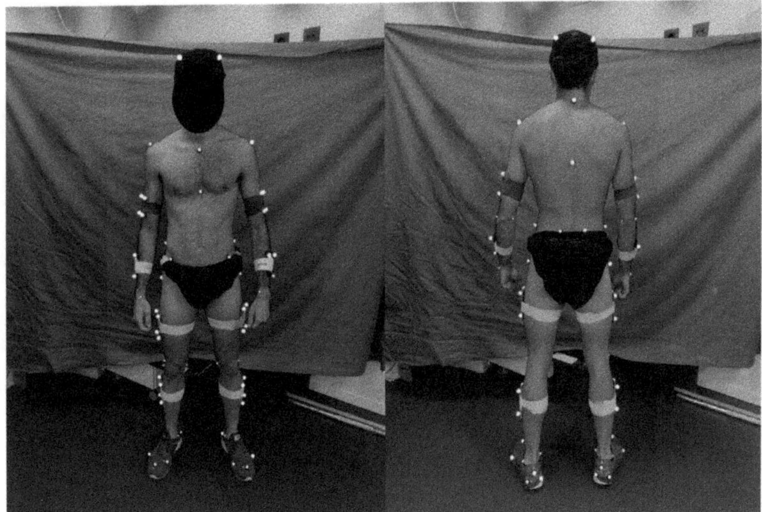

Figure 1. Illustration of the marker set.

All the static and motion tests were exported to Visual 3D (C-Motion Inc, Boyds, ML, USA). The rigid link model created from the static file was then assigned to all the imported motion files. In particular, the pelvic segment was created according to the CODA model [23]. The motion files were filtered with a fourth-order Butterworth low-pass filter with a cut-off frequency of 8 Hz. The x–y–z Cardan sequence was used to calculate joint angles. This sequence corresponds to flexion/extension, abduction/adduction, and axial rotation. Joint angles were not normalized to the static standing test. The laboratory frame followed the right-hand rule and had the positive y-direction oriented in the direction of forward progression, the positive x-direction oriented to the left, and the positive z-direction oriented vertically upward. Vertical oscillation, tilt, obliquity, rotational range of motion, and maximal angular rates were calculated over the entire gait cycle.

The three-sensor Runscribe Sacral Gait LabTM IMU (Scribe Lab. Inc. San Francisco, CA, USA) conducted recording at 500 Hz. This IMU combines an accelerometer, a gyroscope, and a triaxial magnetometer. Following the recommendations of García-Pinillos et al. [18], two RunScribe ™ devices were attached to the laces of the running shoes. A third RunScribe™ device was attached to the waistband of the pants at the height of the sacrum (Figure 2), following the recommendations of the manufacturer. Before data collection, the system was calibrated flat and once mounted according to the manufacturer´s instructions. The pelvic kinematics (i.e., vertical oscillation, tilt, obliquity, rotational range of motion, and maximal angular rates) were collected by the IMU and were then synchronized to the RunScribe platform (https://dashboard.runscribe.com/ accessed on 3 november 2022) where these metrics are reported automatically. From there, they are copied to an Excel sheet for subsequent analysis.

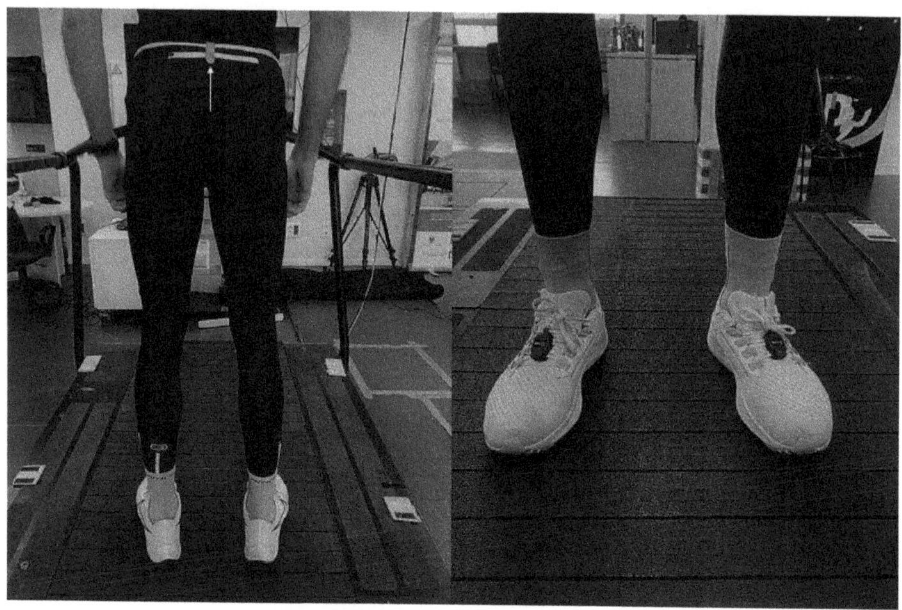

Figure 2. RunScribe sensor placement: **left** panel, sacral sensor and **right** panel, footpods sensors.

2.4. Statistical Analyses

Descriptive data are presented as mean $\pm$ standard deviation (SD) with 95% confidence intervals (CI). The normal distribution of the data and homogeneity of variances were confirmed through the Shapiro–Wilk test and Levene's tests, respectively. The level of agreement of the RunScribe Sacral Gait Lab™ with the optoelectronic motion capture system was examined through systematic bias, Pearson's correlation coefficient (r), the standard error of the estimate (SEE) obtained from the linear regression analysis, and the intraclass correlation coefficient (ICC) using a two-way mixed model for absolute agreement [24]. An acceptable level of agreement was considered if the following criteria were met: a low bias and SEE (<0.2 times the between-subjects differences SD) [25], almost perfect (r > 0.90) [26], and good reliability (ICC > 0.81) [24]. Statistical analyses were performed using the software package SPSS (IBM SPSS version 25.0, Chicago, IL, USA). Statistical significance was set at $p \leq 0.05$.

3. Results

3.1. Vertical Oscillation

Subjects exhibited a vertical oscillation of 4.7 (1.2), 10.0 (1.6), and 8.2 (1.1) cm at 5, 10, and 15 km/h, respectively (Table 1). The IMU system did not reach the validity criteria established at any of the velocities tested.

Table 1. Level of agreement of the pelvic kinematic parameters obtained through the RunScribe Sacral Gait Lab™ and the optoelectronic motion capture system.

	RunScribe	Qualisys	Bias (95% CI)	SEE	SWC	ICC (95% CI)	r
5 km/h							
Vertical oscillation (cm)	25.9 (2.0)	4.7 (1.2)	21.1 (20.0 to 22.3)	4.9	0.2	0.01 (−0.01 to 0.07)	0.458
Obliquity ROM (°)	8.4 (1.8)	9.5 (2.3)	−1.1 (−2.9 to 0.6)	3.2	0.5	−0.10 (−1.97 to 0.63)	−0.056
Tilt ROM (°)	7.0 (1.7)	6.8 (2.4)	0.2 (−1.7 to 2.3)	3.3	0.5	−1.62 (−12.5 to 0.25)	−0.429
Rotation ROM (°)	8.6 (2.9)	7.9 (2.0)	0.7 (−0.4 to 1.7)	1.9	0.4	0.83 (0.46 to 0.95)	0.758 *
Obliquity max rate (°/sec)	75.9 (14.6)	46.2 (13.5)	29.7 (18.8 to 40.4)	35	2.7	0.07 (−0.21 to 0.46)	0.121
Tilt max rate (°/sec)	52.1 (16.6)	49.4 (10.4)	2.7 (−8.9 to 14.4)	18	2.1	0.22 (−2.05 to 0.78)	0.164
Rotation max rate (°/sec)	47.1 (8.7)	67.3 (15.2)	−20.2 (−27.9 to −12.5)	25	3.0	0.34 (−0.24 to 0.75)	0.543
10 km/h							
Vertical oscillation (cm)	9.6 (2.2)	10.0 (1.6)	−0.4 (−1.3 to 0.3)	1.6	0.3	0.81 (0.47 to 0.94)	0.724 *
Obliquity ROM (°)	9.2 (3.4)	14.5 (4.7)	−5.3 (−7.7 to −2.9)	6.8	0.9	0.44 (−0.28 to 0.80)	0.522
Tilt ROM (°)	11.3 (5.6)	6.7 (2.0)	4.6 (1.8 to 7.2)	7.2	0.4	0.28 (−0.36 to 0.69)	0.389
Rotation ROM (°)	11.1 (3.0)	17.2 (5.2)	−6.1 (−11.8 to −4.9)	11	1.0	−0.32 (−1.12 to 0.40)	−0.312
Obliquity max rate (°/sec)	206 (92)	141 (31)	65 (16 to 114)	105	6.2	0.37 (−0.39 to 0.78)	0.533
Tilt max rate (°/sec)	367 (180)	126 (53)	241 (143 to 339)	320	10.6	0.03 (−0.30 to 0.44)	0.085
Rotation max rate (°/sec)	136 (40)	128 (29)	8.1 (−23 to 39)	57	5.8	−0.50 (−4.66 to 0.54)	−0.198
15 km/h							
Vertical oscillation (cm)	5.8 (1.5)	8.2 (1.1)	−2.4 (−3.0 to 1.8)	2.8	9.0	0.37 (−0.18 to 0.77)	0.638 *
Obliquity ROM (°)	16.3 (5.7)	22.4 (7.2)	−6.1 (−9.6 to −2.7)	8.4	1.4	0.577 (−0.22 to 0.86)	0.590 *
Tilt ROM (°)	13.4 (5.3)	10.2 (3.9)	3.2 (−0.2 to 6.6)	7.0	0.8	0.09 (−1.04 to 0.65)	0.056
Rotation ROM (°)	11.7 (3.2)	22.9 (7.8)	−11.2 (−16.6 to −5.8)	15	1.6	−0.30 (−0.79 to 0.39)	−0.481
Obliquity max rate (°/sec)	269 (95)	166 (31)	103 (58 to 148)	138	6.2	0.35 (−0.27 to 0.75)	0.573 *
Tilt max rate (°/sec)	388 (172)	167 (48)	221 (131 to 311)	291	9.6	0.13 (−0.25 to 0.55)	0.333
Rotation max rate (°/sec)	186 (74)	167 (39)	19 (−21.1 to 59.0)	78	7.8	0.39 (−0.77 to 0.79)	0.296

ROM: Range of motion. CI: Confidence interval. SEE: Standard error of the estimate. SWC: Smallest worthwhile change. ICC: Intraclass correlation coefficient. r: Pearson correlation coefficient. *: Correlation is significant at the 0.05 level.

3.2. Pelvic Tilt

Subjects exhibited a pelvic tilt ROM of 6.8 (2.4), 6.7 (2.0), and 10.2 (3.9)° and a maximal tilt rate of 49.4 (10.4), 126 (53), and 167 (48) °/sec at 5, 10, and 15 km/h, respectively. The IMU system did not reach the validity criteria established at any of the velocities tested.

3.3. Pelvic Obliquity

Subjects exhibited a pelvic obliquity ROM of 9.5 (2.3), 14.5 (4.7), and 22.4 (7.2)° and a maximal obliquity rate of 46.2 (13.5), 141 (31), and 166 (31) °/sec at 5, 10, and 15 km/h, respectively. The IMU system did not reach the validity criteria established at any of the velocities tested.

3.4. Pelvic Rotation

Subjects exhibited a pelvic rotation ROM of 7.9 (2.0), 17.2 (5.2), and 22.9 (7.8)° and a maximal rotation rate of 67.3 (15.2), 128 (29), and 167 (39) °/sec at 5, 10, and 15 km/h, respectively. The IMU did not reach the validity criteria established at any of the velocities tested.

4. Discussion

This purpose for this study was to examine the validity of the three-sensor RunScribe Sacral Gait Lab™ to measure pelvic kinematics in terms of vertical oscillation, tilt, obliquity, rotational range of motion, and maximal angular rates during walking and running on a treadmill at 5, 10, and 15 km/h. The results revealed that the IMU did not reach the validity criteria established for any of the variables and velocities tested.

To the authors' knowledge, this is the first study to analyze the validity of the RunScribe Sacral Gait Lab™ IMU for measuring pelvic kinematics during human locomotion. As noted above, the potential of using IMU to assess gait analysis without the limitations of laboratory technology is well-known [18]. This is a step forward in terms of trainers and clinicians being able to measure athletes or clients in a natural environment and in a time-efficient manner. However, this advantage would be useless if the data were invalid The results obtained here indicate that the RunScribe Sacral Gait Lab™ IMU is not valid for measuring pelvic kinematics during walking or running. Similar results have been observed regarding its reliability [27]. However, it is well established that the spatiotemporal parameters reported by this device are valid and reliable [18,19,28].

Several earlier studies have analyzed the validity of IMUs in measuring pelvic kinematics during gait and running and concluded that errors greater than 5° could mislead the clinical and performance interpretation [10,28]. Considering other statistical criteria, Bolink et al. [28] found a reasonably satisfactory agreement of the range of movement measurements between both an IMU and an optoelectronic motion capture system, as the deviations of the measured angles were within the limits of agreement of the Bland–Altman plots. Very large correlations and almost perfect ICCs were obtained during gait with respect to pelvic kinematics (i.e., tilt, obliquity, and rotation). As in the present study, Bugané et al. [29] compared pelvic kinematics in three planes during gait using a sacrum mounted IMU. They obtained an error of less than 3° in both the sagittal and frontal planes and nearly perfect correlations. The validity criteria established here were similar to those reported in previous studies, indicating that the IMU may be a valid device for measuring pelvic kinematics, but further improvements to the RunScribe Sacral Gait Lab™ should be considered to achieve the precision required for its use in clinical and performance settings. In addition, the pelvic kinematics were measured in healthy subjects during walking and running at slow and moderate speeds (i.e., 5, 10, and 15 km/h). These were within the normal kinematic ranges as previously reported [2,30]. However, a different level of accuracy may be required in subjects with clinical conditions or athletes tested at higher running speeds.

Regarding the pelvic angular values provided by the RunScribe™ Sacral Gait Lab™, these are based on the peak values. Therefore, the operating range of the accelerometer could be a potential source of error [31]. Another cause that could explain the low validity and reliability [27] of the device could be that it is attached to the waist rather than sports tights, which could potentially introduce more disturbance to the signal.

The main limitation of the study is that the IMU may have been susceptible to motion artifacts due to the mounting method designed by the manufacturer. Another limitation of the work is the software version used as it was recently updated after this study was conducted. In addition, the present study did not analyze the test-retest reliability, which could be considered as a future line of research. Despite these limitations, the current study examines the validity of a commonly used device by trainers and clinicians to analyze lower limb kinematics during human locomotion, which could provide useful information for these professionals.

5. Conclusions

The results obtained report significant differences between systems for the pelvic kinematic parameters measured during both walking and running. Therefore, the three-sensor RunScribe Sacral Gait Lab™ has shown a questionable validity according to an optoelectronic motion capture system (250 Hz) for measuring pelvic kinematics in terms of vertical oscillation, tilt, obliquity, rotational range of motion, and maximal angular rates at 5, 10, and 15 km/h.

From a practical point of view, the three-sensor RunScribe Sacral Gait Lab™, as stated above, is a valid device for measuring spatiotemporal parameters [18,27] and foot-strike patterns [15] but shows a low level of agreement with the reference system when measuring pelvic kinematics. Therefore, its use for measuring the aforementioned variables is not recommended.

Author Contributions: Conceptualization, F.G.-P., A.M.-M. and S.A.R.-A.; methodology, A.M.-M. and S.A.R.-A.; software, A.M.-M. and V.M.S.-H.; validation, F.G.-P., V.M.S.-H. and A.M.-M.; formal analysis, E.J.R.-M. and S.A.R.-A.; investigation, F.G.-P. and S.A.R.-A.; resources, E.J.R.-M.; data curation, A.M.-M. and S.A.R.-A.; writing—original draft preparation, E.J.R.-M. and S.A.R.-A.; writing—review and editing, A.M.-M. and F.G.-P.; visualization, E.J.R.-M. and S.A.R.-A.; supervision, F.G.-P. and V.M.S.-H.; project administration, F.G.-P. and V.M.S.-H.; funding acquisition, F.G.-P. and V.M.S.-H. All authors have read and agreed to the published version of the manuscript.

Funding: This study was partially supported by the State Research Agency (SRA) and the European Regional Development Fund (ERDF) with the project EDUSPORT (REF: PID2020-115600RB-C21).

Institutional Review Board Statement: The study was conducted in accordance with the Declaration of Helsinki, and approved by the Institutional Review Board (or Ethics Committee) of University of Granada (protocol code 2546/CEIH/2022 and date of approval 18 January 2022).

Informed Consent Statement: Informed consent was obtained from all subjects involved in the study.

Data Availability Statement: Data will be available on request.

Acknowledgments: The authors would like to thank all of the participants.

Conflicts of Interest: The authors declare no conflict of interest.

References

1. Preece, S.J.; Mason, D.; Bramah, C. The coordinated movement of the spine and pelvis during running. *Hum. Mov. Sci.* **2016**, *45*, 110–118. [CrossRef] [PubMed]
2. Schache, A.G.; Blanch, P.; Rath, D.; Wrigley, T.; Bennell, K. Running Lumbal Kinematic. *Hum. Mov. Sci.* **2002**, *21*, 273–293. [CrossRef]
3. Folland, J.P.; Allen, S.J.; Black, M.I.; Handsaker, J.C.; Forrester, S.E. Running Technique is an Important Component of Running Economy and Performance. *Med. Sci. Sports Exerc.* **2017**, *49*, 1412–1423. [CrossRef] [PubMed]
4. Schache, A.G.; Blanch, P.D.; Murphy, A.T. Relation of anterior pelvic tilt during running to clinical and kinematic measures of hip extension. *Br. J. Sports Med.* **2000**, *34*, 279–283. [CrossRef]
5. Foch, E.; Reinbolt, J.A.; Zhang, S.; Fitzhugh, E.C.; Milner, C.E. Associations between iliotibial band injury status and running biomechanics in women. *Gait Posture* **2015**, *41*, 706–710. [CrossRef] [PubMed]
6. Lavine, R. Iliotibial band friction syndrome. *Curr. Rev. Musculoskelet. Med.* **2010**, *3*, 18–22. [CrossRef] [PubMed]
7. Fu, F.H.; Feldman, A. The biomechanics of running: Practical considerations. *Tech. Orthop.* **1990**, *5*, 8–14. [CrossRef]
8. Lu, T.-W.; Chang, C.-F. Biomechanics of human movement and its clinical applications. *Kaohsiung J. Med. Sci.* **2012**, *28* (Suppl. 2), S13–S25. [CrossRef]
9. Topley, M.; Richards, J.G. A comparison of currently available optoelectronic motion capture systems. *J. Biomech.* **2020**, *106*, 109820. [CrossRef]
10. Cuesta-Vargas, A.I.; Galán-Mercant, A.; Williams, J.M. The use of inertial sensors system for human motion analysis. *Phys. Ther. Rev.* **2010**, *15*, 462–473. [CrossRef]
11. Fusca, M.; Negrini, F.; Perego, P.; Magoni, L.; Molteni, F.; Andreoni, G. Validation of a Wearable IMU System for Gait Analysis: Protocol and Application to a New System. *Appl. Sci.* **2018**, *8*, 1167. [CrossRef]
12. Norris, M.; Anderson, R.; Kenny, I.C. Method analysis of accelerometers and gyroscopes in running gait: A systematic review. *Proc. Inst. Mech. Eng. Part P J. Sports Eng. Technol.* **2014**, *228*, 3–15. [CrossRef]
13. Camomilla, V.; Bergamini, E.; Fantozzi, S.; Vannozzi, G. Trends Supporting the In-Field Use of Wearable Inertial Sensors for Sport Performance Evaluation: A Systematic Review. *Sensors* **2018**, *18*, 873. [CrossRef]
14. Zhang, J.-T.; Novak, A.C.; Brouwer, B.; Li, Q. Concurrent validation of Xsens MVN measurement of lower limb joint angular kinematics. *Physiol. Meas.* **2013**, *34*, N63. [CrossRef]
15. Robert-Lachaine, X.; Mecheri, H.; Muller, A.; Larue, C.; Plamondon, A. Validation of a low-cost inertial motion capture system for whole-body motion analysis. *J. Biomech.* **2020**, *99*, 109520. [CrossRef] [PubMed]
16. Diraneyya, M.M.; Ryu, J.; Abdel-Rahman, E.; Haas, C.T. Inertial Motion Capture-Based Whole-Body Inverse Dynamics. *Sensors* **2021**, *21*, 7353. [CrossRef] [PubMed]
17. Ruiz-Malagón, E.J.; Delgado-García, G.; Castro-Infantes, S.; Ritacco-Real, M.; Soto-Hermoso, V.M. Validity and reliability of NOTCH®inertial sensors for measuring elbow joint angle during tennis forehand at different sampling frequencies. *Meas. J. Int. Meas. Confed.* **2022**, *201*, 111666. [CrossRef]
18. García-Pinillos, F.; Chicano-Gutiérrez, J.M.; Ruiz-Malagón, E.J.; Roche-Seruendo, L.E. Influence of RunScribe™ placement on the accuracy of spatiotemporal gait characteristics during running. *Proc. Inst. Mech. Eng. Part P J. Sports Eng. Technol.* **2020**, *234*, 11–18. [CrossRef]
19. DeJong, A.F.; Hertel, J. Validation of Foot-Strike Assessment Using Wearable Sensors During Running. *J. Athl. Train.* **2020**, *55*, 1307–1310. [CrossRef]

20. Lavcanska, V.; Taylor, N.F.; Schache, A.G. Familiarization to treadmill running in young unimpaired adults. *Hum. Mov. Sci.* **2005**, *24*, 544–557. [CrossRef]

21. Schieb, D.A. Kinematic Accommodation of Novice Treadmill Runners. *Res. Q. Exerc. Sport* **1986**, *57*, 1–7. [CrossRef]

22. American College of Sports Medicine. *ACSM's Health-Related Physical Fitness Assessment Manual*; Lippincott Williams & Wilkins: Philadelphia, PA, USA, 2013.

23. Pelvis Overview—Visual3D Wiki Documentation. C-Motion Inc., s. f. Available online: https://www.c-motion.com/v3dwiki/index.php/Pelvis_Overview (accessed on 14 January 2023).

24. Koo, T.K.; Li, M.Y. A Guideline of Selecting and Reporting Intraclass Correlation Coefficients for Reliability Research. *J. Chiropr. Med.* **2016**, *15*, 155–163. [CrossRef] [PubMed]

25. Hopkins, W.G. How to Interpret Changes in an Athletic Performance Test. *Sportscience* **2004**, *8*, 1–7.

26. Hopkins, W.G.; Marshall, S.W.; Batterham, A.M.; Hanin, J. Progressive Statistics for Studies in Sports Medicine and Exercise Science. *Med. Sci. Sports Exerc.* **2009**, *41*, 3–13. [CrossRef] [PubMed]

27. Cartón-Llorente, A.; Roche-Seruendo, L.E.; Jaén-Carrillo, D.; Marcen-Cinca, N.; García-Pinillos, F. Absolute reliability and agreement between Stryd and RunScribe systems for the assessment of running power. *Proc. Inst. Mech. Eng. Part P J. Sports Eng. Technol.* **2021**, *235*, 182–187. [CrossRef]

28. Bolink, S.A.A.N.; Naisas, H.; Senden, R.; Essers, H.; Heyligers, I.C.; Meijer, K.; Grimm, B. Validity of an inertial measurement unit to assess pelvic orientation angles during gait, sit–stand transfers and step-up transfers: Comparison with an optoelectronic motion capture system*. *Med. Eng. Phys.* **2015**, *38*, 225–231. [CrossRef] [PubMed]

29. Buganè, F.; Benedetti, M.G.; D'Angeli, V.; Leardini, A. Estimation of pelvis kinematics in level walking based on a single inertial sensor positioned close to the sacrum: Validation on healthy subjects with stereophotogrammetric system. *Biomed. Eng. Online* **2014**, *13*, 146. [CrossRef]

30. Jamkrajang, P.; Saelee, A.; Suwanmana, S.; Wiltshire, H.; Irwin, G.; Limroongreungrat, W. Temporal Changes of Pelvic and Knee Kinematics during Running. *J. Phys. Educ. Sport* **2022**, *22*, 767–774. [CrossRef]

31. Mitschke, C.; Kiesewetter, P.; Milani, T.L. The Effect of the Accelerometer Operating Range on Biomechanical Parameters: Stride Length, Velocity, and Peak Tibial Acceleration during Running. *Sensors* **2018**, *18*, 130. [CrossRef]

Article

Test–Retest and Between–Device Reliability of *Vmaxpro* IMU at Hip and Ankle for Vertical Jump Measurement

Lamberto Villalon-Gasch, Jose M. Jimenez-Olmedo *, Javier Olaya-Cuartero and Basilio Pueo

Research Group in Health, Physical Activity, and Sports Technology (Health-Tech), Faculty of Education, University of Alicante, San Vicente del Raspeig, 03690 Alicante, Spain
* Correspondence: j.olmedo@ua.es

Abstract: The ability to generate force in the lower body can be considered a performance factor in sports. This study aims to analyze the test–retest and between-device reliability related to the location on the body of the inertial measurement unit *Vmaxpro* for the estimation of vertical jump. Eleven highly trained female athletes performed 220 countermovement jumps (CMJ). Data were simultaneously captured by two *Vmaxpro* units located between L4 and L5 vertebrae (hip method) and on top of the tibial malleolus (ankle method). Intrasession reliability was higher for ankle (ICC = 0.96; CCC = 0.93; SEM = 1.0 cm; CV = 4.64%) than hip (ICC = 0.91; CCC = 0.92; SEM = 3.4 cm; CV = 5.13%). In addition, sensitivity was higher for ankle (SWC = 0.28) than for the hip method (SWC = 0.40). The noise of the measurement (SEM) was higher than the worthwhile change (SWC), indicating lack of ability to detect meaningful changes. The agreement between methods was moderate (r_s = 0.84; ICC = 0.77; CCC = 0.25; SEM = 1.47 cm). Significant differences were detected between methods (−8.5 cm, $p < 0.05$, ES = 2.2). In conclusion, the location of the device affects the measurement by underestimating CMJ on ankle. Despite the acceptable consistency of the instrument, the results of the reliability analysis reveal a significant magnitude of both random and systematic error. As such, the *Vmaxpro* should not be considered a reliable instrument for measuring CMJ.

Keywords: intra-session; intersession; between-session; sensibility; countermovement jump; CMJ; agreement; error

Citation: Villalon-Gasch, L.; Jimenez-Olmedo, J.M.; Olaya-Cuartero, J.; Pueo, B. Test–Retest and Between–Device Reliability of *Vmaxpro* IMU at Hip and Ankle for Vertical Jump Measurement. *Sensors* **2023**, *23*, 2068. https://doi.org/10.3390/s23042068

Academic Editors: Felipe García-Pinillos, Diego Jaén-Carrillo and Alejandro Pérez-Castilla

Received: 31 January 2023
Revised: 9 February 2023
Accepted: 10 February 2023
Published: 12 February 2023

1. Introduction

The ability to generate force in the lower body can be considered a performance factor in sports. One way to assess and monitor this force and power is through the vertical jump (VJ) [1]. The countermovement jump (CMJ) has been used to monitor the fitness of athletes [2–4] and fatigue [5,6]. There are many protocols and instruments used to carry out the analysis and monitoring of VJ [7], the most reliable of which are those that employ the double integration of reaction forces using force platforms and marker tracking with motion capture systems (MoCAP) [8]. However, these instruments are expensive, complex to set up and calibrate, and difficult to operate, and their use is therefore restricted to research use. They also have the added problem that data cannot be collected outdoors or on certain sport-specific surfaces such as grass or sand, which is also a problem with jump mats [9].

Accelerometers represent an alternative method that partially addresses the limitations of force platforms and motion capture (MoCAP) systems. Due to their smaller size and cost-effectiveness, accelerometers have seen a significant increase in usage in recent years for the study of human movement [10,11]. Inertial measurement units (IMUs) are electromechanical measurement systems that can combine time-accurate data acquisition with algorithms collected from all its sensors (i.e., accelerometer, gyroscope, and magnetoscope) to track position in three dimensions, without affecting the natural movements of athletes [12]. As a result, accelerometers are highly ecologically valid instruments, as they allow for data collection in a less obtrusive manner.

The validity of the *Vmaxpro* has been examined in the context of loaded jumps and has been determined to be valid in these studies [13,14]. However, no validation studies have been conducted specifically for the measurement of unloaded (CMJ) [13,14]. In terms of reliability, IMUs typically exhibit values that are considered reliable. This trend is supported by studies such as that of Rago et al. [15], which analyzed the reliability of the *Myotest* IMU. They detected CV values of 4.2%, test–retest ICC of 0.97, and SEM values of 0.5 cm, as well as SWC of 0.8, indicating this IMU to be a valid instrument. Lower reliability indices were reported by Brooks et al. [16], who detected standardized SEM values of 0.3% which they considered moderate, and test–retest ICC of 0.86. However, the authors still considered the IMU to be reliable, despite the standardized SEM exceeding 0.2%.

Many studies place the IMU on the hip during data collection; however, the findings of Spangler et al. [17] suggest that placement of the IMU on the torso does not significantly affect test–retest reliability, as they observed ICC values of 0.85 and CV of 6.7% for the *Catapult GPS* IMU. In contrast, Rantalainen et al. [18] detected lower reliability values (ICC = 0.686 and CV = 8.7%) for the same device on the torso. Similarly, Garnacho-Castaño et al. [19] placed the *Stride* IMU on the ankle and detected values similar to those on the hip (ICC= 0.90; CV = 4.7%).

The surface on which the jumps are performed can also impact the reliability of the measurements, as surfaces such as sand or grass can introduce additional variability and potentially alter the technical execution of the movement [9]. For example, a study by Schleitzer et al. [20] examined the jumping performance of beach volleyball players on a sand surface using the *Movesense* IMU, and detected ICC values of 0.866 [20]. Hence, studies in which the IMU is not placed on the hip, or those conducted on surfaces that are less stable, tend to have lower reliability values compared to studies where the IMU is placed on the hip. The ICC values are usually above 0.95 and CV values are around 3.5% in these studies [15,21–23]. Consequently, IMUs can be considered reliable instruments in the measurement of VJ, although factors such as biological variability, surface, or location of the instrument may influence their reliability [24].

To the best of our knowledge, the reliability of the *Vmaxpro* for measuring VJ has not been evaluated. Additionally, there are currently no studies that have examined the impact of the location of the accelerometer on different body segments for measuring CMJ on reliability. Further research is necessary to determine the reliability of *Vmaxpro* in measuring VJ and to establish whether the instrument can consistently produce measurements that are of sufficient quality to be useful for practitioners. Therefore, the aim of this study is to evaluate the test–retest and inter-device reliability of the *Vmaxpro*, when placed at the ankle or hip, in measuring countermovement jump (CMJ) in highly trained female athletes.

2. Materials and Methods

2.1. Study Design

This observational study utilized a repeated measures design to determine the test–retest and between-device reliability of the IMU-based *Vmaxpro* for VJ measurements. Data were simultaneously collected by two identical specimens of *Vmaxpro*, located on the right ankle, 1 cm above the malleolus of the tibia, and on the back above the hip between the first and fifth lumbar vertebrae. This design allows to compare the results of the jump estimation obtained by both instruments and to study the absolute error, the degree of agreement and the existence of differences between them. The sample size was determined using G*Power (v3.1.9.7, Heinrich-Heine-Universität Düsseldorf, Düsseldorf, Germany), estimating a minimum of 220 jumps for the Wilcoxon test (α = 0.05, two-tailed, ES > 0.25), and for correlations of two paired variables (power of 90%, α = 0.05, two-tailed and ES > 0.4). For this purpose, 11 participants performed 10 jumps with countermovement, resting 2 min between attempts in each session, resulting in a total of 220 valid jumps.

2.2. Procedure

The experimental procedure was conducted in accordance with a randomized, within-subjects design, in which participants were assessed in three separate sessions separated by a seven-day interval. To control for potential effects associated with circadian rhythms, all testing sessions were conducted at the same time of day. During the initial testing session, participants were provided with familiarization of the experimental protocols and anthropometric measurements were collected. In the second and third testing sessions, the same procedures were repeated in the same order: first, a standardized warm-up consisting of five minutes of continuous running was completed, followed by three minutes of dynamic range-of-motion exercises, and then two minutes of familiarization jumps in which subjects were instructed in the initial and final positions of the jump. After the warm-up, a four-minute rest period was implemented during which the inertial device was set up and the jumping protocols were reviewed. Participants then completed 10 CMJ with two minutes of rest between each attempt to control for the effects of fatigue [25].

To avoid displacements in the transverse and frontal plane, take-off and landing jumps were executed completely within the limits. CMJ were executed according to established protocols, with a rapid descent to a depth self-selected by each participant [26,27], followed by an immediate and powerful ascent to achieve take-off [26,27]. All tests were performed with the hands placed on the iliac crests in the Akimbo position [28] to avoid variability generated by the action of the arms. Participants were instructed to jump as high as possible on each attempt, and, in addition, to land on tiptoe imitating the position adopted by the ankle joint in the take-off phase, thus attempting to minimize the error produced by variations in the angle of the ankle flexion in the landing phase [29]. The jumps were monitored by a trained instructor to ensure proper execution, and attempts were deemed invalid if any of particular criteria were not met, namely if the subjects did not land within the established boundaries, if they did not land on the balls of their feet, or if they separated their hands from the iliac crest at any point during the jump. All records were collected simultaneously by two units of the *Vmaxpro*.

2.3. Participants

Eleven highly trained female volleyball players [30] from the Spanish Superliga 2 voluntarily participated in this validation study. All participants met the following criteria to be included as highly trained athletes [30]: (i) competing at the national level, (ii) being part of a team competing in the second division of the Spanish national volleyball league (Superliga 2), (iii) completing structured and periodized training and developing towards (within 20%) of maximal or nearly maximal norms within volleyball, (iv) developing proficiency in skills required to perform volleyball. The descriptive data of the sample can be seen in Table 1.

Table 1. Characteristics of highly trained female volleyball players. Data are presented as M ± SD.

N = 11	Mean	SD
Age (years)	23.10	3.10
Height (m)	1.73	0.05
Body mass (kg)	64.0	7.80
Fat percentage (%)	17.30	2.70
BMI (kg/m^2)	21.30	1.90
Training experience (years)	9.30	1.80

BMI: Body Mass Index; SD: Standard Deviation.

All participants signed an informed consent document informing them of the characteristics of the intervention, as well as the strictly scientific use of the data obtained in the intervention as specified in the World Medical Association (WMA) Declaration of Helsinki; Ethical Principles for Medical Research Involving Human Subjects 1975 (revised in Fortaleza, Brazil in 2013). In addition, this research was approved by the ethics committee of the University of Alicante (UA-2018-11-17).

All participants met the three inclusion criteria for participation in this study: being female, aged over 18 years, having 3 years minimum of training experience in volleyball, being familiar with CMJ. The exclusion criteria included presenting a current or previous pathology that entailed a medical contraindication for physical activity, presenting a previous musculoskeletal injury or one acquired during the experimental phase, not participating in all the interventions included in the study, and ingesting alcohol or drugs in the 48 h before the tests.

2.4. IMU-Based Vmaxpro

The *Vmaxpro* (Blaumann & Meyer-Sports Technology UG, Magdeburg, Germany) consists of a triaxial accelerometer, a gyroscope, and a magnetometer, weighing 16 g and measuring $4.5 \times 2.7 \times 1.2$ cm. It has a sampling rate of 1000 Hz [31] and can be attached to metal surfaces by magnets or placed elsewhere using an elastic strap. This IMU is primarily designed for velocity-based resistance training [32], obtaining data from acceleration integration, so it can provide values for related variables such as peak velocity, average velocity, peak eccentric velocity, average eccentric velocity, percentage of force development, percentage of eccentric force development, average propulsive velocity, distance, and duration. The height of a jump can be calculated based on the velocity of the jumper's center of mass at take-off [33]. By applying the law of conservation of mechanical energy to the flight phase of the jump, a relationship between jump height and take-off velocity can be established. In the case of vertical jumping, air resistance is considered minimal, so the jumper can be treated as a projectile in free flight. Taking into account the changes in kinetic energy and gravitational potential energy from the moment of take-off to the peak of the jump, the jump height reached can be calculated by

$$VJ = \frac{v_0^2}{2g},$$

where v_0 is the take-off velocity. Therefore, by measuring the peak velocity data with the IMU, the vertical jump height can be calculated as this corresponds to the take-off velocity [1]. The data are sent instantly via Bluetooth wireless connection (65 Hz) to a smartphone or tablet device with the *Vmaxpro* app (BM Sports Technology GmbH, Magdeburg, Germany) installed, allowing the data to be viewed instantly and exported to a spreadsheet in CSV format. Before each measurement, each device was calibrated on all six faces by placing it on a completely flat surface for a sufficient period of time, allowing the software to recognize and establish the local three-dimensional coordinates. Once calibrated, the first unit was placed on an elastic band to be as close as possible to the center of mass, at the subject's hip, according to the manufacturer's specifications [34] (hip device). The second unit was placed by securing it with pre-bandage tape on top of the tibial malleolus (ankle device). In this study, a smartphone with the same version of the app was utilized for each sensor to simultaneously collect take-off velocity data on each jump. The paired data were then organized into a spreadsheet format and analyzed using a statistical software package. The complete setup is shown in Figure 1.

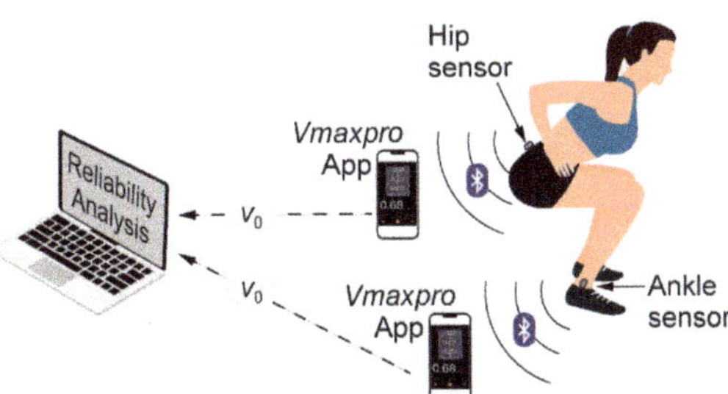

Figure 1. Experimental setup showing the locations of the two *Vmaxpro* devices: hip (close to the center of mass, as indicated by the manufacturer), and ankle (tibial malleolus). Data are sent to smartphones via Bluetooth and the take-off velocity v_0 is used to conduct the reliability analysis.

2.5. Statistical Analysis

Descriptive data are shown as the mean and standard deviation. The Kolmogorov–Smirnov test was used to analyze the normality of the sample, resulting in a non-normal distribution for the *Vmaxpro* data group located at the hip. The reliability of *Vmaxpro* was determined through various tests aimed at estimating the level of agreement and the magnitude of the error in the measure under test–retest and between the different device locations (hip or ankle devices) [35]. Given the non-parametric nature of the sample, the correlation analysis was carried out using the Spearman's coefficients (r_s),

$$r_s = \frac{\mathrm{cov}(\mathrm{R}(X), \mathrm{R}(Y))}{\sigma_{\mathrm{R}(X)}\sigma_{\mathrm{R}(Y)}},$$

where $\mathrm{cov}(\mathrm{R}(X), \mathrm{R}(Y))$ is the covariance and $\sigma_{\mathrm{R}(X)}$ and $\sigma_{\mathrm{R}(Y)}$ are the standard deviations of the rank variables. The intraclass correlation coefficient (ICC) (3,1) was used to determine the intra-session reliability of each of the instruments (consistency),

$$\mathrm{ICC}(3,1) = \frac{MS_R - MS_E}{MS_R + (k-1)MS_E},$$

while ICC (2,*k*) was used to establish the reliability between instruments [36],

$$\mathrm{ICC}(2,k) = \frac{MS_R - MS_E}{MS_R + \frac{MS_C - MS_E}{n}},$$

where MS_R and MS_C are the mean square for data in rows and columns, respectively; MS_E is mean square for error; n is the number of subjects; and k is the number of measurements [36]. The Lin concordance index (CCC) was calculated to determine the degree of agreement between the two device locations,

$$\mathrm{CCC} = \frac{2\rho\sigma_x\sigma_y}{\left(\mu_x - \mu_y\right)^2 + \sigma_x^2 + \sigma_y^2},$$

where ρ is the correlation coefficient, μ and σ^2 are the means and variances for x and y. CCC can be split into two terms $\mathrm{CCC} = \rho \times C_b$, which indicates the degree of similarity between the jump height data obtained from the two devices, where ρ represents the precision of the measurement and C_b represents the accuracy of the measurement. An ideal scenario, where $x = y$, would result in a CCC of 1.0 [37]. The results obtained for the different correlation coefficients were classified as trivial (<0.1), small (0.1–0.29), moderate (0.3–0.49), high (0.5–0.69), very high (0.7–0.89), and practically perfect (>0.9) [38].

Additionally, a linear dependence analysis was performed on the paired observations using the Passing–Bablok linear regression method [39]. This method was used to determine the slope and intercept necessary for obtaining the fitting equation between the two instruments. The standard error of the estimate (SEE) was also determined to evaluate the degree of fit of the data to a linear model,

$$\text{SEE} = \sqrt{\frac{\sum(x - x\prime)^2}{n}},$$

where x is the measured values, x' is the values predicted by the multiple regression model and n is the number of pairs of measures. The magnitude of the error was estimated by calculating the standard error of the measure SEM,

$$\text{SEM} = \frac{Sd_{diff}}{\sqrt{2}},$$

where Sd_{diff} is the standard deviation of the difference [40]. This statistic provides information on the error in absolute terms from the analysis of the dispersion of values around the true value [41]. SEM can also be expressed in its standardized form interpreting those values of SEM as trivial (<0.2), small (0.2–0.59), moderate (0.6–1.19), large (1.2–1.99), and very large (>0.2) [41]. The relative reliability of the measurement was established by calculating the coefficient of variation (CV) as

$$\text{CV} = 100 \cdot \frac{SEM}{\mu},$$

where μ is the mean value. The CV outcomes were then classified according to previous studies [38,42] as follows: low (>10%), moderate (5–10%), good (<5%). To determine if the method is highly reliable, it was established that ICC should be greater than 0.90 and CV should be less than 5% [37,41,43].

Sensitivity of the measurement was evaluated using the smallest worthwhile change (SWC), which allows for determining the minimum improvements that present a practical impact [44],

$$\text{SWC} = 0.2 \cdot \sqrt{2} \cdot \text{SEM},$$

by knowing the SWC, the signal-to-noise ratio can be determined. If the signal-to-noise ratio (SWC/SEM) is greater than unity, the data can be considered reliable [43–45].

To determine the significant differences (systematic bias) in the values shown by the two *Vmaxpro* devices placed on hip and knee, a Wilcoxon test for paired samples was performed, along with the bias-corrected Hedges effect size g (ES) [46]

$$g = \frac{\mu_1 - \mu_2}{\sqrt{\frac{(n_1-1)s_1^2 + (n_2-1)s_2^2}{(n_1-1)+(n_2-1)}}},$$

where μ and s denote the mean and standard deviation of paired samples 1 and 2. This test was used to evaluate the statistical significance and magnitude of the difference between the two devices [46]. The level of significance was established at $p < 0.05$, and the differences expressed as ES were interpreted according to Hopkins et al. [46] as trivial (<0.2), small (0.2–0.59), moderate (0.6–1.19), large (1.2–1.99), very large (0.2–3.99), and huge (>4.0). The degree of agreement between the height data obtained from the two paired devices was evaluated using Bland–Altman plots. These plots allow visualizing the systematic error and the limits of agreement (LoA) for 95%,

$$\text{LoA} = \pm 1.96 \cdot SD_{diff}.$$

The maximum allowed differences were calculated from the CV of each method using the following expression $\sqrt{(\text{CV}^2_{method1} + \text{CV}^2_{method2})}$ [47]. The presence of disagreement between the two methods was determined by analyzing the 95% confidence limits of the upper and lower LoA. If the upper limit is below the minimum allowed difference and the lower limit is above the maximum allowed difference, the methods are considered in agreement [48]. Additionally, the presence of proportional error was identified if the Pearson product–moment correlation coefficient (r^2) is greater than 0.1 [41,49]. This information was used to evaluate the level of agreement and identify any potential sources of error between the two devices.

The level of agreement and potential errors between the two devices were calculated in multiple situations. The reliability between the devices was calculated using paired data from devices located at the hip and ankle, and the intra-session reliability of each device was studied using data from each jump and for each device separately. Additionally, the test–retest reliability between sessions was calculated using data from separate sessions separated by seven days.

Statistical analysis was carried out using the *MedCalc* Statistical Software (v 20.100, MedCalc Software Ltd., Ostend, Belgium) and the validity and reliability analysis spreadsheet available in Sportsciences [50].

3. Results

Table 2 shows the descriptive results of jump heights from both sessions and locations, expressed as mean and 95% confidence intervals. In addition, the differences between sessions and between devices are shown. Statistically significant differences were observed for the values between devices (hip and ankle) with large ES, while no significant differences were observed between sessions with trivial ES.

Table 2. Descriptive results and differences observed between sessions and between devices (hip and ankle).

Device	Total (cm)	Session 1 (cm)	Session 2 (cm)	Mean Diff. between Sessions (cm)	ES (g)
Vmaxpro Hip	27.9	27.1	27.7	−0.1	0.04 (Trivial)
(CI 95%)	(27.1 to 28.65)	(27.2 to 28.6)	(26.9 to 28.5)	(−0.7 to 0.3)	(−0.22 to 0.31)
Vmaxpro Ankle	19.3	19.5	19.4	−0.39	0.07 (Trivial)
(CI 95%)	(18.8 to 19.8)	(18.8 to 20.2)	(18.8 to 19.9)	(−0.9 to 0.1)	(−0.19 to 0.34)
Mean diff. between devices (cm)	−8.5 *	−8.4 *	−8.5 *	–	–
(CI 95%)	(−8.7 to −8.2)	(−8.8 to −7.9)	(−8.8 to −8.1)	–	–
ES (g)	−2.2 (Large)	−2.2 (Large)	−2.2 (Large)	–	–
(CI 95%)	(−2.5 to −1.8)	(−2.5 to −1.9)	(−2.5 to −1.9)	–	–

* Significant difference for 95% confidence interval ($p < 0.001$); CI = confidence interval; ES = Hedge's effect size.

3.1. Intra-Session Test–Retest Reliability

Intra-session test–retest reliability was calculated by using data from the first five jumps obtained in the first session. The results were obtained by pairing consecutive jumps and determining the mean test score for both devices, i.e., with the IMU at the hip and at the ankle. These results are presented in Table 3.

The ICC values indicated near-perfect test–retest correlations in both cases, higher for the ankle (ICC 0.91 and 0.96 for the hip and ankle, respectively). The CCC values indicated greater reliability for the IMU placed at the ankle (CCC = 0.93) compared to the IMU placed at the hip. Both devices displayed near-perfect precision and accuracy as determined by CCC.

The random error or noise of the measure, quantified by SEM, was determined to be 1.41 cm for hip and 1.00 cm for ankle. In both cases, the standardized SEM was greater than 0.2, indicating that the random error was not insignificant. The relative reliability was determined to be consistent for both instruments, with CV values above 5% observed for both devices (5.10% and 5.13%). The sensitivity of the instruments was determined by the SWC, which was 0.40 cm and 0.28 cm for the hip and ankle, respectively. In both cases, the noise was greater than the SWC, resulting in a signal-to-noise ratio of less than 1.

Table 3. Intra-session test–retest reliability (intra-session consistency) for hip and ankle devices.

	Hip					Ankle				
	2–1	3–2	4–3	5–4	Mean	2–1	3–2	4–3	5–4	Mean
Mean change (cm)	0.70	−0.48	−0.37	−0.05	–	0.71	−0.67	0.07	0.25	–
CI-95% lower	−0.51	−1.83	−1.97	−1.19	–	0.05	−1.74	−1.21	−0.37	–
CI-95% upper	1.91	0.87	1.23	1.09	–	1.38	0.39	1.35	0.87	–
ICC	0.92	0.90	0.86	0.93	0.91	0.98	0.95	0.92	0.96	0.96
CI-95% lower	0.72	0.65	0.53	0.75	0.80	0.92	0.82	0.73	0.89	0.89
CI-95% upper	0.98	0.97	0.96	0.98	0.97	0.99	0.99	0.98	0.99	0.99
CCC	0.89	0.88	0.83	0.92	0.88	0.96	0.92	0.89	0.97	0.93
CI-95% lower	0.66	0.62	0.51	0.92	0.68	0.86	0.74	0.67	0.90	0.79
CI-95% upper	0.96	0.99	0.99	0.99	0.98	0.98	0.93	0.99	0.99	0.97
ρ (precision)	0.91	0.89	0.83	0.84	0.90	0.97	0.93	0.91	0.97	0.95
C_b (accuracy)	0.98	0.99	0.97	0.99	0.99	0.98	0.99	0.98	0.99	0.98
SEM (cm)	0.70	1.12	1.35	0.65	1.41	1.19	1.29	1.08	1.15	1.00
CI-95% lower	0.49	0.78	0.94	0.46	1.13	0.90	0.97	0.82	0.87	0.80
CI-95% upper	1.22	1.97	2.37	1.14	1.88	1.80	1.95	1.64	1.74	1.33
SEM_{std}	0.32	0.36	0.44	0.29	0.34	0.17	0.27	0.34	0.17	0.24
CI-95% lower	0.23	0.25	0.31	0.20	0.27	0.12	0.19	0.24	0.12	0.19
CI-95% upper	0.57	0.63	0.77	0.51	0.45	0.30	0.47	0.59	0.29	0.32
CV (%)	4.64	5.04	6.08	4.64	5.10	3.63	5.62	7.00	3.63	5.13
SWC (cm)	0.36	0.40	0.48	0.34	0.40	0.20	0.32	0.38	0.18	0.28
CI-95% lower	0.25	0.28	0.33	0.24	0.32	0.14	0.22	0.27	0.13	0.23
CI-95% upper	0.63	0.71	0.84	0.60	0.53	0.35	0.56	0.67	0.32	0.38

CI = confidence intervals for 95%; ICC = intraclass correlation coefficient; CCC = Lin's concordance coefficient; ρ = CCC-derived precision; C_b = CCC-derived accuracy; SEM = standard error of measurement; SEM_{Std} =standardized standard error of measurement; CV = coefficient of variation; SWC = smallest worthwhile change.

Consistency between jumps was also studied using Bland–Altman plots, as shown in Figure 2. A high degree of concordance was observed in the test–retest analysis performed on the same day, as nearly all pairings were within the bounds of the upper and lower LoA. Additionally, the systematic error, represented by the mean difference, was determined to be low across all charts, with slightly higher values for the hip (−0.7 to 0.3 cm) than for the ankle (−0.2 to 0.5 cm). No significant trends were observed that suggest heteroscedasticity for a particular device or jump pairing, as the slope values, which indicate proportionality of error, ranged from 5×10^{-5} to −0.2. The Bland–Altman plots also reveal wider confidence intervals for the hip, suggesting greater measurement noise in this device.

3.2. Between-Session Test–Retest Reliability

The reliability of *Vmaxpro* was assessed through analysis of results obtained in two sessions with a one-week interval. No significant differences were detected between the two sessions in any of the devices, as indicated by the results of the Wilcoxon test ($p > 0.05$; trivial ES). The SEM values for the hip and ankle were 1.5 cm and 1.7 cm, respectively, with trivial (0.1) and moderate (0.46) SEM_{std}. The between-session correlations were almost perfect for the hip (ICC = 0.98) and high for the ankle (ICC = 0.79).

3.3. Between-Device Reliability (Vmaxpro in Hip vs. Vmaxpro in Ankle)

Table 4 displays the results of the reliability analysis between the devices placed on the hip and ankle. The differences were significant, with a value of −8.5 cm and large ES. The random error was characterized by SEM of 1.47 cm, and a standardized value of 0.4, indicating moderate disagreement (greater than 0.2, the threshold for trivial disagreement). In contrast, the relative reliability was high, with CV values ranging from 14.5% to 19.2% for both groups, and the sensitivity was reflected in the SWC value of 0.4 cm, which represents the minimum jump increment that *Vmaxpro* can detect above the noise of measure.

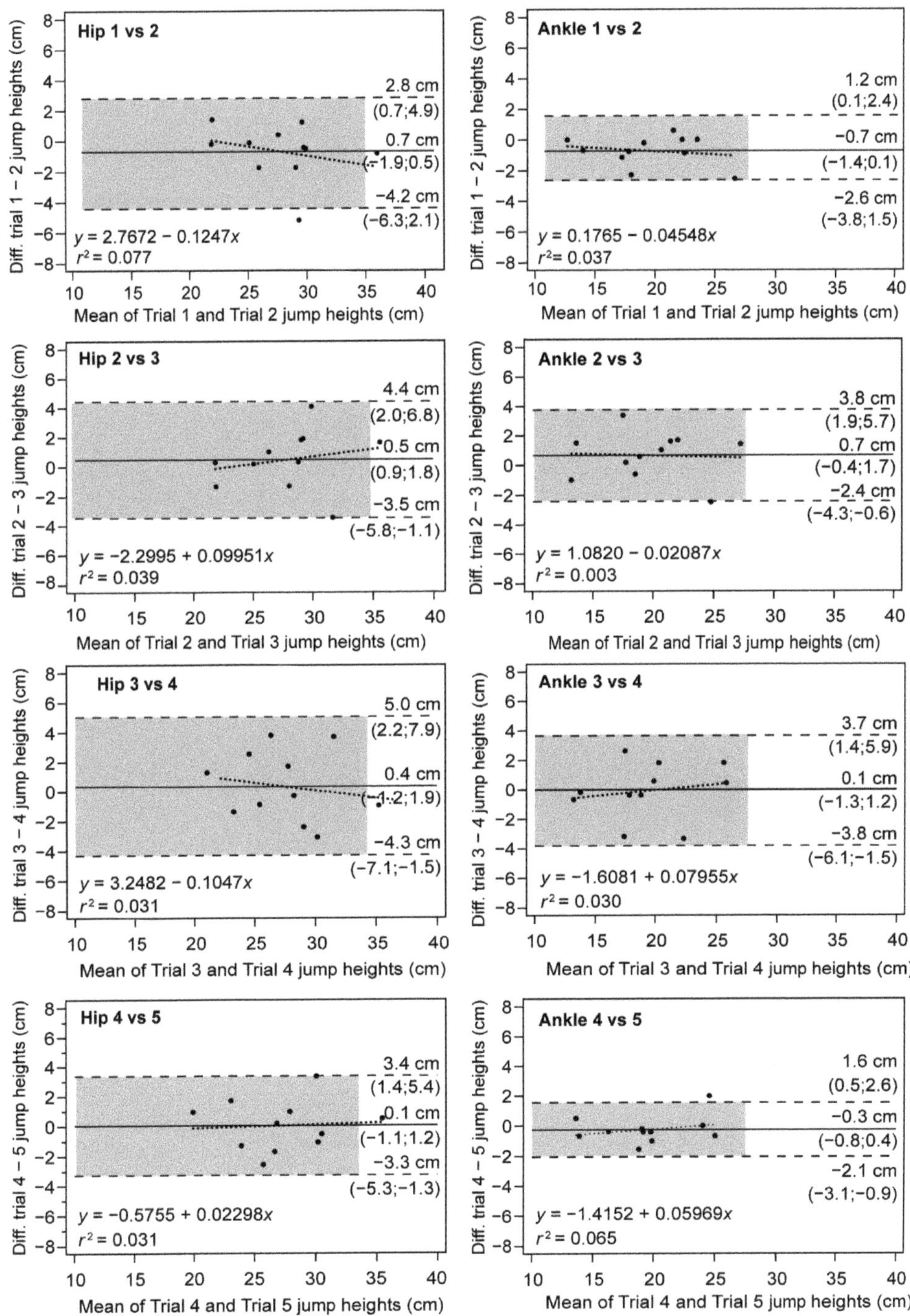

Figure 2. Bland–Altman plots for differences between jumps with the Vmaxpro placed on the hip (**left**) and ankle (**right**). Solid lines: mean differences (systematic error); dashed lines: upper and lower LoA (random error); dotted lines: regression line of the differences between devices.

Table 4. Between device reliability for the *Vmaxpro* inertial measurement unit.

	Ankle vs. Hip Devices	95% CI
Paired differences (cm)	−8.50 *	−8.7 to −8.2
ES (paired)	−2.2	−2.5 to −1.8
ICC	0.77	0.74 to 0.82
CCC	0.25	0.21 to 0.29
ρ (precision)	0.87	–
C_b (accuracy)	0.29	–
SEM (cm)	1.47	1.33 to 1.66
SEM$_{std}$	0.39	0.35 to 0.44
CV$_{hip}$ (%)	14.5	–
CV$_{ankle}$ (%)	19.2	–
SWC (cm)	0.42	0.37 to 0.47
SNR	0.28	0.26 to 0.30

95% CI = confidence intervals for 95%; ES = effect size; ICC = intraclass correlation coefficient; CCC = Lin's coefficient of concordance; SEM = standard error of measurement; SEM$_{Std}$ = standardized SEM; ρ = precision derived from CCC; C_b = accuracy derived from CCC; CV = coefficient of variation; SWC = smallest worthwhile change. SNR = signal to noise ratio; * Statistically significant differences ($p < 0.001$).

The agreement between the two methods was further analyzed using the Passing and Bablok regression (Figure 3) and the Bland–Altman (Figure 4) plots. The Spearman correlation derived from the regression showed high values ($r_s = 0.84$, $p < 0.001$), with with $r^2 = 0.71$. The systematic error between the methods can be quantified using the intercept, which revealed high values of 6.8 cm, a slope of 1.09 (relative error), and random error (SEE) of 1.39 cm. The proportionality of the error was confirmed, as indicated by significant differences in the linearity observed in the Cusum test ($p = 0.44$) and greater dispersion at higher CMJ heights, as visually depicted in Figure 3b.

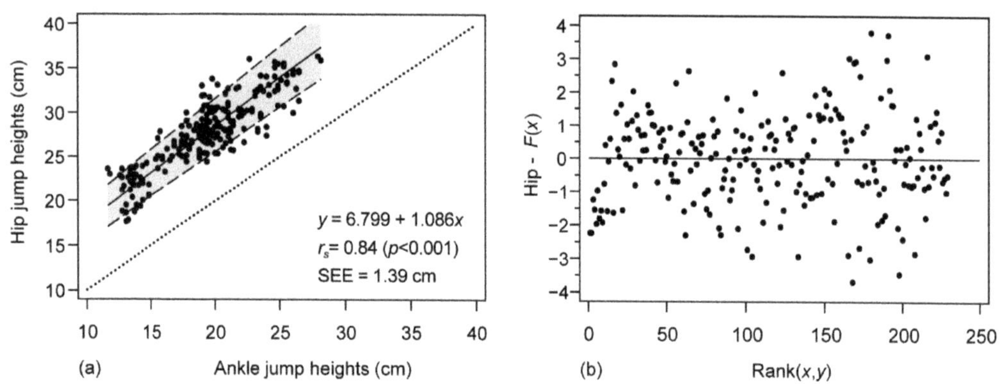

Figure 3. Correlation analysis between the *Vmaxpro* devices placed on hip and ankle through Passing and Bablok regression and residual plot. (a) Regression analysis. Solid line: fitted line; dashed lines: 95% CI of the fitted line; dotted line: perfect agreement line, $x = y$; r_s: Spearman's correlation coefficient; SEE: standard error of the estimate. (b) Residuals plot.

Additionally, the Bland–Altman plot shown in Figure 4 revealed a high systematic error, with a significant difference ($p < 0.001$) of 8.5 cm (95% CI: 8.20 to 8.72 cm) between devices, and LoA of 4.57 to 12.35 cm. The regression equation showed a slope of 0.0851 (95% CI: 0.02 to 0.15), indicating a degree of proportionality (heteroscedasticity). The differences increased with increasing jump values, but a value of $r^2 = 0.03$ (less than 0.1) indicated lack of proportionality in the error. Finally, the maximum and minimum difference allowed was ±24 cm with the limits of agreement included in this range, and only 3.4% of the data was outside the limits of agreement.

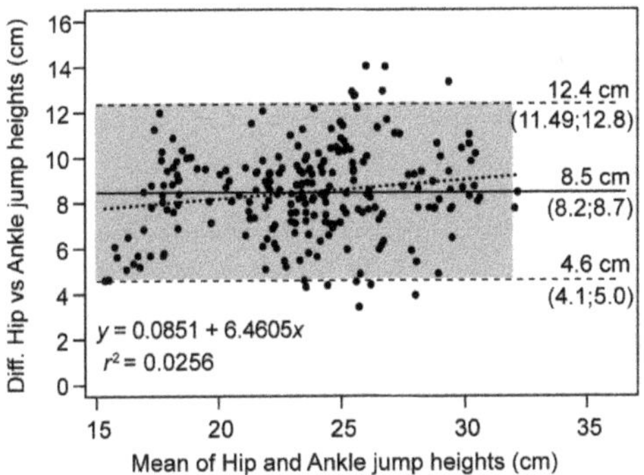

Figure 4. Bland–Altman plot for differences between devices placed on the hip and ankle. Solid lines: mean differences (systematic error); dashed lines: upper and lower LoA (random error); dotted lines: regression line of the differences between devices.

4. Discussion

The aim of the current study was to examine the reliability of the *Vmaxpro* IMU when placed in two different positions, the hip, as recommended by the manufacturer and commonly used in IMU evaluation studies, and the ankle (tibial malleolus), where simultaneous measurements were taken using two devices. The results indicate a variation in the magnitude and consistency of errors between the two placement locations.

The intra-session reliability was assessed through a test–retest design on the same day, analyzing the first five jumps of each device, with the differences between consecutive pairs and the mean of the test being calculated for both hip and ankle devices. Results showed high levels of consistency for the hip with an ICC of 0.91 (ranging from 0.86 to 0.93) and near-perfect reliability for the ankle with an ICC of 0.96 (ranging from 0.92 to 0.98). Additionally, CCC was also high, with values of 0.88 (ranging from 0.831 to 0.92) for the hip and 0.96 (ranging from 0.89 to 0.97) for the ankle. The accuracy values obtained from Lin's concordance correlation coefficient (*Cb*) were almost perfect in both locations, with a score of 0.99. Precision was determined to be the most significant factor in the final accuracy index, with a score of 0.90 for the hip and 0.95 for the ankle devices. This would mean that the consistency of both devices is affected by the accuracy of the device in making repeated measurements, this phenomenon being more noticeable in the hip placement of the device. The observed consistency values agree with those obtained by Montalvo et al. [21] who determined near-perfect values for the CMJ using the *Push Band* 2.0 IMU with ICC values of 0.98 (95% CI: 0.97 to 0.99). Additionally, the results of other studies indicate higher levels of reliability compared to the *Vmaxpro*. Rago et al. [15] reported ICC values of 0.97 (95% CI: 0.92 to 0.99) for the *Myotest Pro* device in their analysis of six CMJ jumps. The CV values obtained in our study, 5.1% for hip and ankle, are considered moderate and align with the results of other similar studies, where CVs ranged from 4.2% to 7.1% for the CMJ [15,21]. On the other hand, the results of Buchheit et al. [51] showed similar reliability values to the *Vmaxpro* when the device was placed on the tibia, with ICC values of 0.83 and a CV of 5.4%, which are considered reliable.

Regarding the measurement error and sensitivity of the instrument, *Vmaxpro* showed inconsistent values compared to the study by Rago et al. [15], which determined that the device is sensitive enough for vertical jump measurement (SEM = 0.5 cm, SWC = 0.8 cm). However, *Vmaxpro* displayed higher noise values in our study: SEM = 1.4 and 1.0 cm for hip and ankle, respectively, higher than the minimum practically significant value (SWC = 0.4 and 0.3 cm). In contrast, Buchheit et al. [51] determined low standardized SEM values (0.44) and SWC = 3% for ankle device placement, values that align more with *Vmaxpro* (SEMstd = 0.34 to 0.24). The data suggest that small changes can be obscured by the noise and thus affect the instrument's reliability, so only moderate or large changes can be detected with a single jump. Sensitivity is also much lower than that determined by Rago et al. [15], at 2.8 cm. The standardized SEM values of 0.3 for hip and 0.2 for ankle indicated errors greater than 0.2 and cannot be considered trivial. In general terms, agreement and magnitude values were higher for ankle vs. hip placement (higher ICC, CCC, lower CV, and SEM, lower SWC values).

The Bland–Altman plots confirmed the trends seen in the reliability values. The systematic error low for both the ankle and hip placements, with smaller systematic error observed for the hip (-0.7 to 0.7 cm for the ankle and 0.1 to 0.7 cm for the hip). However, there is a larger dispersion of the data (random error) for the hip, which can be seen in a wider range of agreement in all comparisons to the ankle. Heteroscedasticity was not observed in any method, as all r^2 values are below 0.1. Thus, while the reliability values showed high correlation, the observed noise was still high.

This study highlights that various factors can contribute to measurement noise in IMU assessments. Factors such as the method used to attach the device to the body and variations in detecting the exact take-off moment can affect accuracy and alter the sensitivity of the device [52]. We used an elastic band to attach the IMU at the hip and a tape bandage at the ankle, which may have resulted in more instability at the hip and contributed to the differences in reliability between the two instruments [53]. To improve reliability, it is crucial to control IMUs to minimize fluctuations and avoid disturbing elements that generate noise. Small fluctuations in measurement can lead to significant random error that can compromise reliability.

Reliability of the instrument has been analyzed in a test–retest design in various studies [54]. For *Vmaxpro*, reliability on the hip was determined to be similar (ICC = 0.98; CV = 6.1%) compared to other studies where the device was placed on the hip (ICCs range: 0.86 to 0.98; CVs range: 3.1% to 10.7%) [15,16,19,21,55]. Placing the IMU on the forefoot showed better reliability figures (ICC = 0.89 to 0.90; CV = 4.1% to 4.3%) [18,56] than the *Vmaxpro* on the ankle in this study (ICC = 0.79; CV = 8.1%). The studies that placed the IMU on the torso showed lower reliability values compared to those that placed it on the hip and forefoot (ICC = 0.69 to 0.85; CV = 8.7% to 6.7%) [17,18].

However, the reliability of the instrument should not be solely blamed as the error observed between test sessions may not be entirely due to the instrument. The time period between sessions can cause biological variability that should not be ignored. Such fluctuations in jumping performance can result from changes in physical (fitness, fatigue, learning), psychological (stress, motivation, etc.), and biomechanical (variability in performance technique) factors [54]. These factors can be so significant that they might mask the variability of the instrument. To mitigate this, averaging multiple jumps in the data analysis instead of using raw data can help avoid uncertainty [55].

Once the consistency of the device was evaluated, the agreement between devices was analyzed. A paired jumping study was performed between the two units of the *Vmaxpro* with the device placed at the hip, which is the standardized method, and the ankle. The agreement was considered high with a relative reliability of ICC = 0.77 (95% CI 0.74 to 0.82). However, CCC values were low (CCC = 0.25; 95% CI 0.21 to 0.29) due to lack of accuracy ($C_b = 0.29$) despite having high accuracy ($\rho = 0.87$). The coefficients of variation were lower for hip (14.5%) compared to ankle (19.2%).

It is observed that studies comparing the reliability of IMU placement during vertical jumping are lacking. However, studies on IMU-collected flight time showed that CVs are lower for placement at the hip (CV < 5.2%) compared to the ankle (11.6%) [15,23,56,57] or torso (CV 6.7% to 7.8%) [17,18]. Placing the device on the forefoot had the lowest CV of 4.7% as observed by Garnacho-Castaño et al. [19], which was lower than the tibia placement above the ankle for *Vmaxpro* but higher than the values observed by Montoro-Bombú et al. of 2.5% [41]. Nevertheless, all placement methods (ankle, hip, torso, and forefoot) were considered reliable [57].

The agreement between devices was assessed through a paired jumping study using two *Vmaxpro* IMU identical units. The paired difference analysis revealed the presence of a high systematic error, leading to a lack of accuracy. The *Vmaxpro* device at the ankle consistently underestimated the hip measurement by approximately 9 cm ($p < 0.001$, ES = 2.2). A previous study by Montoro-Bombú et al. [42] investigated the systematic bias in drop jumps and determined a smaller underestimation of 4.5 cm. The random error (SEM) was determined to be 1.5 cm with a standardized value of 0.4, above the trivial level (0.2), resulting in a sensitivity of 4.1 cm and SWC of 0.4 cm. The Passing–Bablok regression analysis confirmed the trends observed, with a high correlation ($r_s = 0.84$) and linearity between the two methods. The systematic error was estimated to be 6.7 cm and the slope was 1.09, while the random error was quantified as 1.4 cm using the standard error of the estimate. Although a correlation between the two instruments was indicated, the presence of large systematic and random errors was noted. The Bland–Altman graph showed a systematic error of 8.5 cm, with limits of agreement of 4.6 cm for the lower and 12.4 cm for the upper limit. This implies an underestimation by the ankle instrument and a high degree of dispersion. A proportional error (slope = 0.08) was also observed, but considered trivial ($r^2 < 0.1$).

The findings of the study indicate a strong linear dependence between the ankle and hip methods, as evidenced by the high correlation results. However, the level of systematic error and noise detected may raise concerns about the reliability of the *Vmaxpro* device in both locations. To the best of our knowledge, there are no studies that specifically examine the reliability of ankle–hip positioning for IMUs. Previous studies on the validity of accelerometer positions on the body have concluded that both ankle and hip positions are acceptable [58]. However, these studies focused on variables related to range of motion rather than dynamic variables like CMJ. In contrast, Althouse [59] conducted a study to determine which body segments or combinations of segments yield the most accurate data for CMJ estimation. Results showed that the root-mean-square error increased as the IMUs were positioned further away from the hypothetical center of mass, with the largest errors observed for accelerations measured in the feet and tibias (15.1 m/s^2 and 9.0 m/s^2) compared to those located in the hip or trunk (3.0 m/s^2). These results differ from those obtained for the *Vmaxpro*, where the magnitude of error was greater for the device located at the hip.

In the design of a reliability study, it is important to recognize that error can stem from two sources: biological variation among subjects and technological variation among items [43]. The objective of these studies is often to compare technological variation; therefore, it is desirable to minimize biological variation. One approach to minimize biological variation is to utilize athletes as subjects, as they tend to exhibit higher reliability compared to non-athletes, regardless of gender. Our study specifically utilized female athletes as subjects, as they meet the criteria of being athletes, and therefore, in our view, testing male participants was not deemed necessary. However, it is recommended that future studies also include male athletes to provide further confirmation of our findings. This would provide a more comprehensive understanding of the relationship between athletic status and reliability in the context of a reliability study.

5. Conclusions

The results of this study indicate that the consistency of the *Vmaxpro* shows acceptable values, but the magnitude of systematic and random error observed in the test–retest reliability analysis in the measuring of vertical jump in highly trained female athletes is significant. The location of the device on various body segments impacts the accuracy of the measurement, leading to statistically significant differences when the IMU is placed on the ankle compared to its standard position at the hip. Thus, the inter-device reliability is affected by the placement of the instrument on the body, resulting in an underestimation of the measurement when placed on the ankle.

Author Contributions: Conceptualization, J.M.J.-O. and B.P.; Data curation, J.O.-C.; Formal analysis J.M.J.-O.; Investigation, L.V.-G. and J.O.-C.; Methodology, L.V.-G. and J.M.J.-O.; Project administration B.P.; Supervision, B.P.; Validation, L.V.-G. and J.O.-C.; Writing—original draft, L.V.-G. and B.P.; Writing—review and editing, J.M.J.-O. and J.O.-C. All authors have read and agreed to the published version of the manuscript.

Funding: This work was supported by Generalitat Valenciana (grant number GV/2021/098).

Institutional Review Board Statement: The study was conducted in accordance with the Declaration of Helsinki and approved by the Institutional Review Board of the University of Alicante (IRB No UA-2018-11-17)

Informed Consent Statement: Informed consent was obtained from all subjects involved in the study.

Data Availability Statement: The data presented in this study are available on reasonable request from the corresponding author.

Conflicts of Interest: The authors declare no conflict of interest.

References

1. Rantalainen, T.; Finni, T.; Walker, S. Jump Height from Inertial Recordings: A Tutorial for a Sports Scientist. *Scand. J. Med. Sci. Sports* **2020**, *30*, 38–45. [CrossRef] [PubMed]
2. Cronin, J.; Hansen, K. Strength and Power Predictors of Sports Speed. *J. Strength Cond. Res.* **2005**, *19*, 349–357. [PubMed]
3. Pálinkás, G.; Béres, B.; Tróznai, Z.; Utczás, K.; Petridis, L. The Relationship of Maximal Strength with the Force-Velocity Profile in Resistance Trained Women. *Acta Polytech. Hung.* **2021**, *18*, 173–185. [CrossRef]
4. Washif, J.A.; Kok, L.-Y. Relationships between Vertical Jump Metrics and Sprint Performance, and Qualities That Distinguish Between Faster and Slower Sprinters. *J. Sci. Sport Exerc.* **2021**, *4*, 135–144. [CrossRef]
5. Alba-Jiménez, C.; Moreno-Doutres, D.; Peña, J. Trends Assessing Neuromuscular Fatigue in Team Sports: A Narrative Review. *Sports* **2022**, *10*, 33. [CrossRef]
6. Gathercole, R.J.; Sporer, B.C.; Stellingwerff, T.; Sleivert, G.G. Comparison of the Capacity of Different Jump and Sprint Field Tests to Detect Neuromuscular Fatigue. *J. Strength Cond. Res.* **2015**, *29*, 2522–2531. [CrossRef]
7. McMahon, J.J.; Suchomel, T.J.; Lake, J.P.; Comfort, P. Understanding the Key Phases of the Countermovement Jump Force-Time Curve. *Strength Cond. J.* **2018**, *40*, 96–106. [CrossRef]
8. Aragón, L.F. Evaluation of Four Vertical Jump Tests: Methodology, Reliability, Validity, and Accuracy. *Meas. Phys. Educ. Exerc. Sci.* **2000**, *4*, 215–228. [CrossRef]
9. Villalon-Gasch, L.; Penichet-Tomás, A.; Jimenez-Olmedo, J.M.; Espina-agulló, J.J. Reliability of a Linear Sprint Test on Sand in Elite Female Beach Handball Players. *J. Phys. Educ. Sport (JPES)* **2022**, *22*, 1246–1251. [CrossRef]
10. Olaya-Cuartero, J.; Cejuela, R. Influence of Biomechanical Parameters on Performance in Elite Triathletes along 29 Weeks of Training. *Appl. Sci.* **2021**, *11*, 1050. [CrossRef]
11. Glatthorn, J.F.; Gouge, S.; Nussbaumer, S.; Stauffacher, S.; Impellizzeri, F.M.; Maffiuletti, N.A. Validity and Reliability of Optojump Photoelectric Cells for Estimating Vertical Jump Height. *J. Strength Cond. Res.* **2011**, *25*, 556–560. [CrossRef] [PubMed]
12. Marković, S.; Dopsaj, M.; Tomažič, S.; Umek, A. Potential of IMU-Based Systems in Measuring Single Rapid Movement Variables in Females with Different Training Backgrounds and Specialization. *Appl. Bionics Biomech.* **2020**, *2020*, 1–7. [CrossRef] [PubMed]
13. Fritschi, R.; Seiler, J.; Gross, M. Validity and Effects of Placement of Velocity-Based Training Devices. *Sports* **2021**, *9*, 123. [CrossRef] [PubMed]
14. Olovsson Ståhl, E.; Öhrner, P. Concurrent Validity of an Inertial Sensor for Measuring Muscle Mechanical Properties. 2020. Available online: http://urn.kb.se/resolve?urn=urn:nbn:se:umu:diva-173423 (accessed on 30 January 2023).
15. Rago, V.; Brito, J.; Figueiredo, P.; Carvalho, T.; Fernandes, T.; Fonseca, P.; Rebelo, A. Countermovement Jump Analysis Using Different Portable Devices: Implications for Field Testing. *Sports* **2018**, *6*, 91. [CrossRef]
16. Brooks, E.R.; Benson, A.C.; Bruce, L.M. Novel Technologies Found to Be Valid and Reliable for the Measurement of Vertical Jump Height with Jump-and-Reach Testing. *J. Strength Cond. Res.* **2018**, *32*, 2838–2845. [CrossRef]
17. Spangler, R.; Rantalainen, T.; Gastin, P.B.; Wundersitz, D. Inertial Sensors Are a Valid Tool to Detect and Consistently Quantify Jumping. *Int. J. Sports Med.* **2018**, *39*, 802–808. [CrossRef]
18. Rantalainen, T.; Gastin, P.B.; Spangler, R.; Wundersitz, D. Concurrent Validity and Reliability of Torso-Worn Inertial Measurement Unit for Jump Power and Height Estimation. *J. Sports Sci.* **2018**, *36*, 1937–1942. [CrossRef]
19. Garnacho-Castaño, M.v.; Faundez-Zanuy, M.; Serra-Payá, N.; Maté-Muñoz, J.L.; López-Xarbau, J.; Vila-Blanch, M. Reliability and Validity of the Polar V800 Sports Watch for Estimating Vertical Jump Height. *J. Sports Sci. Med.* **2021**, *20*, 149–157. [CrossRef]
20. Schleitzer, S.; Wirtz, S.; Julian, R.; Eils, E. Development and Evaluation of an Inertial Measurement Unit (IMU) System for Jump Detection and Jump Height Estimation in Beach Volleyball. *Ger. J. Exerc. Sport Res.* **2022**, *52*, 228–236. [CrossRef]
21. Montalvo, S.; Gonzalez, M.P.; Dietze-Hermosa, M.S.; Eggleston, J.D.; Dorgo, S. Common Vertical Jump and Reactive Strength Index Measuring Devices: A Validity and Reliability Analysis. *J. Strength Cond. Res.* **2021**, *35*, 1234–1243. [CrossRef] [PubMed]

22. Pino-Ortega, J.; García-Rubio, J.; Ibáñez, S.J. Validity and Reliability of the WIMU Inertial Device for the Assessment of the Vertical Jump. *PeerJ* **2018**, *6*, e4709. [CrossRef] [PubMed]

23. Watkins, C.M.; Maunder, E.; van den Tillaar, R.; Oranchuk, D.J. Concurrent Validity and Reliability of Three Ultra-Portable Vertical Jump Assessment Technologies. *Sensors* **2020**, *20*, 7240. [CrossRef] [PubMed]

24. Clemente, F.; Badicu, G.; Hasan, U.C.; Akyildiz, Z.; Pino-Ortega, J.; Silva, R.; Rico-González, M. Validity and Reliability of Inertial Measurement Units for Jump Height Estimations: A Systematic Review. *Hum. Mov.* **2022**, *23*, 1–20. [CrossRef]

25. Read, M.M. *The Effects of Varied Rest Interval Lengths on Depth Jump Performance*; San Jose State University: San Jose, CA, USA, 1997; Volume 15.

26. Mandic, R.; Jakovljevic, S.; Jaric, S. Effects of Countermovement Depth on Kinematic and Kinetic Patterns of Maximum Vertical Jumps. *J. Electromyogr. Kinesiol.* **2015**, *25*, 265–272. [CrossRef]

27. Pérez-Castilla, A.; Rojas, F.J.; Gómez-Martínez, F.; García-Ramos, A. Vertical Jump Performance Is Affected by the Velocity and Depth of the Countermovement. *Sports Biomech.* **2019**, *20*, 1015–1030. [CrossRef]

28. López, J.M.; López, J.L. Relevance of the Technique of Immobilizing Arms for the Kinetic Variables in the Countermovement Jump Test. *Cult. Cienc. Y Deporte* **2012**, *7*, 173–178. [CrossRef]

29. Wade, L.; Lichtwark, G.A.; Farris, D.J. Comparisons of Laboratory-Based Methods to Calculate Jump Height and Improvements to the Field-Based Flight-Time Method. *Scand. J. Med. Sci. Sports* **2020**, *30*, 31–37. [CrossRef]

30. McKay, A.K.A.; Stellingwerff, T.; Smith, E.S.; Martin, D.T.; Mujika, I.; Goosey-Tolfrey, V.L.; Sheppard, J.; Burke, L.M. Defining Training and Performance Caliber: A Participant Classification Framework. *Int. J. Sports Physiol. Perform.* **2022**, *17*, 317–331. [CrossRef]

31. Held, S.; Rappelt, L.; Deutsch, J.-P.; Donath, L.; Jimenez-Olmedo, M. Valid and Reliable Barbell Velocity Estimation Using an Inertial Measurement Unit. *Int. J. Environ. Res. Public Health* **2021**, *18*, 9170. [CrossRef]

32. Jimenez Olmedo, J.M.; Olaya-cuartero, J.; Villalon-gasch, L.; Penichet-tomás, A. Validity and Reliability of the VmaxPro IMU for Back Squat Exercise in Multipower Machine. *J. Phys. Educ. Sport (JPES)* **2022**, *22*, 2920–2926. [CrossRef]

33. Linthorne, N.P. Analysis of Standing Vertical Jumps Using a Force Platform. *Am. J. Phys.* **2001**, *69*, 1198–1204. [CrossRef]

34. Linnecke, T. How to Perform Jump Tests with Vmaxpro. Available online: https://vmaxpro.de/how-to-perform-jump-tests-with-vmaxpro/ (accessed on 19 September 2022).

35. Courel-Ibáñez, J.; Martínez-Cava, A.; Morán-Navarro, R.; Escribano-Peñas, P.; Chavarren-Cabrero, J.; González-Badillo, J.J.; Pallarés, J.G. Reproducibility and Repeatability of Five Different Technologies for Bar Velocity Measurement in Resistance Training. *Ann. Biomed. Eng.* **2019**, *47*, 1523–1538. [CrossRef] [PubMed]

36. Shrout, P.E.; Fleiss, J.L. Intraclass Correlations: Uses in Assessing Rater Reliability. *Psychol. Bull.* **1979**, *86*, 420–428. [CrossRef] [PubMed]

37. Lin, L.; Hedayat, A.S.; Sinha, B.; Yang, M. Statistical Methods in Assessing Agreement: Models, Issues, and Tools. *J. Am. Stat. Assoc.* **2002**, *97*, 257–270. [CrossRef]

38. Lake, J.; Augustus, S.; Austin, K.; Mundy, P.; McMahon, J.; Comfort, P.; Haff, G. The Validity of the Push Band 2.0 during Vertical Jump Performance. *Sports* **2018**, *6*, 140. [CrossRef]

39. Passing, H.; Bablok, W. Comparison of Several Regression Procedures for Method Comparison Studies and Determination of Sample Sizes. Application of Linear Regression Procedures for Method Comparison Studies in Clinical Chemistry, Part II. *Clin. Chem. Lab. Med.* **1984**, *22*, 431–445. [CrossRef]

40. Portney, L.G. *Foundations of Clinical Research: Applications to Evidence-Based Practice*, 4th ed.; Dean Emerita MGH Institute of Health Professions School of Health and Rehabilitation Sciences Boston, M., F.A. Davis Company, Eds.; F.A. Davis Company: Philadelphia, PA, USA, 2020; ISBN 9780803661134.

41. Atkinson, G.; Nevill, A.M. Statistical Methods for Assessing Measurement Error (Reliability) in Variables Relevant to Sports Medicine. *Sport. Med.* **1998**, *26*, 217–238. [CrossRef]

42. Montoro-Bombú, R.; Field, A.; Santos, A.C.; Rama, L. Validity and Reliability of the Output Sport Device for Assessing Drop Jump Performance. *Front. Bioeng. Biotechnol.* **2022**, *10*, 1015526. [CrossRef]

43. Hopkins, W.G. Measures of Reliability in Sports Medicine and Science. *Sport. Med.* **2000**, *30*, 1–15. [CrossRef]

44. Bernards, J.R.; Sato, K.; Haff, G.G.; Bazyler, C.D. Current Research and Statistical Practices in Sport Science and a Need for Change. *Sports* **2017**, *5*, 87. [CrossRef]

45. Pueo, B.; Lopez, J.J.; Mossi, J.M.; Colomer, A.; Jimenez-Olmedo, J.M. Video-Based System for Automatic Measurement of Barbell Velocity in Back Squat. *Sensors* **2021**, *21*, 925. [CrossRef] [PubMed]

46. Hopkins, A.G.; Marshall, S.W.; Batterham, A.M.; Hanin, J. Progressive Statistics for Studies in Sports Medicine and Exercise Science. *Med. Sci. Sports Exerc.* **2009**, *41*, 3–12. [CrossRef] [PubMed]

47. Petersen, P.H.; Fraser, C.G. Strategies to Set Global Analytical Quality Specifications in Laboratory Medicine: 10 Years on from the Stockholm Consensus Conference. *Accredit. Qual. Assur.* **2010**, *15*, 323–330. [CrossRef]

48. Stöckl, D.; Rodríguez Cabaleiro, D.; van Uytfanghe, K.; Thienpont, L.M. Interpreting Method Comparison Studies by Use of the Bland-Altman Plot: Reflecting the Importance of Sample Size by Incorporating Confidence Limits and Predefined Error Limits in the Graphic. *Clin. Chem.* **2004**, *50*, 2216–2218. [CrossRef]

49. Bartlett, J.W.; Frost, C. Reliability, Repeatability and Reproducibility: Analysis of Measurement Errors in Continuous Variables. *Ultrasound Obstet. Gynecol.* **2008**, *31*, 466–475. [CrossRef] [PubMed]

50. Hopkins, W.G. A Spreadsheet to Compare Means of Two Groups. *Sportscience* **2015**, *19*, 36–44.
51. Buchheit, M.; Lacome, M.; Cholley, Y.; Simpson, B.M. Neuromuscular Responses to Conditioned Soccer Sessions Assessed via GPS-Embedded Accelerometers: Insights into Tactical Periodization. *Int. J. Sports Physiol. Perform.* **2018**, *13*, 577–583. [CrossRef]
52. Pérez-Castilla, A.; Fernandes, J.F.T.; Rojas, F.J.; García-Ramos, A. Reliability and Magnitude of Countermovement Jump Performance Variables: Influence of the Take-off Threshold. *Meas. Phys. Educ. Exerc. Sci.* **2021**, *25*, 227–235. [CrossRef]
53. Weenk, D.; Stevens, A.G.; Koning, B.; Van Bernhard, B.-J.; Hermanus, J.H.; Veltink, P.H. A Feasibility Study in Measuring Soft Tissue Artifacts on the Upper Leg Using Inertial and Magnetic Sensors. In Proceedings of the 4th Dutch Bio-Medical Engineering Conference sensors, Egmond aan Zee, The Netherlands, 24–25 January 2013; p. 154.
54. Claudino, J.G.; Cronin, J.; Mezêncio, B.; McMaster, D.T.; McGuigan, M.; Tricoli, V.; Amadio, A.C.; Serrão, J.C. The Countermovement Jump to Monitor Neuromuscular Status: A Meta-Analysis. *J. Sci. Med. Sport* **2017**, *20*, 397–402. [CrossRef]
55. Pueo, B.; Hopkins, W.; Penichet-Tomas, A.; Jimenez-Olmedo, J. Accuracy of Flight Time and Countermovement-Jump Height Estimated from Videos at Different Frame Rates with MyJump. *Biol. Sport* **2023**, *40*, 595–601. [CrossRef]
56. García-Pinillos, F.; Roche-Seruendo, L.E.; Marcén-Cinca, N.; Marco-Contreras, L.A.; Latorre-Román, P.A. Absolute Reliability and Concurrent Validity of the Stryd System for the Assessment of Running Stride Kinematics at Different Velocities. *J. Strength Cond. Res.* **2021**, *35*, 78–84. [CrossRef] [PubMed]
57. Gindre, C.; Lussiana, T.; Hebert-Losier, K.; Morin, J.B. Reliability and Validity of the Myotest ® for Measuring Running Stride Kinematics. *J. Sports Sci.* **2015**, *34*, 664–670. [CrossRef] [PubMed]
58. Al-Amri, M.; Nicholas, K.; Button, K.; Sparkes, V.; Sheeran, L.; Davies, J. Inertial Measurement Units for Clinical Movement Analysis: Reliability and Concurrent Validity. *Sensors* **2018**, *18*, 719. [CrossRef] [PubMed]
59. Althouse, D. Effects of IMU Sensor Location and Number on the Validity of Vertical Acceleration Time-Series Data in Countermovement Jumping. All Graduate Plan B and other Reports. 2022. Available online: https://digitalcommons.usu.edu/gradreports/1657/ (accessed on 30 January 2023).

Review

Internal and External Load Profile during Beach Invasion Sports Match-Play by Electronic Performance and Tracking Systems: A Systematic Review

Pau Vaccaro-Benet [1], Carlos D. Gómez-Carmona [2,3,*], Joaquín Martín Marzano-Felisatti [4] and José Pino-Ortega [1,2]

1 Department of Physical Activity and Sport, Faculty of Sport Sciences, University of Murcia, 30720 Murcia, Spain; p.vaccarobenet@um.es (P.V.-B.); josepinoortega@um.es (J.P.-O.)
2 BioVetMed & SportSci Research Group, University of Murcia, 30100 Murcia, Spain
3 Training Optimization and Sports Performance Research Group (GOERD), Department of Didactics of Music Plastic and Body Expression, Faculty of Sport Science, University of Extremadura, 10003 Caceres, Spain
4 Research Group in Sports Biomechanics (GIBD), Department of Physical Education and Sports, Faculty of Physical Activity and Sport Sciences, Universitat de València, 46010 Valencia, Spain; joaquin.marzano@uv.es
* Correspondence: carlosdavid.gomez@um.es

Abstract: Beach variants of popular sports like soccer and handball have grown in participation over the last decade. However, the characterization of the workload demands in beach sports remains limited compared to their indoor equivalents. This systematic review aimed to: (1) characterize internal and external loads during beach invasion sports match-play; (2) identify technologies and metrics used for monitoring; (3) compare the demands of indoor sports; and (4) explore differences by competition level, age, sex, and beach sport. Fifteen studies ultimately met the inclusion criteria. The locomotive volumes averaged 929 ± 269 m (average) and 16.5 ± 3.3 km/h (peak) alongside 368 ± 103 accelerations and 8 ± 4 jumps per session. The impacts approached 700 per session. The heart rates reached 166–192 beats per minute (maximal) eliciting 60–95% intensity. The player load was 12.5 ± 2.9 to 125 ± 30 units. Males showed 10–15% higher external but equivalent internal loads versus females. Earlier studies relied solely on a time–motion analysis, while recent works integrate electronic performance and tracking systems, enabling a more holistic quantification. However, substantial metric intensity zone variability persists. Beach sports entail intermittent high-intensity activity with a lower-intensity recovery. Unstable surface likely explains the heightened internal strain despite moderately lower running volumes than indoor sports. The continued integration of technology together with the standardization of workload intensity zones is needed to inform a beach-specific training prescription.

Keywords: athlete monitoring; global positioning systems; time–motion analysis; high-intensity interval training; speed; team sports

Citation: Vaccaro-Benet, P.; Gómez-Carmona, C.D.; Marzano-Felisatti, J.M.; Pino-Ortega, J. Internal and External Load Profile during Beach Invasion Sports Match-Play by Electronic Performance and Tracking Systems: A Systematic Review. *Sensors* **2024**, *24*, 3738. https://doi.org/10.3390/s24123738

Academic Editors: Felipe García-Pinillos, Alejandro Pérez-Castilla and Diego Jaén-Carrillo

Received: 12 May 2024
Revised: 28 May 2024
Accepted: 6 June 2024
Published: 8 June 2024

1. Introduction

Beach sports have also experienced an increase in participation during the last decade [1]. Following the sports modalities classification realized by Read and Edwards [2], beach sports could be classified as: (a) invasion sports (e.g., soccer and handball), (b) net and wall sports (e.g., volleyball and pickleball), and (c) striking/fielding games (e.g., softball and baseball). Specifically, invasion sports consist of invading the opponent's territory and scoring a goal or point depending on time [3]. Whilst conventional invasion team sports like soccer, rugby, or handball have been highly monitored and described in the sport sciences area [4–6], beach sports variants presented a lack of research. In this sense, precisely quantifying the external and internal loads that athletes are exposed to during the training and competition context has demonstrated their fundamental utility for injury prevention and performance optimization in sports settings [7]. This highlights the need for workload monitoring also in beach sports disciplines, which entail unique demands compared to their indoor equivalents [8].

The workload imposed on athletes can be broadly categorized into two components: external load and internal load [9]. External load represents the mechanical and locomotor actions completed during training or competition, which is quantified through variables like total distance, accelerations/decelerations, jumps, and impacts [10,11]. On the other hand, internal load refers to the relative physiological and psychological stress elicited by the activity, commonly measured via metrics including heart rate, blood lactate concentration, and subjective ratings of perceived exertion scales in different formats (CR-10, CR-100, or 6–20) [12,13].

In beach sports, the time–motion analysis, microtechnologies, or heart rate monitors have been utilized for workload monitoring purposes [14,15]. The most commonly used technology for external workload quantification is global navigation satellite systems (GNSSs), particularly global positioning systems (GPSs) of the American government [16]. Subsequently, local positioning systems (LPSs) were developed to improve the GNSS signal where the satellite coverage is deficient, changing the satellites by antennas around the court [17]. Moreover, microtechnologies (accelerometers, gyroscopes, and magnetometers) have been incorporated to improve the workload monitoring in high-intensity actions with no locomotion (e.g., jumps and collisions) [11].

In this sense, new devices called electronic performance and tracking systems (EPTSs) have been developed that include tracking technologies, microtechnologies, and wireless technologies (e.g., Ant+, Bluetooth, and Wi-Fi) to connect external sensors (e.g., heart rate, muscle oximeters, and blood lactate) and provide a holistic view of internal and external workload demands on athletes [18]. These devices allow the gathering of external load metrics, including the total distance, distance in different speed zones, accelerations/decelerations in different intensity zones, peak speeds, and number of jumps or impacts exceeding given g-forces, amongst others [5]. When coupled with the simultaneous heart rate or rate of perceived exertion values, they permit quantifying both external and internal loads of beach sports matches and training drills in a holistic and non-invasive way [13].

Beach sports present specific characteristics compared to indoor modalities like an unstable surface (sand), variable environmental elements (e.g., wind, temperature), reduced gameplay and players that condition decision-making, and physical and physiological demands on athletes [8]. When compared with indoor modalities, beach sports involved a lower external workload (total distance and at lower speed, changes of speed, and impacts) but produced a higher internal workload (heart rate and rate of perceived exertion) [14,19]. Movements in sand require high levels of strength and speed due to the reduction in the applied energy in each instance of ground-to-ground contact [20]. Therefore, beach sports are demanding activities, with numerous moderate-to-high-intensity displacements and actions that are distributed intermittently throughout the game with less intense periods to facilitate recovery [15,21].

While extensive research has profiled the workload demands of invasion team sports, the literature focusing specifically on beach sports disciplines remains comparatively limited [8]. Yet, quantifying the precise external and internal loads imposed on beach athletes can, in turn, enable the individualization of training prescription and recovery, thereby, ultimately, enhancing performance and preventing overuse injuries [13]. Incorporating new technologies could enhance the analysis granularity beyond traditional monitoring approaches [7]. Therefore, the aims of this systematic review are threefold: (a) to characterize the internal and external workload demands of invasion beach sports based on competition level, age group, and sex; (b) to identify the different technologies and specific variables utilized to quantify internal and external loads in beach sports research; and (c) to report and compare the intensity zones that have been established for the various internal and external load metrics registered in beach athletes. Findings will highlight monitoring best practices to inform individualized beach training design.

2. Materials and Methods

2.1. Study Design

This manuscript is a systematic review [22] of scientific articles related to the analysis of internal and external load in invasion beach sports. The methodological procedures outlined in the Preferred Reporting Items for Systematic Reviews and Meta-Analyses (PRISMA) guidelines were followed for the development of this systematic review [23], as well as the standards for conducting systematic reviews in sports sciences [24].

2.2. Search Strategy and Study Eligibility

The following databases were used to search for relevant publications on 25 May 2024, after completing the registry protocol: Web of Science (Web of Science Core Collection, MEDLINE, Current Contents Connect, Derwent Innovations Index, KCI-Korean Journal Database, Russian Science Citation Index, and Scielo Citation Index), PubMed Electronics, and Scopus Electronic. The search strategy utilized to identify relevant studies with topics related to the study aims in the title, abstract, or keywords was: ("beach") AND ("sport" OR "sports" OR "physical activity") AND ("local positioning system" OR "LPS" OR "ultra-wideband" OR "UWB" OR "global positioning system" OR "GPS" OR "global navigation satellite system" OR "GNSS" OR "wearable" OR "inertial measurement units" OR "IMUs") AND ("demands" OR "training load" OR "match" OR "energy expenditure" OR "internal load" OR "external load" OR "heart rate" OR "player load").

An author (P.V.-B.) performed an electronic search to identify potentially eligible studies for this systematic review, and extracted data in an unblended, standardized manner. Then, two authors (P.V.-B and C.D.G.-C.) independently reviewed the titles, abstracts, and reference lists of retrieved studies to identify potentially relevant papers. Additionally, they evaluated the full texts of included articles to confirm those meeting the predetermined eligibility criteria. Disagreements regarding study eligibility were resolved by discussion and consensus between the two reviewers, with arbitration by a third author (J.P.-O.) when needed to resolve.

Finally, study eligibility was based on the PICOS framework as per the PRISMA guidelines [23]. The "Comparison" (C) and "Study design" (S) parameters were not considered for the inclusion/exclusion criteria as they were not critical for this systematic review. Studies were excluded if the type of document was case studies, doctoral thesis, books or book chapters, conference papers, patents, or reviews. Table 1 shows the eligibility criteria.

Table 1. Eligibility criteria.

PICOS	Search Words	Inclusion Criteria	Exclusion Criteria
Population	Beach, sport	Athletes that participate on invasion beach sports.	Athletes that participate in conventional sports or other non-invasion beach sports.
Intervention/ Exposure	Local Positioning System or LPS, Ultra-Wide Band or UWB, Global Positioning System or GPS, Global Navigation Satellite Systems or GNSS, Wearable, Inertial Measurement Units or IMUs.	Using one of the non-invasive and portable technologies for monitoring internal or external workload.	Not using one of the non-invasive and portable technologies for monitoring internal or external workload.
Outcomes	Demands, Training Load, Match Performance, Energy Expenditure, Internal Load, External Load, Heart Rate, or Player Load.	Studies should refer to load monitoring (internal or external load) and register specific variables in training and competition contexts.	Studies that did not register internal and external workload variables, and realize non-ecological assessments out of the training and competition contexts.

2.3. Data Extraction and Analysed Variables

The Cochrane Consumers and Communication Review Group data extraction protocol [25] was utilized to extract the following information from studies analyzing internal and external load in beach sports: (1) authors, (2) publication year, (3) sport, (4) competition level, (5) sample characteristics, (6) instruments, (7) internal and external workload variables, (8) intensity zones, and (9) referential values.

Data extraction from the included studies was performed independently by two researchers to minimize bias and error. One researcher extracted the relevant data, and the second researcher independently checked the extracted information for accuracy and completeness. Any disagreements between the two reviewers regarding the extracted data were resolved through discussion and consensus. The search results were exported as a comma-separated values (CSV) file using Windows 10 operating system. The exported data were then organized into a Microsoft Excel spreadsheet (Microsoft Corporation, Redmond, WA, USA) to systematically categorize the identified studies.

2.4. Quality of the Studies

The methodological quality of the included studies was evaluated using the Methodological Index for Non-Randomized Studies (MINORS) [26], which is a widely accepted and validated assessment tool for non-randomized studies. It includes 8 items for non-comparative studies and 4 additional items for comparative studies. The eight items for non-comparative studies are: (1) clearly stated aim, (2) inclusion of consecutive patients, (3) prospective data collection, (4) endpoints appropriate to study aim, (5) unbiased assessment of study endpoint, (6) follow-up period appropriate to study aim, (7) <5% lost to follow-up, and (8) prospective calculation of study size. The four additional items for comparative studies are: (9) adequate control group, (10) contemporary groups, (11) baseline equivalence of groups, and (12) adequate statistical analyses. Each item is scored as (0) not reported, (1) reported but inadequate, or (2) reported and adequate, obtaining a maximum score of 16 points for non-comparative and 24 points for comparative study designs. The MINORS quality assessment was realized by two reviewers independently, and interrater reliability was assessed by the intraclass correlation coefficient.

3. Results

3.1. Search Results

Seventy-nine studies were identified from the database search on Web of Science ($n = 23$), Scopus ($n = 26$), and PubMed ($n = 30$). In addition, four additional studies were identified through the list of references and other sources, being a total of 83 articles. The Zotero reference manager software (version 6, Corporation for Digital Scholarship, Vienna, VI, USA) was used to import and eliminate any duplicates (25 studies). Then, 26 records were excluded from screening due to the type of document (four books or book chapters, one patent, and two conference papers) and being out of the sport context ($n = 19$). From the remaining 32 studies, 12 did not fulfill the inclusion criteria after the revision of the full text due to: (a) specific evaluation tests ($n = 2$), (b) match or training demands in non-invasion beach sports ($n = 11$), (c) match or training demands in conventional sports (n= 2), (d) the notational analysis of matches ($n = 1$), and (e) referees ($n = 1$). Finally, 15 studies that evaluate the internal and external workload in beach invasion sports were included in this systematic review: (a) beach handball ($n = 10$) [15,19,27–35] and (b) beach soccer ($n = 4$) [14,21,36,37]. None of the studies assessed beach rugby. A detailed representation of the selection process is illustrated in the flow diagram depicted in Figure 1.

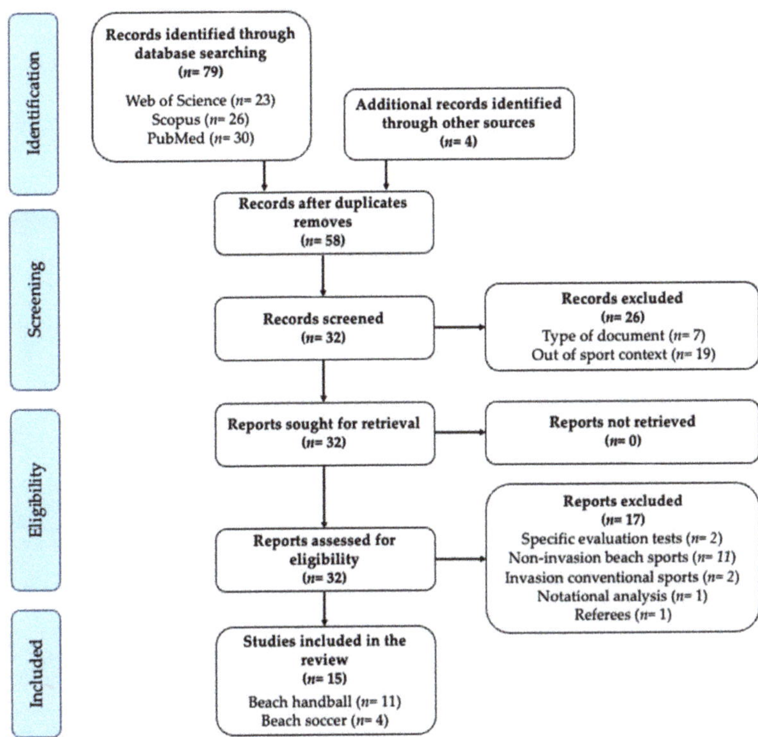

Figure 1. PRISMA flow diagram of study selection process.

3.2. Quality of the Studies

In order to evaluate the quality of the selected studies, the MINORS scale was employed [26]. Prior to the quality assessment, an inter-coder reliability analysis was conducted, yielding a value of 0.95, indicating a high level of agreement between observers (95% confidence interval: 0.93 to 0.97). The principal findings of the quality indicators for the chosen studies were as follows: (1) all studies obtained a B score with an average methodological quality of 12.93/16 (80.83%); (2) one study attained 15/16 points [15]; (3) two studies obtained 14/16 points [14,30], (4) seven studies obtained 13/16 points [19,21,27,32–34,37], (4) five studies achieved 12/16 points [28,29,31,35,36], and (5) no study received a score below 12 points that correspond to the C score (insufficient methodological quality) (see Table 2 for more details).

Four key aspects were primarily associated with methodological deficiencies in the selected studies: (1) Criterion 8, where 100% of studies did not report appropriately the prospective calculation of study size; (2) Criterion 2, where 73.3% of articles did not clearly acknowledge the inclusion of consecutive patients; (3) Criterion 6, where 60.0% did not report appropriately the follow-up period to study aim; and (4) Criterion 7, where 40.0% did not clearly report the <5% lost to follow-up.

Table 2. Methodological quality of selected studies.

Selected Studies (Authors and Year)	MINORS Criteria								Total Score
	1	2	3	4	5	6	7	8	
Castellano and Casamichana (2010) [21]	2	1	2	2	2	1	2	1	13/16
Scarfone et al. (2015) [37]	2	2	2	2	1	1	2	1	13/16
Bozdogan (2017) [36]	2	1	2	2	1	2	1	1	12/16
Pueo et al. (2017) [19]	2	1	2	2	1	2	2	1	14/16
Gutiérrez-Vargas et al. (2019) [29]	2	1	2	2	2	1	1	1	12/16
Gómez-Carmona et al. (2020) [28]	2	1	2	2	2	1	1	1	12/16
Mancha-Triguero et al. (2020) [35]	2	1	2	2	2	1	1	1	12/16
Zapardiel and Asín-Izquierdo (2020) [34]	2	2	2	2	2	1	1	1	13/16
Iannaccone et al. (2021) [30]	2	2	2	2	1	2	2	1	14/16
Sánchez-Sáez et al. (2021) [32]	2	1	2	2	1	2	2	1	13/16
Costa et al. (2022) [14]	2	1	2	2	2	2	2	1	14/16
Müller et al. (2022) [31]	2	1	2	2	2	1	1	1	12/16
Gómez-Carmona et al. (2023) [15]	2	2	2	2	2	2	2	1	15/16
Cobos et al. (2023) [27]	2	1	2	2	2	1	2	1	13/16
Zapardiel et al. (2023) [33]	2	1	2	2	2	1	2	1	13/16

Note. MINORS criteria: (1) clearly stated aim, (2) inclusion of consecutive patients, (3) prospective data collection, (4) endpoints appropriate to study aim, (5) unbiased assessment of study endpoint, (6) follow-up period appropriate to study aim, (7) <5% lost to follow-up, and (8) prospective calculation of study size.

3.3. Research Evolution, Competition Level, and Characterization of Beach Sports Athletes

Table 3 shows the research evolution (authors and year of publication), competition-level athletes' characterization, internal and external load variables registered, and tools per beach sport. Publication dates ranged from 2010 to 2023, indicating increasing research attention on these sports from 2020 to the present (studies < 2020: $n = 5$; studies $\geq$ 2020: $n = 10$). The included studies examined beach soccer and beach handball players ranging from amateur to professional international levels. In beach soccer, athletes competed at the amateur [37] and national level [14,21,36]. For beach handball, regional- [28,35], national- [15,29], European- [33,34], and international-level players [19,27,30–32] were analyzed. This indicates the research has progressed from recreational to elite settings.

The beach athletes present the following characteristics: (a) age—in beach soccer, ranging from 23.6 ± 4.4 to 29.4 ± 6.9 years, and, in beach handball, ranging from 20.1 ± 4.9 to 26.3 ± 4.8 years; (b) height—in beach soccer, ranging from 1.82 ± 0.06 to 1.77 ± 0.05 m, and, in beach handball, ranging from 1.87 ± 0.09 to 1.78 ± 0.04 m; and (c) body mass—in beach soccer, ranging from 79.3 ± 9.1 to 71.8 ± 3.8 kg, and, in beach handball, ranging from 86.9 ± 9.5 to 77.6 ± 13.4 kg. Beach handball players were younger, taller, and heavier than beach soccer players. Regarding sex differences in beach handball, male players were taller (male vs. female: 1.87–1.78 vs. 1.70–1.66 m) and heavier (male vs. female: 86.9–77.6 vs. 70.5–60.0 kg) than female players.

A variety of technologies were used, primarily GPS, UWB systems, accelerometers, and heart rate monitors. Earlier works relied more on video analysis ($n = 1$) [37], while recent studies integrated EPTS units ($n = 13$; e.g., WIMU PRO, SPI Pro X, Optimeye S5) to capture the external load [14,15,19,21,27–35]. The internal load assessment has been carried out via a heart rate band ($n = 11$; Polar Electro and coded T14 systems) [15,19,21,27,29,30,32–34,36,37], a lactate meter ($n = 2$) [36,37] or psychological rated effort ($n = 4$) [14,30,33,34].

External load variables focused on locomotive demands like total distance or per different speed zones ($n = 10$) [14,15,19,21,27,29,31–34], accelerations/decelerations or per intensities ($n = 7$) [15,19,27,30–34], jumps ($n = 7$) [15,28,30,31,33–35], impacts or per intensities ($n = 5$) [15,19,28,31,35], steps ($n = 3$) [15,28,35], changes of direction ($n = 2$) [30,31], player load ($n = 7$) [15,28,30,31,33–35], or body load ($n = 2$) [19,29]. For the internal load, average, maximum, and minimum HR ($n = 10$) continue to be primary measures [15,19,21,29,32–34,36,37], supplemented by time in HR zones ($n = 5$) [19,21,27,32,33], blood lactate ($n = 2$) [36,37], RPE ($n = 2$) [14,30], and TRIMP ($n = 2$) [33,34] more recently.

Table 3. Research evolution, competition level, athletes' characteristics, assessment instruments, and registered variables per beach sport.

Selected Studies (Authors and Year)	Players' Level	Athletes' Characteristics						Instruments		Variables	
		N°	Gender	Age (Years)	Height (Meters)	Body Mass (kg)	External Load Tool, Technology (Company)	Internal Load Tool, Variable (Company)		External Load	Internal Load
Beach soccer											
Castellano and Casamichana (2010) [21]	National (Spanish League)	10	male	25.5 ± 0.5	1.80 ± 0.08	78.2 ± 5.6	MinimaxX 2.0, GPS (Catapult Sports)	Team Sport, HR (Polar)		TD and per speed zones, $Speed_{max}$	HR_{avg}, HR_{min}, HR_{max} and $\%HR_{max}$ per zones
Scarfone et al. (2015) [37]	Amateur (3 years' experience)	10	male	23.6 ± 4.4	1.77 ± 0.05	71.8 ± 3.8	Hi8 Pro Camera, Video (SONY)	Team Sport, HR (Polar) Accusport, Lactate (Roche)		% time per locomotive category	HR_{avg}, HR_{max}, HR_{min} Blood lactate
Bozdogan (2017) [36]	National (Turkish League)	12	male	28.33 ± 3.70	1.79 ± 0.08	79.3 ± 9.1		No model, HR (Polar) No model, Lactate (Scout)			HR_{avg}, HR_{max} Blood lactate
Costa et al. (2022) [14]	International (Portugal National Team)	11	male	29.4 ± 6.9	1.82 ± 0.06	74.1 ± 7.9	Apex, GPS (STATSports)	Hooper Index, RPE		TD and per speed zones	Hooper Index
Beach handball											
Pueo et al. (2017) [19]	International (Spanish National Team)	12 12	male female	26.3 ± 4.8 23.7 ± 4.8	1.87 ± 0.09 1.68 ± 0.05	84.5 ± 12.1 62.4 ± 4.6	SPI Pro X, GPS (GPSport)	Electro, HR (Polar)		TD and per speed zones, $Speed_{avg}$, Imp, Acc per intensities, BL	HR_{avg}, HR_{max} and $\%HR_{max}$ per zones
Gutiérrez-Vargas et al. (2019) [29]	National (Costa Rican Tournament)	8 8	male female	25.6 ± 9.0 26.0 ± 7.0	1.78 ± 0.04 1.67 ± 0.08	78.1 ± 6.5 70.5 ± 12.7	SPI Pro X II, GPS (GPSport)	Coded T14, HR (Polar)		TD, $Speed_{avg}$, $Speed_{max}$, BL, Imp	HR_{avg}
Gómez-Carmona et al. (2020) [28]	Regional (Senior Extremadura Championship)	20 25	male female	23.92 ± 0.05 19.54 ± 3.72	1.79 ± 0.08 1.66 ± 0.06	80.71 ± 13.83 58.65 ± 8.46	WIMU PRO, Acel (RealTrack)			Total Imp and per intensities, PL, steps, jumps	

Table 3. *Cont.*

Selected Studies (Authors and Year)	Players' Level	Athletes' Characteristics						Instruments		Variables	
		N°	Gender	Age (Years)	Height (Meters)	Body Mass (kg)	External Load Tool, Technology (Company)	Internal Load Tool, Variable (Company)	External Load	Internal Load
Mancha-Triguero et al. (2020) [35]	Regional (U-16 Extremadura Championship)	20	male	15.6 ± 0.3	1.72 ± 0.06	65.52 ± 9.94	WIMU PRO, Acel (RealTrack)		Total Imp and per intensities, Acc/Dec, PL, steps, jumps	
Zapardiel and Asín-Izquierdo (2020) [34]	International (European Championship 2017)	25 32	male female	25.3 ± 4.8 25.3 ± 4.8	1.87 ± 0.07 1.68 ± 0.04	86.9 ± 9.5 60.7 ± 3.8	OptimEye S5, GPS (Catapult)		TD, Acc/Dec, Jumps, PL	HR_{avg}, TRIMPS
Iannaccone et al. (2021) [30]	International (U17 Lithuanian National Team)	13	male	15.9 ± 0.3	1.80 ± 0.10	67.4 ± 6.8	Clearsky, GPS (Catapult)	H10 Electro, HR (Polar)	PL, Acc/Dec, CoD, Jumps, Events per intensities	SHRZ score, RPE
Sánchez-Sáez et al. (2021) [32]	International (Spanish National Team)	9	female	24.6 ± 4.0	1.68 ± 0.06	62.4 ± 4.6	SPI Pro X, GPS (GPSport)	Electro, HR (Polar)	TD and per speed zones, $Speed_{avg}$, $Speed_{max}$	HR_{avg}, HR_{max} and $\%HR_{max}$ per zones
Müller et al. (2022) [31]	International (German National Team)	69	male	20.1 ± 4.9	1.87 ± 0.09	83.9 ± 11.5	Clearsky, GPS (Catapult)		TD, $Speed_{max}$, Max Acc/Dec, Total Imp, EE, PL, CoD, Jumps	
Gómez-Carmona et al. (2023) [15]	National (Brazilian League)	54 38	male female	22.1 ± 2.6 24.4 ± 5.5	1.80 ± 0.05 1.70 ± 0.05	77.6 ± 13.4 67.5 ± 6.5	WIMU PRO, UWB (RealTrack)	No model, HR (Garmin)	TD and per speed zones, $Speed_{max}$, Total Acc/Dec and per intensities, Total Imp and per intensities, PL, steps, jumps	HR_{avg}
Cobos et al. (2023) [27]	International (Spanish National Team)	14	female	24.6 ± 4.0	1.69 ± 0.06	60.0 ± 4.1	SPI HPU, GPS (GPSport)	Electro, HR (Polar)	TD, $Speed_{max}$, Acc/Dec per intensities	$\%HR_{max}$ per zones
Zapardiel et al. (2023) [33]	International (Europe Championship)	25 32	male female	25.38 ± 4.82 25.38 ± 4.82	1.87 ± 0.07 1.68 ± 0.04	86.96 ± 9.53 60.78 ± 3.87	OptimEye S5, GPS (Catapult)	Electro, HR (Polar)	TD and per speed zones, $Speed_{max}$, Acc/Dec, Jumps, PL	HR_{avg}, HR_{max}, HR_{min}, $\%HR_{max}$ per zones, TRIMPS

Note. N: sample; GPS: Global Positioning Systems, UWB: Ultra-Wide Band; Acel: Accelerometer; HR: Heart Rate; TD: Total Distance; $Speed_{max}$: Maximum Speed; $Speed_{avg}$: Average Speed; Acc: Accelerations; Dec: Decelerations; Imp: Impacts; PL: Player Load; BL: Body Load; EE: Energy Expenditure; MP: Metabolic Power; CoD: Changes of Direction; HR_{AVG}: Average Heart Rate; HR_{max}: Maximum Heart Rate; HR_{avg}: Average Heart Rate; HR_{min}: Minimum Heart Rate; TRIMP: Training Impulse.

3.4. External Workload Demands and Intensity Zones

The external workload demands and intensity zones profile in beach sports players are showed in Table 4. Players covered 929.5 ± 269.5 m (566-to-1606 m.) [14,15,19,27,29,31–34] at an average speed of 3.1 ± 0.95 km/h (2.5-to-4.2 km/h) [19,29,32], and reached a maximum speed of 16.5 ± 3.3 km/h (11.9-to-21.7 km/h) [15,21,27,29,31–34]. Athletes performed 368.2 ± 103.4 (268-to-533) accelerations over 0 m/s^2 [15,34,35] or 41.1 ± 13.3 (19-to-53) accelerations over 1 m/s^2 [19,27,30,31,33], and suffered 718-to-1251 impacts over 0 g [29], 477-to-572 impacts over 2 g [28] or 78-to-95 impacts over 5 g [19]. In addition, beach players realized 8.3 ± 4.1 (4-to-14) jumps [28,30,31,33–35] and 815.3 ± 45.1 (765-to-852) steps [28,35] that entail a player load of 12.5 ± 2.9 a.u. (8.8-to-16.2 a.u.) by RealTrack Systems or 125.1 ± 29.7 a.u. (88-to-162 a.u.) by Catapult Sports [15,28,30,31,33–35] and a body load of 18.8 ± 5.9 a.u (11.3-to-24.4) [19,29].

Intensity zones vary according to selected studies in distance, accelerations, and impacts. Regarding the distance covered, two procedures to fix intensity zones were identified: (1) video analysis with five zones (standing, walking, jogging, running, and sprinting) considering the intensity of movements by coders [37], and (2) tracking technologies (GPS or UWB) with five or six zones considering the speed of displacements [14,15,19,21,27,32–34]. With five intensity zones, two classifications were found: (1) Z1 0–4 km/h, Z2 4–7 km/h, Z3 7–13 km/h, Z4 13–18 km/h, and Z5 > 18 km/h [21]; and (2) Z1 1–5.9 km/h, Z2 6–8.9 km/h, Z3 9–11.9 km/h, Z4 12–14.9 km/h, and Z5 >15 km/h [33,34]. In addition, three studies modified the classification utilized by Castellano and Casamichana [21] with six intensity zones dividing the lowest intensity zone into two zones: standing (0–0.4 km/h) and walking (0.4–4 km/h) [15,19,32]. Finally, Lara-Cobos et al. [27] utilized a different classification system based on the maximum speed of the session dividing the distance covered in six zones: Z1 (<10% S_{max}), Z2 (10–29% S_{max}), Z3 (30–49% S_{max}), Z4 (50–79% S_{max}), Z5 (80–95% S_{max}), and Z6 (>95% S_{max}). The majority of displacements (85–90%) were realized <13 km/h [14,15,19,21,27,32–34].

Accelerations were classified in three zones (Z1: 1–2 m/s^2, Z2: 2–3 m/s^2, and Z3: >3 m/s^2) [19,27] or four zones, adding a highest intensity zone (Z4: >4 m/s^2) [15]. One study also counted only accelerations over 2,5 m/s^2 [34]. Almost all acceleration were performed >2 m/s^2 [15,19,27]. On the other hand, three different settings of impact intensity zones were found: (1) three zones (0–5 g, 5–8 g, and >8 g) [35], (2) five zones (2–4 g, 4–6 g, 6–8 g, 8–10 g, and >10 g) [28], or (3) six zones (5–6 g, 6–6.5 g, 6.5–7 g, 7–8 g, 8–10 g, and >10 g) [19]. The greatest number of impacts suffered are of low or very low intensity (<6 g) [19,28,35].

When comparing beach sports, beach soccer covered a greater total distance (1370 vs. 860 m) and a greater distance over 13 km/h (140 vs. 45 m) [14,15,19,27,29,31–34], and achieved a higher maximum speed (21.7 vs. 15.9 km/h) [15,21,27,29,31–34]. No data were available to compare the accelerations, impacts, jumps, steps, player load, or body load between beach handball and beach soccer. In terms of sex-related external workload during beach handball, male players covered a greater total distance (+200–300 m) and a greater distance over 13 km/h (+15–25 m), reached a higher maximum speed (+2–3 km/h), performed more total accelerations (+30–60), and experienced more impacts (+100–300). These higher external workload demands resulted in a higher player load (2–4 a.u. calculated by RealTrack; +20–40 a.u. calculated by Catapult Sports) and body load (+3–4 a.u.) [15,19,28,29,33,34].

Table 4. External workload demands (mean ± SD) and intensity zones (cursive text) in beach sports.

Selected Studies (Authors and Year)	Players' Level N/Sex	Distance Total	Distance Z1	Distance Z2	Distance Z3	Distance Z4	Distance Z5	Speed AVG	Speed MAX	Accelerations Total	Accel. Z1	Accel. Z2	Accel. Z3	Accel. Z4	Impacts Total	Impacts Z1	Impacts Z2	Impacts Z3	Impacts Z4	Impacts Z5	Jumps n ± SD	Steps n ± SD	PL/BL a.u. ± SD
Beach soccer																							
Castellano and Casamichana (2010) [21]	National 10/male	1135 ± 27	*0–3.9* 249 ± 25	*4–6.9* 297 ± 57	*7–12.9* 422 ± 132	*13–17.9* 110 ± 38	*>18* 30 ± 28		21.7 ± 4.5														
Scarfone et al. (2015) [37]	Amateur 10/male		*standing* 35 ± 6	*walking* 46 ± 5	*jogging* 5 ± 1	*running* 12 ± 4	*sprinting* 2 ± 1																
Bozdogan (2017) [36]	National 12/male																						
Costa et al. (2022) [14]	International 11/male	1606 ± 88				*>13* 66 ± 18																	
Beach handball																							
Pueo et al. (2017) [19]	International 12/male	1235 ± 192	*0–4* 398 ± 64	*4.1–7* 433 ± 103	*7.1–13* 356 ± 101	*13.1–18* 46 ± 32	*>18* 2 ± 4	4.2 ± 0.6		53 ± 17	*1–2* 43 ± 12	*2–3* 9 ± 5	*>3* 1 ± 1		78 ± 25	*5–6* 40 ± 14	*6–7* 21 ± 5	*7–8* 5 ± 3	*8–10* 4 ± 4	*>10* 8 ± 4			*BL* 22.7 ± 9.2
	12/female	1118 ± 222	409 ± 65	371 ± 94	318 ± 109	20 ± 16	0	3.9 ± 0.8		45 ± 16	40 ± 13	4 ± 3	0		95 ± 29	51 ± 15	21 ± 6	10 ± 4	6 ± 4	6 ± 5			24.4 ± 9.5
Gutiérrez-Vargas et al. (2019) [29]	National 8/male	939 ± 212						2.8 ± 0.6	15.9 ± 2.1	1251 ± 30													*BL* 16.7 ± 7.4
	8/female	613 ± 145						1.8 ± 0.4	13.6 ± 2.2	718 ± 138													11.3 ± 4
Gómez-Carmona et al. (2020) [28]	Regional 20/male														572 ± 266	*2–4* 467 ± 175	*4–6* 125 ± 105	*6–8* 31 ± 20	*8–10* 7 ± 5	*>10* 4 ± 5	6 ± 5	765 ± 309	*PL* 15.1 ± 5.7
	25/female														477 ± 218	381 ± 168	77 ± 47	14 ± 10	3 ± 3	2 ± 3	4 ± 3	852 ± 294	14.4 ± 4.4
Mancha-Triguero et al. (2020) [35]	Regional/ 20/male									533 ± 309					476 ± 192		*0–5* 334 ± 130	*5–3* 117 ± 60	*>8* 25 ± 13		5 ± 3	829 ± 267	*PL* 17.4 ± 4.8
Zapardiel and Asín-Izquierdo (2020) [34]	International 25/male	891 ± 313	*1–5.9* 581 ± 244	*6–8.9* 133 ± 100	*9–11.9* 76 ± 80	*12–14.9* 18 ± 24	*>15* 2 ± 2		20.5 ± 4.3	396 ± 110		*>2.5* 9 ± 4									12 ± 5		*PL* 10.3 ± 3.4
	32/female	739 ± 317	440 ± 200	145 ± 89	63 ± 60	21 ± 25	2 ± 5		18.4 ± 0.4	338 ± 105		5 ± 6									5 ± 5		8.8 ± 3.6
Iannaccone et al. (2021) [30]	International 13/male									26 ± 14											14 ± 10		*PL* 10.9 ± 4.2
Sánchez-Sáez et al. (2021) [32]	International 9/female	898 ± 216	*0–4* 485 ± 132	*4.1–7* 262 ± 77	*7.1–13* 123 ± 39	*>13.1* 27 ± 25		2.5 ± 0.6	14.9 ± 2.3														
Müller et al. (2022) [31]	International 69/male	806 ± 214						16.5 ± 2.0	19 ± 9												11 ± 7		*PL* 9.2 ± 2.8

Table 4. *Cont.*

Selected Studies (Authors and Year)	Players' Level N/Sex	Distance (Zones in km/h; e.g., 13–17.9 or *Running*) Meters/% Time ± SD						Speed (km/h)			Accelerations (Zones in m/s^2; e.g., 2–3) n ± SD					Impacts (Zones in g Force; e.g., 8–10) n ± SD					Jumps n ± SD	Steps n ± SD	PL/BL a.u. ± SD
		Total	Z1	Z2	Z3	Z4	Z5	AVG	MAX	Total	Z1	Z2	Z3	Z4	Total	Z1	Z2	Z3	Z4	Z5			
Gómez-Carmona et al. (2023) [15]	National 54/male 38/female	754 ± 250 566 ± 252	*0–4* ND ND	*4–7* 226 ± 82 164 ± 80	*7–13* ND ND	*13–18* 81 ± 38 58 ± 30	*>18* 2 ± 10 4 ± 4		20.7 ± 8.1 15.8 ± 2.8	306 ± 72 268 ± 92	*1–2* ND ND	*2–3* ND ND	*3–4* ND ND	*>4* 40 ± 60 20 ± 20										*PL* 16.2 ± 5.1 12.7 ± 6.0
Cobos et al. (2023) [27]	International 14/female	702 ± 251	*0–4.7* 313	*4.7–8* 91	*8–12* 225	*12–14.7* 41	*>14.7* 2		15.5 ± 1.9	59 ± 37	*1–2* 36 ± 16	*2–3* 20 ± 12	*>3* 2 ± 7											
Zapardiel et al. (2023) [33]	International 25/male 32/female	1059 ± 420 882 ± 380	*1–5.9* 595 ± 241 506 ± 239	*6–8.9* 214 ± 145 204 ± 116	*9–11.9* 134 ± 116 82 ± 73	*12–14.9* 31 ± 52 36 ± 75	*>15* 3 ± 11 5 ± 13		11.9 ± 2.6 12.1 ± 2.9	47 ± 10 39 ± 13												13 ± 10 2 ± 5		*PL* 12.0 ± 4.3 10.6 ± 4.6

Note. N: sample; *n*: count; SD: standard deviation; PL: player load; BL: body load; AVG: average; MAX: maximum; Z1, Z2, Z3, Z4, and Z5: Zone 1, 2, 3, 4, and 5; ND: no data.

3.5. Internal Workload Demands and Intensity Zones

Table 5 represents the internal workload demands and intensity zones profile in beach sports players. Athletes demonstrated an average heart rate between 137–170 bpm, a minimum heart rate between 117–129 bpm, and a maximum heart rate between 166–192 bpm [15,19,21,29,32–34,36,37]. During the intermittent activity profile of beach sports, players developed anaerobic and aerobic fitness due to the heart rate demands ranging from 60% to 95% of the maximum heart rate [19,21,32–34,37].

Heart rate zones were classified according to the percentage of the maximum heart rate (%HRmax) in four zones (Z1: <75%, Z2: 76–84%, Z3: 85–89%, and Z4: >90%) [21], five zones (Z1: <65%, Z2: 66–75%, Z3: 76–85%, Z4: 86–95%, and Z5: >95%) [37], or six zones (Z1: <60%, Z2: 60–70%, Z3: 70–80%, Z4: 80–90%, Z5: 90–95%, and Z6: >95%) [15,19,27,32–34]. In addition, players performed a TRIMPS of 57.8-to-69.4 a.u. or a sHRZ of 77.4 ± 26.5, and reported a Hooper index of 8/27 a.u. and a RPE of 6–8/10 [14,30,33,34]. Finally, 6.7 to 7.0 mmoL of blood lactate have been registered during beach sports [36,37].

When comparing beach handball and beach soccer, both modalities produced similar heart rate demands (average, maximum, minimum, and intensity zones) and rate of perceived exertion. In the same way, no sex-related differences were found on internal workload demands between male and female handball players [15,19,28,29,33,34].

Table 5. Internal workload demands (mean ± SD) and intensity zones (cursive text) in beach sports.

Selected Studies (Authors and Year)	Players' Level N/sex	HR_{avg} (bpm)	HR_{min} (bpm)	HR_{max} (bpm)	HR Zones (Zones in %HR_{max}; e.g., 90–95%) % Time ± SD						TRIMPS/sHRZ (a.u. ± SD)	Hooper Index/RPE (1–10) (a.u. ± SD)	Blood Lactate (mmol ± SD)
					Z1	Z2	Z3	Z4	Z5	Z6			
Beach soccer													
Castellano and Casamichana (2010) [21]	National 10/male	165 ± 20	121 ± 5	188 ± 6		*<75* 18.8	*76–84* 8.8	*85–89* 12	*>90* 59				
Scarfone et al. (2015) [37]	Amateur 10/male	166 ± 16	117 ± 17	188 ± 11	*<65* 8 ± 8	*66–75* 11 ± 7	*76–85* 27 ± 4	*86–95* 35 ± 12		*>95* 19 ± 7			6.7 ± 3.8
Bozdogan (2017) [36]	National 12/male	158 ± 11		181 ± 9									7.0 ± 2.6
Costa et al. (2022) [14]	International 11/male											*HI/RPE* 8 ± 2/7 ± 2	
Beach handball													
Pueo et al. (2017) [19]	International 12/male 12/female	137 ± 12 138 ± 18		174 ± 15 178 ± 15	*<60* 19 ± 17 27 ± 22	*61–70* 26 ± 12 16 ± 11	*71–80* 26 ± 12 18 ± 13	*81–90* 20 ± 14 29 ± 20	*91–95* 7 ± 10 8 ± 12	*>95* 2 ± 4 2 ± 6			
Gutiérrez-Vargas et al. (2019) [29]	National 8/male 8/female	156 ± 18 158 ± 20											
Zapardiel and Asín-Izquierdo (2020) [34]	International 25/male 32/female	150 ± 10 145 ± 18	129 ± 12 120 ± 23	164 ± 11 164 ± 16	*<60* ND ND	*60–70* ND ND	*70–80* ND ND	*80–90* 36 ± 27 36 ± 26	*>90* 1 ± 2 4 ± 7	*>95* ND ND	*TRIMPS* 65.6 ± 18.8 57.8 ± 20.4		
Iannaccone et al. (2021) [30]	International 13/male										*sHRZ* 77.4 ± 26.5	*RPE* 6 ± 1	
Sánchez-Sáez et al. (2021) [32]	International 9/female	170 ± 15		192 ± 13	*<60* 3 ± 7	*60–70* 9 ± 14	*70–80* 18 ± 15	*80–90* 33 ± 18	*90–95* 19 ± 15	*>95* 18 ± 24			
Gómez-Carmona et al. (2023) [15]	National 54/male 38/female	162 ± 23 150 ± 25			*<60* ND ND	*60–70* ND ND	*70–80* ND ND	*80–90* ND ND	*90–95* ND ND	*>95* ND ND			
Cobos et al. (2023) [27]	International 14/female				*<60* 26	*60–70* 10	*70–80* 17	*80–90* 29	*90–95* 12	*>95* 6			
Zapardiel et al. (2023) [33]	International 25/male 32/female	148 ± 14 147 ± 22	123 ± 21 122 ± 32	166 ± 14 168 ± 18	*<60* ND ND	*60–70* ND ND	*70–80* ND ND	*80–90* 35 ± 20 39 ± 17	*>90* 1 ± 2 2 ± 4	*>95* ND ND	*TRIMPS* 69.4 ± 19.9 58.9 ± 24.0		

Note. SD: standard deviation; bpm: beats per minute; a.u.: arbitrary units; HR: heart rate; HR_{avg}: average heart rate; HR_{min}: minimum heart rate; HR_{max}: maximum heart rate; Z: Zone; TRIMPS: training impulses; sHRZ: summated heart rate zones; RPE: rate of perceived exertion (1–10 scale); HI: Hooper index; ND: no data.

4. Discussion

Participation in beach sports has increased over the last decade. However, the research characterizing the workload demands in beach sports is comparatively limited compared to indoor modalities. Precisely quantifying external and internal loads in training and competition enables the individualization of training prescription and recovery for performance optimization and injury prevention. Therefore, this systematic review aimed to (a) characterize the workload demands of invasion beach sports, (b) identify technologies and variables utilized to quantify loads, and (c) report intensity zones for load metrics.

4.1. External Workload Demands

Beach soccer and beach handball, as beach invasion sports, present unique external load demands due to their specific playing conditions. In beach soccer, players cover more total distance (1606 ± 88 m) and achieve higher maximum speeds (21.7 ± 4.5 km/h), as well as more meters in high-speed zones (110 ± 38 in Z4, 30 ± 28 in Z5) [14,21]. Beach handball, on the other hand, involves more explosive and high-intensity actions (533 ± 309 accelerations), including rapid accelerations, decelerations, frequent jumps (12 ± 5), and a higher impact (117 ± 60 in Z3, 25 ± 13 in Z4) [33,35]. These differences can be explained by factors such as: court size, sports rules, or game dynamics [38–40]. Concerning the court size, beach soccer presents a bigger playing area, allowing soccer players to cover longer distances, reaching higher speeds more frequently [38]. In relation to sport rules, beach handball players can be substituted several times during the game, reducing the volume load values and increasing potential explosive efforts [39]. Furthermore, the unique dynamics of beach handball involve a predominant stationary attack vs. defense interaction near the goal area, where explosive movements such as changes in direction, jumps, and "in-flight" acrobatic moves are necessary in order to generate scoring opportunities [39,40]. As a result, beach handball players experience a greater number of accelerations and decelerations, as well as frequent jumps and impacts in comparison to beach soccer.

4.2. Internal Workload Demands

Regarding the internal workload, both beach soccer and beach handball players display significant cardiovascular demands, as shown by their average heart rates ranging from 137–170 bpm [19,32]. The minimum and maximum values also reflect the intense nature of these sports, ranging from 117–129 bpm to 164–192 bpm, respectively [32,34,37]. Moreover, players can reach 20% of total game time in the maximum heart rate zone (95% HR_{max}), indicating the considerable demands imposed on the cardio-vascular system during the match [32,37]. Moreover, metrics like RPE and blood lactate can also assess the internal load in beach invasion sports. RPE scores typically range from 6 to 8 out of 10, indicating a high level of exertion experienced by players during matches [14,30]. Concurrently, blood lactate concentrations ranging from 6.7 to 7.0 mmol/L underscore the metabolic stress and anaerobic energy system [36,37]. In addition, the internal load demands in beach soccer and beach handball have direct implications for training and performance optimization. The substantial time spent in higher heart rate zones requires targeted endurance and high-intensity interval training. Moreover, the high RPE and lactate levels suggest the need for recovery strategies and metabolic conditioning. Understanding these internal load metrics is essential in order for coaches and technical staffs to design specific training sessions that enhance performance while minimizing the risk of overtraining and injury. This approach ensures that players achieve the physiological demands of match-play in beach invasion sports, leading to optimal performance outcomes.

4.3. Intensity Zones

Determining performance intensities zones based on internal and external load variables can be challenging due to different methodologies and technologies used. Various devices and brands, and the author's decision to divide the reference intensity zones considering a proposed criterion lead to variability. This variability makes it difficult to compare data across studies and sports [21,27,34,37]. Reference studies have highlighted the need for individualized intensity reference zones based on athletes characteristics for better player load management [41,42]. However, a common criterion for establishing intensity zones in beach sports would facilitate data integration and interpretation [43,44]. For this purpose, the evaluation of raw data of the magnitude of speeds, accelerations, decelerations, and impacts, and the classification by Gaussian distributions, k-means clustering, or spectral clustering could provide objective intensity zones to classify external workload demands [45,46]. From these objective zones, a unified approach would help in comparing and generalizing across studies, which can be useful in developing sport-specific training and recovery protocols. Therefore, it is necessary to carry out extensive methodological studies led by experts to establish reference intensity zones for internal and external load variables according to each sport and playing context's requirements.

4.4. Sex-Related Differences in Beach Sports

Male players exhibit a higher external workload in beach handball compared to female players. This is characterized by a greater total distance covered (additional 200–300 m), a greater distance over 13 km/h (extra 15–25 m), a higher maximum speed (additional 2–3 km/h), increased total accelerations (additional 30–60), and more impacts (additional 100–300) [15,19,28,29,33,34]. These differences in external workload can be explained by considering differences in physical capacities, strength, and types of effort exerted during the game. Male beach handball players have higher values of weight, body mass index, anthropometric characteristics, and strength, that allow them to move larger distances and at a higher intensity during competition [19,28,34].

Moreover, despite the pronounced differences in external workload, studies indicate no significant sex-related differences in internal workload demands between male and female handball players [19,33,34]. This suggests that, although male players may engage in a more intense external load, the internal physiological response and exertion experienced by female players are comparable. Factors such as heart rate, perceived exertion, and metabolic stress do not show marked differences between sexes, indicating that females experience the same internal load at lower external demands in beach sports.

4.5. Workload Demands for Indoor vs. Beach Sports

Comparative studies have found that indoor sports like handball and soccer require higher external workload demands, while beach sports require a significantly higher internal workload [19,21,47,48]. This difference is mainly attributed to the nature of the playing surfaces [49]. Beach sports, with their softer and unstable sand surface, require more energy expenditure and muscle engagement for movement, which increases the internal load despite lower external load metrics like the total distance covered and speed [19,21,47,48]. The unstable and yielding nature of the beach surface plays a crucial role in influencing the demands of beach sports [40]. Activities like running, jumping, and changing direction are more physically demanding on sand compared to the solid surface of indoor courts [49]. Additionally, variable outdoor conditions such as wind and temperature further contribute to the increased physical demands on beach sports athletes [49]. Findings suggest that training in beach conditions, characterized by challenging environments like sand, significantly impacts an athlete's aerobic capacity and running economy [50]. The distinct workload profiles of indoor and beach sports require custom training approaches for athletes. For beach sports, training should emphasize developing the strength and endurance required to overcome the challenges of the sand surface and variable conditions. In contrast, indoor sports training would focus on repeated speed and agility exercises, to better adapt to the higher external load demands of these sports [47,48]. Understanding these workload demands is crucial for optimizing athletic performance and formulating effective injury prevention strategies [40,49].

4.6. Considerations on Electronic Performance and Tracking Systems (EPTSs) in Beach Sports

The use of EPTSs in beach sports requires the careful consideration of various factors that can influence the accuracy, reliability, and validity of the collected data. These factors include tracking positioning, microtechnology data accuracy, device attachment and comfort, and data transmission methods.

Tracking, Microtechnology, and Precision: EPTS devices are composed of tracking technologies such as GPS in outdoor environments and LPS in indoor environments to register player movements and positions during beach sports matches and training sessions. However, the accuracy and precision of these systems can be affected by several factors, including satellite availability, environmental conditions (e.g., wind, temperature, and humidity), and the presence of obstacles or obstructions in the playing area [51,52]. In beach sports, the open and unobstructed playing environment may provide better satellite visibility and signal reception, potentially improving the accuracy of GPS-based tracking systems. However, the presence of sand and other environmental factors may introduce signal interference or multipath effects, which can degrade the positioning precision [51,52].

In addition to tracking systems, EPTSs often incorporate microtechnologies such as accelerometers, gyroscopes, and magnetometers to capture high-intensity actions like jumps, changes in direction, and impacts [11,53]. The precision of these microtechnologies can be influenced by factors such as sensor quality, sampling rates, and data processing algorithms [11,53]. It is essential that we consider the specifications and limitations of the microtechnologies used in EPTSs to ensure the accurate and reliable measurement of these high-intensity actions, which are particularly relevant in beach sports due to different movements without displacement (e.g., vertical jumps and impacts).

Device Attachment, Comfort, and Validity: The attachment of EPTS devices to the athletes' body is commonly realized by a customized vest at the interscapular level. Its attachment is a critical factor that can influence the comfort and precision of the collected data [53]. In beach sports, athletes typically wear minimal clothing which can pose challenges in securely attaching EPTS devices while maintaining athlete comfort and minimizing potential movement artifacts. Additionally, the presence of sand and sweat may affect the adhesion and stability of the devices, especially of heart rate bands, potentially compromising data validity [54]. Researchers and practitioners should carefully consider the attachment methods, device placement, and athlete feedback to ensure optimal data collection while maintaining athlete comfort and performance.

Data Transmission and Real-Time Monitoring: The transmission of data from EPTS devices to external systems is another important aspect to consider in beach sports. Real-time monitoring and data transmission can provide valuable insights for coaches and support staff during matches or training sessions, enabling them to make informed decisions and adjustments [55]. However, the beach environment may introduce challenges in wireless data transmission, potentially affecting the reliability and latency of real-time data streaming [56]. Alternative methods, such as post-session data retrieval, may be necessary in certain situations, which could impact the ability to make immediate adjustments during the activity.

To address these considerations, researchers and practitioners should carefully evaluate the capabilities and limitations of the EPTSs used in beach sports, and consider the specific environmental and practical challenges associated with these sports. Collaboration between sports scientists, engineers, and manufacturers is essential in order to develop tailored solutions that ensure accurate and reliable data collection while maintaining athlete comfort and performance. Additionally, conducting validation and reliability studies in beach sports settings can provide valuable insights and inform best practices for the effective use of EPTSs in these unique environments.

4.7. Limitations and Future Research

While this systematic review provides valuable insights into the workload demands of beach invasion sports, some limitations should be acknowledged based on the included studies. The majority of research has focused on beach handball, with relatively fewer studies examining beach soccer. This limits the ability for us to draw comprehensive comparisons between the two sports. Additionally, most studies analyzed elite or professional athletes, leaving a gap in the understanding of the demands across different competitive levels, such as amateur or recreational participants. Furthermore, several studies did not report detailed information on participant characteristics, such as years of experience or playing position, which could influence the interpretation of workload profiles. Lastly, the variability in technologies used and the lack of standardized intensity zones for external load metrics hindered direct comparisons across studies and sports, highlighting the need for consistent methodological approaches in future research.

Standardizing methodological approaches is a pressing need for future research in beach sports. Establishing consensus on intensity zones for external load metrics, such as the distance covered, accelerations, and impacts, would facilitate meaningful comparisons across studies and sports. Additionally, efforts should be made to integrate emerging wearable technologies that can comprehensively assess internal workload demands. In particular, incorporating breathing pattern monitoring could complement the traditionally used measures of heart rate and lactate, providing a more holistic understanding of the physiological strain experienced by athletes during beach sports competitions and training.

Furthermore, investigations should explore the potential influence of environmental factors, such as wind, temperature, and humidity, on the workload demands of beach sports. These unique conditions may significantly impact both external and internal loads, necessitating tailored training and recovery strategies. Research should also examine potential differences in workload profiles based on playing positions, enabling the development of position-specific training protocols for optimal performance and injury prevention. Expanding the scope of research to include a broader range of competition levels, from recreational to elite, would provide a more comprehensive understanding of the sport-specific demands and inform appropriate training programs across various athlete populations.

5. Conclusions

This systematic review characterized the internal and external workload demands of beach invasion sports across competition levels, ages, and sexes. Key findings demonstrate beach sports are intermittent in nature, entailing moderate-to-high-intensity bouts interspersed with lower-intensity activity for recovery. Average and peak heart rates align with high-intensity interval training principles for both aerobic and anaerobic adaptations. The unstable sand surface and variable outdoor conditions elicit a greater perceptual effort despite lower external load volumes versus indoor equivalents. Male players showed higher locomotive demands, but no internal workload differences exist between sexes. Finally, acceleration, impact, and internal load intensity zones vary substantially in the limited beach sports research, highlighting an urgent need for standardization to enable comparisons and inform training design.

Practical Applications

The data on internal and external workload demands in beach sports can be leveraged for individualized training prescription and recovery through several approaches. Firstly, by integrating the external load metrics (e.g., distances covered, accelerations, and impacts) with individual athlete characteristics such as age, sex, playing position, and training age, coaches can tailor training drills and intensities to optimally prepare each player for the specific demands of competition. Secondly, the internal load data, particularly the heart rate responses and perceived exertion, can be used to monitor individual athletes' physiological and psychological readiness, enabling appropriate adjustments to training loads and recovery strategies.

For this purpose, standardized external load zones are needed, while internal load zones by percentage of maximal heart rate appear reasonably consistent. Coaches should recognize that an unstable surface accentuates internal stress, enabling high cardio respiratory training stimuli despite deceptively lower running volumes. Training should prepare players physically and perceptually for intermittent effort profiles. Finally, male and female beach players likely require similar cardio conditioning, but more position-specific locomotive development.

Moreover, the integration of EPTSs with athlete monitoring systems can facilitate real-time feedback and online monitoring processes. Through real-time data, adjustments can be made during training sessions or matches based on individual athletes' external and internal load responses. This real-time monitoring can help prevent excessive fatigue, overload, or underload, ultimately reducing the risk of injury and optimizing performance. Furthermore, by combining the workload data with additional personal biological parameters, such as hormonal profiles, sleep quality, and nutrient intake, a more comprehensive understanding of each athlete's readiness and recovery status can be achieved. This holistic approach can inform personalized periodization strategies, tailoring training phases, intensities, and recovery interventions to individual needs.

Author Contributions: Conceptualization, P.V.-B. and J.P.-O.; methodology, P.V.-B. and C.D.G.-C.; software, C.D.G.-C. and J.M.M.-F.; validation, C.D.G.-C., J.M.M.-F. and J.P.-O.; formal analysis, P.V.-B. and C.D.G.-C.; investigation, P.V.-B. and J.P.-O.; resources, C.D.G.-C. and J.M.M.-F.; data curation, C.D.G.-C. and J.P.-O.; writing—original draft preparation, P.V.-B., C.D.G.-C. and J.M.M.-F.; writing—review and editing, C.D.G.-C. and J.P.-O.; visualization, C.D.G.-C., J.M.M.-F. and J.P.-O.; supervision, J.P.-O.; project administration, J.P.-O. All authors have read and agreed to the published version of the manuscript.

Funding: The author Joaquín Martín Marzano-Felisatti was supported by a pre-doctoral grant from the Ministry of Universities of Spain (Grant number: FPU20/01060).

Institutional Review Board Statement: Not applicable.

Informed Consent Statement: Not applicable.

Data Availability Statement: Not applicable.

Conflicts of Interest: The authors declare no conflicts of interest. The funders had no role in the design of the study; in the collection, analyses, or interpretation of the data; in the writing of the manuscript; or in the decision to publish the results.

References

1. Bělka, J.; Hůlka, K.; Šafář, M.; Weisser, R.; Chadimova, J. Beach Handball and Beach Volleyball as Means Leading to Increasing Physical Activity of Recreational Sportspeople—Pilot Study. *J. Sports Sci.* **2015**, *3*, 165–170.
2. Read, B.; Edwards, P. Blue Section. Introducing Formal Games. In *Teaching Children to Play Games*; White Line Publishing Services: Leeds, UK, 1992; pp. 61–65.
3. Read, B.; Edwards, P. Blue Section. Invasion Games. In *Teaching Children to Play Games*; White Line Publishing Services: Leeds, UK, 1992; pp. 91–139.
4. Hausler, J.; Halaki, M.; Orr, R. Application of Global Positioning System and Microsensor Technology in Competitive Rugby League Match-Play: A Systematic Review and Meta-Analysis. *Sports Med.* **2016**, *46*, 559–588. [CrossRef] [PubMed]
5. Miguel, M.; Oliveira, R.; Loureiro, N.; García-Rubio, J.; Ibáñez, S.J. Load Measures in Training/Match Monitoring in Soccer: A Systematic Review. *Int. J. Environ. Res. Public Health* **2021**, *18*, 2721. [CrossRef] [PubMed]
6. Petway, A.J.; Freitas, T.T.; Calleja-González, J.; Leal, D.M.; Alcaraz, P.E. Training Load and Match-Play Demands in Basketball Based on Competition Level: A Systematic Review. *PLoS ONE* **2020**, *15*, e0229212. [CrossRef] [PubMed]
7. Bourdon, P.C.; Cardinale, M.; Murray, A.; Gastin, P.; Kellmann, M.; Varley, M.C.; Gabbett, T.J.; Coutts, A.J.; Burgess, D.J.; Gregson, W.; et al. Monitoring Athlete Training Loads: Consensus Statement. *Int. J. Sports Physiol. Perform.* **2017**, *12*, S2-161–S2-170. [CrossRef] [PubMed]
8. Achenbach, L. Beach Sports. In *Injury and Health Risk Management in Sports: A Guide to Decision Making*; Krutsch, W., Mayr, H.O., Musahl, V., Della Villa, F., Tscholl, P.M., Jones, H., Eds.; Springer: Berlin/Heidelberg, Germany, 2020; pp. 749–753; ISBN 978-3-662-60752-7.
9. Impellizzeri, F.M.; Marcora, S.M.; Coutts, A.J. Internal and External Training Load: 15 Years on. *Int. J. Sports Physiol. Perform.* **2019**, *14*, 270–273. [CrossRef] [PubMed]
10. Buchheit, M.; Lacome, M.; Cholley, Y.; Simpson, B.M. Neuromuscular Responses to Conditioned Soccer Sessions Assessed via GPS-Embedded Accelerometers: Insights into Tactical Periodization. *Int. J. Sports Physiol. Perform.* **2018**, *13*, 577–583. [CrossRef] [PubMed]

11. Gómez-Carmona, C.D.; Bastida-Castillo, A.; Ibáñez, S.J.; Pino-Ortega, J. Accelerometry as a Method for External Workload Monitoring in Invasion Team Sports. A Systematic Review. *PLoS ONE* **2020**, *15*, e0236643. [CrossRef] [PubMed]

12. Foster, C.; Florhaug, J.A.; Franklin, J.; Gottschall, L.; Hrovatin, L.A.; Parker, S.; Doleshal, P.; Dodge, C. A New Approach to Monitoring Exercise Training. *J. Strength Cond. Res.* **2001**, *15*, 109–115.

13. Halson, S.L. Monitoring Training Load to Understand Fatigue in Athletes. *Sports Med.* **2014**, *44*, 139–147. [CrossRef]

14. Costa, J.A.; Figueiredo, P.; Prata, A.; Reis, T.; Reis, J.F.; Nascimento, L.; Brito, J. Associations between Training Load and Well-Being in Elite Beach Soccer Players: A Case Report. *Int. J. Environ. Res. Public Health* **2022**, *19*, 6209. [CrossRef] [PubMed]

15. Gómez-Carmona, C.D.; Rojas-Valverde, D.; Rico-González, M.; De Oliveira, V.; Lemos, L.; Martins, C.; Nakamura, F.; Pino-Ortega, J. Crucial Workload Variables in Female-Male Elite Brazilian Beach Handball: An Exploratory Factor Analysis. *Biol. Sport* **2023**, *40*, 345–352. [CrossRef] [PubMed]

16. Scott, M.T.U.; Scott, T.J.; Kelly, V.G. The Validity and Reliability of Global Positioning Systems in Team Sport: A Brief Review. *J. Strength Cond. Res.* **2016**, *5*, 1470–1490. [CrossRef] [PubMed]

17. Bastida Castillo, A.; Gómez Carmona, C.D.; De la Cruz Sánchez, E.; Pino Ortega, J. Accuracy, Intra- and Inter-Unit Reliability, and Comparison between GPS and UWB-Based Position-Tracking Systems Used for Time–Motion Analyses in Soccer. *Eur. J. Sport Sci.* **2018**, *18*, 450–457. [CrossRef]

18. Pino Ortega, J.; Rico González, M. Standarization of Electronic Performance and Tracking Systems. In *The Use of Applied Technology in Team Sport*; Routledge/Taylor & Francis Group: Abingdon, UK, 2021; pp. 3–10. ISBN 978-1-003-15700-7.

19. Pueo, B.; Jimenez-Olmedo, J.M.; Penichet-Tomas, A.; Becerra, M.O.; Agullo, J.J.E. Analysis of Time-Motion and Heart Rate in Elite Male and Female Beach Handball. *J. Sport Sci. Med.* **2017**, *16*, 450–458.

20. Bishop, D. A Comparison between Land and Sand-Based Tests for Beach Volleyball Assessment. *J. Sports Med. Phys. Fit.* **2003**, *43*, 418–423.

21. Castellano, J.; Casamichana, D. Heart Rate and Motion Analysis by GPS in Beach Soccer. *J. Sports Sci. Med.* **2010**, *9*, 98.

22. Ato, M.; López-García, J.J.; Benavente, A. Un Sistema de Clasificación de Los Diseños de Investigación en Psicología. *An. Psicol.* **2013**, *29*, 1038–1059. [CrossRef]

23. Page, M.J.; McKenzie, J.E.; Bossuyt, P.M.; Boutron, I.; Hoffmann, T.C.; Mulrow, C.D.; Shamseer, L.; Tetzlaff, J.M.; Akl, E.A.; Brennan, S.E. The PRISMA 2020 Statement: An Updated Guideline for Reporting Systematic Reviews. *Int. J. Surg.* **2021**, *88*, 105906. [CrossRef]

24. Rico-González, M.; Pino-Ortega, J.; Clemente, F.; Los Arcos, A. Guidelines for Performing Systematic Reviews in Sports Science. *Biol. Sport* **2022**, *39*, 463–471. [CrossRef]

25. Moher, D.; Shamseer, L.; Clarke, M.; Ghersi, D.; Liberati, A.; Petticrew, M.; Shekelle, P.; Stewart, L.A. Preferred Reporting Items for Systematic Review and Meta-Analysis Protocols (PRISMA-P) 2015 Statement. *Syst. Rev.* **2015**, *4*, 1. [CrossRef] [PubMed]

26. Slim, K.; Nini, E.; Forestier, D.; Kwiatkowski, F.; Panis, Y.; Chipponi, J. Methodological Index for Non-randomized Studies (*MINORS*): Development and Validation of a New Instrument. *ANZ J. Surg.* **2003**, *73*, 712–716. [CrossRef] [PubMed]

27. Cobos, D.L.; Ortega-Becerra, M.; Daza, G.; Sánchez-Sáez, J.A. Internal and External Load in International Women's Beach Handball: Official and Unofficial Competition. *Apunts Educ. Física Esports* **2023**, *151*, 79–87.

28. Gómez-Carmona, C.D.; García-Santos, D.; Mancha-Triguero, D.; Antúnez, A.; Ibáñez, S.J.; Gómez-Carmona, C.D.; García-Santos, D.; Mancha-Triguero, D.; Antúnez, A.; Ibáñez, S.J. Analysis of Sex-Related Differences in External Load Demands on Beach Handball. *Braz. J. Kinanthropometry Hum. Perform.* **2020**, *22*, 1–13. [CrossRef]

29. Gutiérrez-Vargas, R.; Gutiérrez-Vargas, J.C.; Ugalde-Ramírez, J.A.; Rojas-Valverde, D. Kinematics and Thermal Sex-Related Responses during an Official Beach Handball Game in Costa Rica: A Pilot Study. *Arch. Med. Deporte* **2019**, *36*, 13–18.

30. Iannaccone, A.; Fusco, A.; Skarbalius, A.; Kniubaite, A.; Cortis, C.; Conte, D. Relationship between External and Internal Load Measures in Youth Beach Handball. *Int. J. Sports Physiol. Perform.* **2021**, *17*, 256–262. [CrossRef] [PubMed]

31. Müller, C.; Willberg, C.; Reichert, L.; Zentgraf, K. External Load Analysis in Beach Handball Using a Local Positioning System and Inertial Measurement Units. *Sensors* **2022**, *22*, 3011. [CrossRef] [PubMed]

32. Sánchez-Sáez, J.A.; Sánchez-Sánchez, J.; Martínez-Rodríguez, A.; Felipe, J.L.; García-Unanue, J.; Lara-Cobos, D. Global Positioning System Analysis of Physical Demands in Elite Women's Beach Handball Players in an Official Spanish Championship. *Sensors* **2021**, *21*, 850. [CrossRef]

33. Zapardiel, J.C.; Paramio, E.M.; Ferragut, C.; Vila, H.; Asin-Izquierdo, I. Comparison of Internal and External Load Metrics between Won and Lost Game Segments in Elite Beach Handball. *Hum. Mov.* **2023**, *24*, 85–94. [CrossRef]

34. Zapardiel, J.C.; Asín-Izquierdo, I. Conditional Analysis of Elite Beach Handball According to Specific Playing Position through Assessment with GPS. *Int. J. Perform. Anal. Sport* **2020**, *20*, 118–132. [CrossRef]

35. Mancha-Triguero, D.; González-Espinosa, S.; Córdoba, L.G.; García-Rubio, J.; Feu, S. Differences in the Physical Demands between Handball and Beach Handball Players. *Rev. Bras. Cineantropometria Desempenho Hum.* **2020**, *22*, e72114. [CrossRef]

36. Bozdogan, T.K. Evaluate the Physical and Physiological Characteristics of Turkish National Beach Soccer Players. *Acta Sci. Intellectus* **2017**, *3*, 58–67.

37. Scarfone, R.; Tessitore, A.; Minganti, C.; Capranica, L.; Ammendolia, A. Match Analysis Heart-Rate and CMJ of Beach Soccer Players during Amateur Competition. *Int. J. Perform. Anal. Sport* **2015**, *15*, 241–253. [CrossRef]

38. Fédération Internationale de Football Association Beach Soccer Laws of the Game 2023-2024 2023.

39. International Handball Federation Rules of the Game for Beach Handball 2021.

40. Achenbach, L.; Loose, O.; Laver, L.; Zeman, F.; Nerlich, M.; Angele, P.; Krutsch, W. Beach Handball Is Safer than Indoor Team Handball: Injury Rates during the 2017 European Beach Handball Championships. *Knee Surg. Sports Traumatol. Arthrosc.* **2018**, *26*, 1909–1915. [CrossRef] [PubMed]

41. Reardon, C.; Tobin, D.P.; Delahunt, E. Application of Individualized Speed Thresholds to Interpret Position Specific Running Demands in Elite Professional Rugby Union: A GPS Study. *PLoS ONE* **2015**, *10*, e0133410. [CrossRef] [PubMed]

42. Abt, G.; Lovell, R. The Use of Individualized Speed and Intensity Thresholds for Determining the Distance Run at High-Intensity in Professional Soccer. *J. Sports Sci.* **2009**, *27*, 893–898. [CrossRef] [PubMed]

43. Clemente, F.; Ramirez-Campillo, R.; Beato, M.; Moran, J.; Kawczynski, A.; Makar, P.; Sarmento, H.; Afonso, J. Arbitrary Absolute vs. Individualized Running Speed Thresholds in Team Sports: A Scoping Review with Evidence Gap Map. *Biol. Sport* **2023**, *40*, 919–943. [CrossRef] [PubMed]

44. Polglaze, T.; Hogan, C.; Dawson, B.; Buttfield, A.; Osgnach, C.; Lester, L.; Peeling, P. Classification of Intensity in Team Sport Activity. *Med. Sci. Sports Exerc.* **2018**, *50*, 1487–1494. [CrossRef] [PubMed]

45. Sweeting, A.J.; Cormack, S.J.; Morgan, S.; Aughey, R.J. When Is a Sprint a Sprint? A Review of the Analysis of Team-Sport Athlete Activity Profile. *Front. Physiol.* **2017**, *8*, 432. [CrossRef]

46. Ibáñez, S.J.; Gómez-Carmona, C.D.; Mancha-Triguero, D. Individualization of Intensity Thresholds on External Workload Demands in Women's Basketball by K-Means Clustering: Differences Based on the Competitive Level. *Sensors* **2022**, *22*, 324. [CrossRef]

47. Spyrou, K.; Freitas, T.T.; Marín-Cascales, E.; Alcaraz, P.E. Physical and Physiological Match-Play Demands and Player Characteristics in Futsal: A Systematic Review. *Front. Psychol.* **2020**, *11*, 569897. [CrossRef] [PubMed]

48. García-Sánchez, C.; Navarro, R.M.; Karcher, C.; de la Rubia, A. Physical Demands during Official Competitions in Elite Handball: A Systematic Review. *Int. J. Environ. Res. Public Health* **2023**, *20*, 3353. [CrossRef] [PubMed]

49. Eils, E.; Wirtz, S.; Brodatzki, Y.; Zentgraf, K.; Büsch, D.; Szwajca, S. Optimizing the Transition from the Indoor to the Beach Season Improves Motor Performance in Elite Beach Handball Players. *Ger. J. Exerc. Sport Res.* **2022**, *52*, 637–646. [CrossRef]

50. Balasas, D.; Vamvakoudis, E.; Christoulas, K.; Stefanidis, P.; Prantsidis, D.; Evangelia, P. The Effect of Beach Volleyball Training on Running Economy and VO2max of Indoor Volleyball Players. *J. Phys. Educ. Sport* **2013**, *13*, 33–38. [CrossRef]

51. Malone, J.J.; Lovell, R.; Varley, M.C.; Coutts, A.J. Unpacking the Black Box: Applications and Considerations for Using GPS Devices in Sport. *Int. J. Sports Physiol. Perform.* **2017**, *12*, S2-18–S2-26. [CrossRef] [PubMed]

52. Gómez Carmona, C.D.; Pino Ortega, J.; Rico González, M.; Bastida Castillo, A. Global Navigation Satellite Systems. In *The Use of Applied Technology in Team Sport*; Routledge/Taylor & Francis Group: Abingdon, UK, 2021; pp. 20–38; ISBN 978-1-003-15700-7.

53. Edwards, S.; White, S.; Humphreys, S.; Robergs, R.; O'Dwyer, N. Caution Using Data from Triaxial Accelerometers Housed in Player Tracking Units during Running. *J. Sports Sci.* **2019**, *37*, 810–818. [CrossRef] [PubMed]

54. Camomilla, V.; Bergamini, E.; Fantozzi, S.; Vannozzi, G. Trends Supporting the In-Field Use of Wearable Inertial Sensors for Sport Performance Evaluation: A Systematic Review. *Sensors* **2018**, *18*, 873. [CrossRef] [PubMed]

55. Aughey, R.J.; Falloon, C. Real-Time versus Post-Game GPS Data in Team Sports. *J. Sci. Med. Sport* **2010**, *13*, 348–349. [CrossRef]

56. Chambers, R.; Gabbett, T.J.; Cole, M.H.; Beard, A. The Use of Wearable Microsensors to Quantify Sport-Specific Movements. *Sports Med.* **2015**, *45*, 1065–1081. [CrossRef]

MDPI AG
Grosspeteranlage 5
4052 Basel
Switzerland
Tel.: +41 61 683 77 34

Sensors Editorial Office
E-mail: sensors@mdpi.com
www.mdpi.com/journal/sensors